AP® PHYSICS 2 PREMIUM PREP

11th Edition

The Staff of The Princeton Review

PrincetonReview.com

Penguin Random House

The Princeton Review
110 East 42nd Street, 7th Floor
New York, NY 10017
princetonreview.com
penguinrandomhouse.com

Published in the United States by Penguin Random House, LLC, New York.

Terms of Service: The Princeton Review Online Companion Tools ("Student Tools") for retail books are available for only the two most recent editions of that book. Student Tools may be activated only once per eligible book purchased for a total of 24 months of access. Activation of Student Tools more than once per book is in direct violation of these Terms of Service and may result in discontinuation of access to Student Tools Services.

ISBN: 979-8-217-11910-3
eBook ISBN: 979-8-217-11911-0
ISSN: 3067-4638

The material in this book is up-to-date at the time of publication. However, changes may have been instituted by the testing body in the test after this book was published.

If there are any important late-breaking developments, changes, or corrections to the materials in this book, we will post that information online in the Student Tools. Register your book and check your Student Tools to see if there are any updates posted there.

Editor: Selena Coppock
Production Editors: Liz Dacey, Heidi Torres
Production Artist: Deborah Weber

Manufactured in the United States of America

10 9 8 7 6 5 4 3 2 1

11th Edition

EU Contact:
Penguin Random House Ireland
32 Nassau Street
Dublin D02 YH68
https://eu-contact.penguin.ie

For customer service, please contact **editorialsupport@review.com**, and be sure to include:

- full title of the book
- ISBN
- page number

Acknowledgments

The Princeton Review would like to extend special thanks to Tim Amen for his contributions to the 11th edition of this book. We are also, as always, very appreciative of the time and attention given to each page by Liz Dacey, Heidi Torres, and Deborah Weber.

Contents

Get More (Free) Content at PrincetonReview.com/prep

As easy as 1·2·3

1 Go to PrincetonReview.com/prep or scan the **QR code** and enter the following ISBN to register your book:

9798217119103

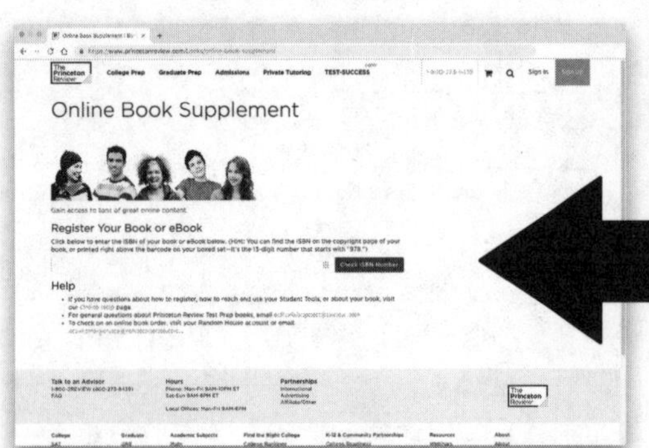

2 Answer a few simple questions to set up an exclusive Princeton Review account. *(If you already have one, you can just log in.)*

3 Enjoy access to your **FREE** content!

Access Your Online Test Practice

Your AP Prep book comes with new interactive practice exams to help you prepare for digital test-taking! Find these tests in your online Student Tools, provided in at least one of the following formats:

- Digital versions with a timer option to simulate the hybrid exam experience
- Downloadable interactive PDFs with digital features like clicking your answer in Section I

Check back often as we continue to update the included AP Student Tools.

PLUS, IN YOUR ACCOUNT YOU CAN:

- Get valuable advice about the college application process, tips for essay writing, and financial aid info
- Use our searchable rankings to learn more about your dream school
- Access comprehensive study guides, digital flashcards, and formula sheets
- Check whether there have been any updates or corrections to this edition

Need to report a potential **content** issue?

Contact **EditorialSupport@review.com** and include:

- full title of the book
- ISBN
- page number

Need to report a **technical** issue?

Contact **TPRStudentTech@review.com** and provide:

- your full name
- email address used to register the book
- full book title and ISBN
- Operating system (Mac/PC) and browser (Chrome, Firefox, Safari, etc.)

Look For These Icons Throughout The Book

 APPLIED STRATEGIES

 ONLINE ARTICLES

 PROVEN TECHNIQUES

 MORE GREAT BOOKS

 TIME-SAVING TIP

 STUDY BREAK

 ONLINE VIDEO

 CALCULATOR

Part I
Using This Book to Improve Your AP Score

- Preview: Your Knowledge, Your Expectations
- Your Guide to Using This Book
- How to Begin

PREVIEW: YOUR KNOWLEDGE, YOUR EXPECTATIONS

Your route to a high score on the AP Physics 2 Exam will depend on how you plan to use this book. Please respond to the following questions.

1. Rate your level of confidence about your knowledge of the content tested by the AP Physics 2 Exam.

   A. Very confident—I know it all
   B. I'm pretty confident, but there are topics for which I could use help
   C. Not confident—I need quite a bit of support
   D. I'm not sure

2. If you have a goal score in mind, circle your goal score for the AP Physics 2 Exam.

   5 4 3 2 1 I'm not sure yet.

3. What do you expect to learn from this book? Circle all that apply to you.

   A. A general overview of the test and what to expect
   B. Strategies for how to approach the test
   C. The content tested by this exam
   D. I'm not sure yet

Stay Up to Date!
For late-breaking information about test dates, exam formats, and any other changes pertaining to AP Physics 2, make sure to check the College Board's website at apstudents.collegeboard.org/courses/ap-physics-2-algebra-based

YOUR GUIDE TO USING THIS BOOK

This book is organized to provide as much—or as little—support as you need, so you can use this book in whatever way will be most helpful for improving your score on the AP Physics 2 Exam.

- The remainder of **Part I** provides guidance on how to use this book and helps you determine your strengths and weaknesses.
- **Part II** of this book contains your first practice test, its Diagnostic Answer Key, its answers and explanations, and a scoring guide. (Bubble sheets can be found in the very back of the book for easy tear-out.) Begin your test preparation with this practice test in order to realistically determine:
  - your starting point right now
  - which question types you're ready for and which you might need to practice
  - which content topics you are familiar with and which you will want to carefully review

  Once you have nailed down your strengths and weaknesses with regard to this exam, you can focus your test preparation, build a study plan, and be efficient with your time. Our Diagnostic Answer Key will assist you with this process.

- **Part III** of this book will:
    - provide information about the structure, scoring, and content of the AP Physics 2 Exam
    - help you to make a study plan
    - point you toward additional resources
- **Part IV** of this book will explore the following strategies:
    - how to attack multiple-choice questions
    - how to approach free-response questions
    - how to manage your time to maximize the number of points available to you
- **Part V** of this book covers the content you need for the AP Physics 2 Exam.
- **Part VI** of this book contains Practice Test 2, its answers and explanations, and a scoring guide. (Bubble sheets can be found in the very back of the book for easy tear-out.) If you skipped Practice Test 1, we recommend that you do both (with at least a day or two between them) so that you can compare your progress between the two.

    Practice Tests 3 and 4 can be found online in your Student Tools, along with their answers and explanations, and a scoring guide. We recommend leaving at least a few days or, ideally, a few weeks between practice tests.

    Additionally, taking all four practice tests will help to identify any external issues: if you get a certain type of question wrong multiple times, you probably need to review it. If you get it wrong only once, you may have run out of time or been distracted by something. In either case, completing all three practice tests will allow you to focus on the factors that caused the discrepancy in scores and to be as prepared as possible on the day of the test.
- **Online Resources** contain Practice Tests 3 and 4 and their answers and explanations, a score conversion chart, and bubble sheets.

Don't Forget!
When you register your book at PrincetonReview.com/prep following the instructions on pages x–xi, you'll gain access to a wealth of other helpful Student Tools, including study guides and printable bubble sheets.

You may choose to use some parts of this book over others, or you may work through the entire book. Your approach will depend on your needs and how much time you have. Let's now look at how to make this determination.

HOW TO BEGIN

1. **Take Practice Test 1**

    Before you can decide how to use this book, you need to take a practice test. Doing so will give you insight into your strengths and weaknesses, and the test will also help you make an effective study plan. If you're feeling test-phobic, remind yourself that a practice test is a tool for diagnosing yourself—it's not how well you do that matters but how you use information gleaned from your performance to guide your preparation.

So, before you read further, take AP Physics 2 Practice Test 1 starting at page 9 of this book. Be sure to do so in one sitting, following the instructions that appear before the test.

Scoring Worksheets
We've included a scoring worksheet for each of the Practice Tests. Remember that these worksheets are meant to serve as a rough guideline only. AP Exam scores are weighted according to a statistical process that varies slightly every year based on how students perform on the exam. But you can use the worksheets to approximate your score!

2. Check Your Answers

Using the Diagnostic Answer Key on page 33, follow our three-step process to identify your strengths and weaknesses with regard to the tested topics. This will help you determine which content review chapters to prioritize when studying this book. Don't worry about the explanations for now, and don't worry about why you missed questions. We'll get to that soon.

3. Reflect on the Test

After you take your first test, respond to the following questions:

- How much time did you spend on the multiple-choice questions?
- How much time did you spend on each free-response question?
- How many multiple-choice questions did you miss?
- Do you feel you had the knowledge to address the subject matter of the free-response questions?

4. Read Part III of this Book and Complete the Self-Evaluation

Part III will provide information on how the test is structured and scored. It will also set out areas of content that are tested.

As you read Part III, reevaluate your answers to the questions above. At the end of Part III, you will revisit the questions above and refine your answers to them. You will then be able to make a study plan, based on your needs and time available, that will allow you to use this book most effectively.

5. Engage with Parts IV and V as Needed

AP Exams Go Digital
As of May 2025, some AP Exams are fully digital and others are hybrid digital. For the latest info regarding specific AP subjects and testing accommodations, visit apcentral.collegeboard.org/exam-administration-ordering-scores/digital-ap-exams

Notice the word *engage.* You'll get more out of this book if you use it intentionally than if you read it passively, hoping for an improved score through osmosis.

The strategy chapters in Part IV will help you think about your approach to the question types on this exam. This part opens with a reminder to think about how you approach questions now and then closes with a reflection section asking you to think about how/whether you will change your approach in the future.

The content chapters in Part V are designed to provide a review of the content tested on the AP Physics 2 Exam, including the level of detail you need to know and how the content is tested. You will have the opportunity to assess your proficiency in the content of each chapter through test-appropriate questions.

6. **Take Practice Tests 2, 3, and 4 and Assess Your Performance**

Once you feel you have developed the strategies you need and gained the knowledge you lacked, you should take Practice Test 2. You should do so in one sitting, following the instructions at the beginning of the test.

When you are done, check your answers to the multiple-choice sections. See if a teacher will read your free-response answers and provide feedback.

Once you have taken the test, reflect on what areas you still need to work on, and revisit the chapters in this book that address those deficiencies. Through this type of reflection and engagement, you will continue to improve.

After a few weeks or a few days, take Practice Test 3 and assess your performance on that one as well. Then after a few more days or weeks, take Practice Test 4 and assess your performance once more.

Got a Question?
For answers to test-prep questions for all your tests and additional test-taking tips, subscribe to our YouTube channel at YouTube.com/ThePrincetonReview

7. **Keep Working**

After you have revisited certain chapters in this book, continue the process of testing, reflecting, and engaging with Practice Tests 2, 3, and 4. Consider what additional work you need to do and how you will change your strategic approach to different parts of the test. You can continue to explore areas that can stand to be improved and engage in those areas right up to the day of the test. As we will discuss in Part III, there are other resources available to you, including a wealth of information online at AP Students.

Part II
Practice Test 1

- Practice Test 1
- Practice Test 1: Diagnostic Answer Key and Explanations
- Practice Test 1: Free-Response Questions Grading Rubrics
- How to Score Practice Test 1

Practice Test 1

AP® Physics 2 Exam

SECTION I: Multiple-Choice Questions

At a Glance

Total Time
80 minutes
Number of Questions
40
Percent of Total Grade
50%

DISCLAIMER: The official AP Physics 2 Exam will be administered in a hybrid digital format. Instructions for the digital exam may differ from this practice test.

Instructions

Section I of this examination contains 40 multiple-choice questions. Fill in only the ovals for numbers 1 through 40 on your answer sheet.

CALCULATORS MAY BE USED IN BOTH SECTIONS OF THE EXAMINATION.

Indicate all of your answers to the multiple-choice questions on the answer sheet. Give only one answer to each question.

Use your time effectively, working as quickly as you can without losing accuracy. Do not spend too much time on any one question. Go on to other questions and come back to the ones you have not answered if you have time. It is not expected that everyone will know the answers to all the multiple-choice questions.

GO ON TO THE NEXT PAGE.

ADVANCED PLACEMENT PHYSICS 2 TABLE OF INFORMATION

CONSTANTS AND CONVERSION FACTORS	
Avogadro's number,	$N_0 = 6.02 \times 10^{23}\ \text{mol}^{-1}$
Universal gas constant,	$R = 8.31\ \text{J/(mol}\cdot\text{K)}$
Boltzmann's constant,	$k_B = 1.38 \times 10^{-23}\ \text{J/K}$
1 atmosphere of pressure,	$1\ \text{atm} = 1.0 \times 10^5\ \text{N/m}^2 = 1.0 \times 10^5\ \text{Pa}$
Coulomb's law constant,	$k = \dfrac{1}{4\pi\varepsilon_0} = 9.0 \times 10^9\ \dfrac{\text{N}\cdot\text{m}^2}{\text{C}^2}$
Proton mass,	$m_p = 1.67 \times 10^{-27}\ \text{kg}$
Neutron mass,	$m_n = 1.67 \times 10^{-27}\ \text{kg}$
Electron mass,	$m_e = 9.11 \times 10^{-31}\ \text{kg}$
Elementary charge,	$e = 1.60 \times 10^{-19}\ \text{C}$
Vacuum permittivity,	$\varepsilon_0 = 8.85 \times 10^{-12}\ \text{C}^2/(\text{N}\cdot\text{m}^2)$
Vacuum permeability,	$\mu_0 = 4\pi \times 10^{-7}\ (\text{T}\cdot\text{m})/\text{A}$
1 electron volt,	$1\ \text{eV} = 1.60 \times 10^{-19}\ \text{J}$
Planck's constant,	$h = 6.63 \times 10^{-34}\ \text{J}\cdot\text{s} = 4.14 \times 10^{-15}\ \text{eV}\cdot\text{s}$
	$hc = 1.99 \times 10^{-25}\ \text{J}\cdot\text{m} = 1.24 \times 10^3\ \text{eV}\cdot\text{nm}$
Speed of light,	$c = 3.00 \times 10^8\ \text{m/s}$
Wien's constant,	$b = 2.90 \times 10^{-3}\ \text{m}\cdot\text{K}$
Stefan-Boltzmann constant	$\sigma = 5.67 \times 10^{-8}\ \text{W/(m}^2\cdot\text{K}^4)$
1 unified atomic mass unit,	$1\ \text{u} = 1.66 \times 10^{-27}\ \text{kg} = 931\ \text{MeV}/c^2$

Universal gravitational constant, $G = 6.67 \times 10^{-11}\ \text{m}^3/(\text{kg}\cdot\text{s}^2) = 6.67 \times 10^{-11}\ \text{N}\cdot\text{m}^2/\text{kg}^2$
Magnitude of the acceleration due to gravity at Earth's surface, $g = 9.8\ \text{m/s}^2$
Magnitude of the gravitational field strength at the Earth's surface, $g = 9.8\ \text{N/kg}$

UNIT SYMBOLS	
ampere,	A
coulomb,	C
degree Celsius,	°C
electron volt,	eV
farad,	F
hertz,	Hz
joule,	J
kelvin,	K
kilogram,	kg
meter,	m
mole,	mol
newton,	N
ohm,	Ω
pascal,	Pa
second,	s
tesla,	T
volt,	V
watt,	W

PREFIXES

Factor	Prefix	Symbol
10^{12}	tera	T
10^9	giga	G
10^6	mega	M
10^3	kilo	k
10^{-2}	centi	c
10^{-3}	milli	m
10^{-6}	micro	μ
10^{-9}	nano	n
10^{-12}	pico	p

VALUES OF TRIGONOMETRIC FUNCTIONS FOR COMMON ANGLES

θ	0°	30°	37°	45°	53°	60°	90°
$\sin\theta$	0	1/2	3/5	$\sqrt{2}/2$	4/5	$\sqrt{3}/2$	1
$\cos\theta$	1	$\sqrt{3}/2$	4/5	$\sqrt{2}/2$	3/5	1/2	0
$\tan\theta$	0	$\sqrt{3}/3$	3/4	1	4/3	$\sqrt{3}$	∞

The following conventions are used in this exam.

- The frame of reference of any problem is assumed to be inertial unless otherwise stated.
- Air resistance is assumed to be negligible unless otherwise stated.
- In all situations, positive work is defined as work done **ON** a system.
- Springs and strings are assumed to be ideal unless otherwise stated.
- The electric potential is zero at an infinite distance from an isolated point charge.
- The direction of current is the direction in which positive charges would drift.
- All batteries and meters are assumed to be ideal unless otherwise stated.

GO ON TO THE NEXT PAGE.

ADVANCED PLACEMENT PHYSICS 2 EQUATIONS

ELECTRICITY

$$\left|\vec{F}_E\right| = \frac{1}{4\pi\varepsilon_0}\frac{|q_1 q_2|}{r^2} = k\frac{|q_1 q_2|}{r^2}$$

$$\vec{E} = \frac{\vec{F}_E}{q}$$

$$\left|\vec{E}\right| = \frac{1}{4\pi\varepsilon_0}\frac{|q|}{r^2} = k\frac{|q|}{r^2}$$

$$U_E = \frac{1}{4\pi\varepsilon_0}\frac{q_1 q_2}{r} = k\frac{q_1 q_2}{r}$$

$$\Delta V = \frac{\Delta U_E}{q}$$

$$V = \frac{1}{4\pi\varepsilon_0}\sum_i \frac{q_i}{r_i}$$

$$\left|\vec{E}\right| = \left|\frac{\Delta V}{\Delta r}\right|$$

$$C = \frac{Q}{\Delta V}$$

$$C = \kappa\varepsilon_0\frac{A}{d}$$

$$E_C = \frac{Q}{\kappa\varepsilon_0 A}$$

$$U_C = \frac{1}{2}Q\Delta V$$

$$I = \frac{\Delta Q}{\Delta t}$$

$$R = \frac{\rho\ell}{A}$$

$$P = I\Delta V$$

$$I = \frac{\Delta V}{R}$$

$$R_{eq,s} = \sum_i R_i$$

$$\frac{1}{R_{eq,p}} = \sum_i \frac{1}{R_i}$$

$$\frac{1}{C_{eq,s}} = \sum_i \frac{1}{C_i}$$

$$C_{eq,p} = \sum_i C_i$$

$$\tau = R_{eq}C_{eq}$$

A = area
C = capacitance
d = distance
E = electric field
F = force
I = current
ℓ = length
P = power
q = charge
Q = charge
r = distance, radius, or position
R = resistance
t = time
U = potential energy
V = electric potential
κ = dielectric constant
ρ = resistivity
τ = time constant

MAGNETISM

$$F_B = qvB\sin\theta$$

$$B = \frac{\mu_0 I}{2\pi r}$$

$$F_B = I\ell B\sin\theta$$

$$\Phi_B = \vec{B}\cdot\vec{A}$$

$$\Phi_B = \left|\vec{B}\right|\cos\theta\left|\vec{A}\right|$$

$$|\varepsilon| = \left|\frac{\Delta\Phi_B}{\Delta t}\right|$$

$$\varepsilon = B\ell v$$

A = area
B = magnetic field
F = force
I = current
ℓ = length
q = charge
r = distance, radius, or position
t = time
v = velocity or speed
ε = emf
θ = angle
Φ = flux

THERMAL PHYSICS

$$P = \frac{F_\perp}{A}$$

$$K_{avg} = \frac{3}{2}k_B T = \frac{1}{2}mv_{rms}^2$$

$$\frac{Q}{\Delta t} = \frac{kA\Delta T}{L}$$

$$PV = nRT = Nk_B T$$

$$U = \frac{3}{2}nRT = \frac{3}{2}Nk_B T$$

$$W = -P\Delta V$$

$$\Delta U = Q + W$$

$$Q = mc\Delta T$$

A = area
c = specific heat
F = force
k = thermal conductivity
K = kinetic energy
L = length
m = mass
n = number of moles
N = number of atoms
P = pressure
Q = energy transferred to a system by heating
t = time
T = temperature
U = internal energy
v = velocity or speed
V = volume
W = work done on a system

GO ON TO THE NEXT PAGE.

ADVANCED PLACEMENT PHYSICS 2 EQUATIONS

WAVES, SOUND, AND OPTICS

$$\lambda = \frac{v}{f}$$

$$n = \frac{c}{v}$$

$$n_1 \sin\theta_1 = n_2 \sin\theta_2$$

$$\frac{1}{s_i} + \frac{1}{s_o} = \frac{1}{f}$$

$$|M| = \left|\frac{h_i}{h_o}\right| = \left|\frac{s_i}{s_o}\right|$$

$$\Delta D = m\lambda$$

$$\Delta D = a\sin\theta$$

$$a\left(\frac{y_{min}}{L}\right) \approx m\lambda$$

$$\Delta D = d\sin\theta$$

$$d\left(\frac{y_{max}}{L}\right) \approx m\lambda$$

$$v_{string} = \sqrt{\frac{F_T}{m/\ell}}$$

$$T = \frac{1}{f}$$

$$x(t) = A\cos(\omega t) = A\cos(2\pi ft)$$

$$y(x) = A\cos\left(2\pi\frac{x}{\lambda}\right)$$

$$|f_{beat}| = |f_1 - f_2|$$

a = width
A = amplitude
d = separation
D = path length
f = frequency or focal length
F = force
h = height
ℓ = length
L = distance
m = order or mass
M = magnification
n = index of refraction
s = position
t = time
T = period
v = speed
x = position
y = position
λ = wavelength
θ = angle
ω = angular frequency

MODERN PHYSICS

$$E = hf$$

$$\lambda = \frac{h}{p}$$

$$\lambda = \frac{c}{f}$$

$$\lambda_{max} = \frac{b}{T}$$

$$P = A\sigma T^4$$

$$K_{max} = hf - \phi$$

$$\Delta\lambda = \frac{h}{m_e c}(1 - \cos\theta)$$

$$E = mc^2$$

$$N = N_0 e^{-\lambda t}$$

$$\lambda = \frac{\ln 2}{t_{1/2}}$$

A = area
b = constant
E = energy
f = frequency
K = kinetic energy
m = mass
N = number of particles
p = momentum
P = power
t = time
T = absolute temperature
θ = angle
λ = wavelength or decay constant
σ = constant
ϕ = work function

GEOMETRY AND TRIGONOMETRY

Rectangle
$A = bh$

Triangle
$A = \frac{1}{2}bh$

Circle
$A = \pi r^2$
$C = 2\pi r$
$s = r\theta$

Rectangular Solid
$V = \ell wh$

Cylinder
$V = \pi r^2 \ell$
$S = 2\pi r\ell + 2\pi r^2$

Sphere
$V = \frac{4}{3}\pi r^3$
$S = 4\pi r^2$

s, r, θ

A = area
b = base
C = circumference
h = height
ℓ = length
r = radius
s = arc length
S = surface area
V = volume
w = width
θ = angle

Right Triangle

$a^2 + b^2 = c^2$

$\sin\theta = \frac{a}{c}$

$\cos\theta = \frac{b}{c}$

$\tan\theta = \frac{a}{b}$

c, a, b, θ, 90°

GO ON TO THE NEXT PAGE.

ADVANCED PLACEMENT PHYSICS 2 EQUATIONS

MECHANICS AND FLUIDS

$v_x = v_{x0} + a_x t$

$x = x_0 + v_{x0}t + \frac{1}{2}a_x t^2$

$v_x^2 = v_{x0}^{\ 2} + 2a_x(x - x_0)$

$\vec{x}_{cm} = \frac{\sum m_i \vec{x}_i}{\sum m_i}$

$\vec{a}_{sys} = \frac{\sum \vec{F}}{m_{sys}} = \frac{\vec{F}_{net}}{m_{sys}}$

$\left|\vec{F}_g\right| = G\frac{m_1 m_2}{r^2}$

$\left|\vec{F}_f\right| \le \left|\mu \vec{F}_n\right|$

$\vec{F}_S = -k\Delta\vec{x}$

$a_c = \frac{v^2}{r}$

$K = \frac{1}{2}mv^2$

$W = F_{\parallel}d = Fd\cos\theta$

$\Delta K = \sum W_i = \sum F_{\parallel,i}d_i$

$\Delta U_S = \frac{1}{2}k(\Delta x)^2$

$U_G = -\frac{Gm_1 m_2}{r}$

$\Delta U_g = mg\Delta y$

$P_{avg} = \frac{W}{\Delta t} = \frac{\Delta E}{\Delta t}$

$P_{inst} = F_{\parallel}v = Fv\cos\theta$

$\vec{p} = m\vec{v}$

$\vec{F}_{net} = \frac{\Delta\vec{p}}{\Delta t} = m\frac{\Delta\vec{v}}{\Delta t} = m\vec{a}$

$\vec{J} = \vec{F}_{avg}\Delta t = \Delta\vec{p}$

$\vec{v}_{cm} = \frac{\sum \vec{p}_i}{\sum m_i} = \frac{\sum m_i \vec{v}_i}{\sum m_i}$

a = acceleration
d = distance
E = energy
F = force
J = impulse
k = spring constant
K = kinetic energy
m = mass
p = momentum
P = power
r = radius, distance, or position
t = time
U = potential energy
v = velocity or speed
W = work
x = position
y = height
θ = angle
μ = coefficient of friction

$\omega = \omega_0 + \alpha t$

$\theta = \theta_0 + \omega_0 t + \frac{1}{2}\alpha t^2$

$\omega^2 = \omega_0^2 + 2\alpha(\theta - \theta_0)$

$v = r\omega$

$a_T = r\alpha$

$\tau = r_{\perp}F = rF\sin\theta$

$I = \sum m_i r_i^2$

$I' = I_{cm} + Md^2$

$\alpha_{sys} = \frac{\sum \tau}{I_{sys}} = \frac{\tau_{net}}{I_{sys}}$

$K = \frac{1}{2}I\omega^2$

$W = \tau\Delta\theta$

$L = I\omega$

$L = rmv\sin\theta$

$\Delta L = \tau\Delta t$

$\Delta x_{cm} = r\Delta\theta$

$T = \frac{1}{f}$

$T_s = 2\pi\sqrt{\frac{m}{k}}$

$T_P = 2\pi\sqrt{\frac{\ell}{g}}$

$x = A\cos(2\pi ft)$
$x = A\sin(2\pi ft)$

$\rho = \frac{m}{V}$

$P = \frac{F_{\perp}}{A}$

$P = P_0 + \rho gh$

$P_{gauge} = \rho gh$

$F_b = \rho Vg$

$A_1 v_1 = A_2 v_2$

$P_1 + \rho gy_1 + \frac{1}{2}\rho v_1^2 = P_2 + \rho gy_2 + \frac{1}{2}\rho v_2^2$

a = acceleration
A = amplitude or area
d = distance
f = frequency
F = force
h = height
I = rotational inertia
k = spring constant
K = kinetic energy
ℓ = length
L = angular momentum
m = mass
M = mass
P = pressure
r = radius, distance, or position
t = time
T = period
v = velocity or speed
V = volume
W = work
x = position
y = vertical position
α = angular acceleration
θ = angle
ρ = density
τ = torque
ω = angular speed

GO ON TO THE NEXT PAGE.

AP PHYSICS 2
SECTION I
Time—80 minutes
40 Questions

Note: To simplify calculations, you may use $g = 10 \text{ m/s}^2$ in all problems.

Directions: Each of the questions or incomplete statements below is followed by four suggested answers or completions. Select the one that is best in each case and mark it on your sheet.

1 Mark for Review

A green light ($\lambda = 550$ nm) is focused on a material with a work function of $\varphi = 3$ eV. What is the maximum kinetic energy of a photoelectron emitted from the material?

(A) 0.74 eV

(B) 2.3 eV

(C) Electrons will not be emitted in this scenario.

(D) Cannot be determined without additional information

2 Mark for Review

Two different ideal gases are brought to the same temperature. The molecules of the first gas are four times more massive compared to the second gas. How do the average speeds of the molecules of the two gases compare?

(A) The average speeds are the same.

(B) The average speed of the molecules of the first gas are higher by a factor of 2.

(C) The average speed of the molecules of the second gas are higher by a factor of 2.

(D) The average speed of the molecules of the first gas are higher by a factor of 4.

Questions 3 through 4 refer to the following.

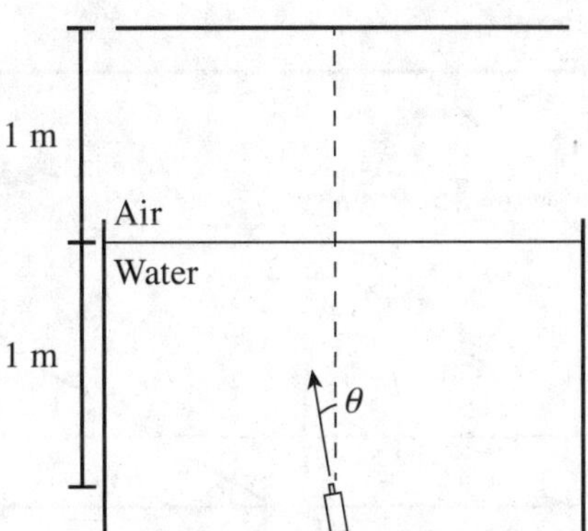

A laser pointer is placed in a tank of water with no lid so that the light is emitted at a depth of 1 m from the surface. 1 m above the surface, a projection screen is positioned so that the center of the screen is directly above the laser pointer. The pointer is tilted at an angle, θ, from the vertical.

3 Mark for Review

If the angle, θ, is 20 degrees, and given that the index of refraction of water is $n = 1.3$, how far from the center would the laser hit the screen?

(A) 79 cm from the center

(B) 86 cm from the center

(C) 184 cm from the center

(D) Cannot be determined without more information

GO ON TO THE NEXT PAGE.

4 Mark for Review

If a plane of glass (n = 1.5) is used as a lid, with no gap between the water and the glass, what will happen to the position of the laser on the screen?

(A) There will be no change in position.

(B) The laser will hit farther from the center.

(C) The laser will hit closer to the center.

(D) Cannot be determined without knowing the thickness of the glass

5 Mark for Review

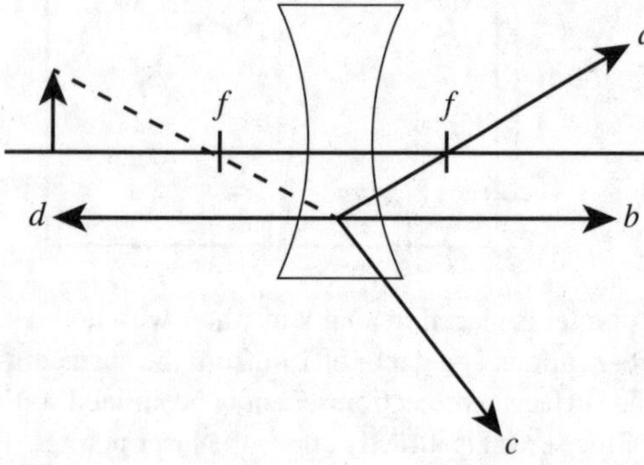

In the figure above, a ray of light hits an object and travels through the focal point of a concave lens as shown by the dotted line. Which line best shows the correct continuation of the ray after it hits the lens?

(A) *a*

(B) *b*

(C) *c*

(D) *d*

6 Mark for Review

An experiment is performed where standing sound waves are produced in open-ended cylinders of varying lengths. The frequencies of the waves are adjusted until the first harmonic is reached in a 1 m cylinder, the second harmonic is reached in a 2 m cylinder, and the third harmonic is reached in a 3 m cylinder. If the frequencies are f_1, f_2, and f_3, respectively, what is the relationship between these frequencies?

(A) $f_1 = f_2 = f_3$

(B) $f_1 < f_2 < f_3$

(C) $f_1 > f_2 > f_3$

(D) $f_1 = f_2 < f_3$

7 Mark for Review

An ideal gas undergoes an isothermal compression. What best describes the heat flow during this process?

(A) No heat flows.

(B) Heat flows out of the gas.

(C) Heat flows into the gas.

(D) Heat initially flows out of the gas, but the same amount of heat then flows into the gas.

GO ON TO THE NEXT PAGE.

Questions 8 through 10 refer to the following.

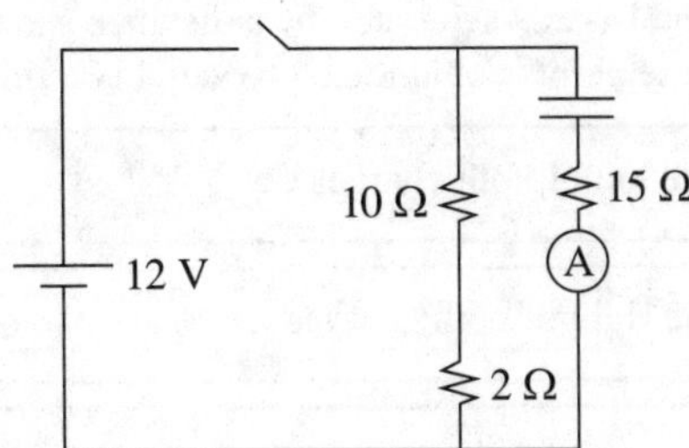

8 Mark for Review

In the above circuit, the capacitor is initially uncharged and the switch is closed. What best describes the current through the ammeter after the switch is closed?

(A) The current starts at a maximum and decreases to zero.

(B) The current starts at a maximum and decreases to a non-zero amount.

(C) The current starts at zero and increases to a set amount.

(D) Cannot be determined without knowing the capacitance of the capacitor

9 Mark for Review

After a long period of time, the switch is opened. What is the current through the ammeter immediately after the switch is opened?

(A) 0 A

(B) 0.44 A

(C) 0.8 A

(D) 1.8 A

10 Mark for Review

The capacitor is replaced with a piece of wire with negligible resistance. If the switch is again closed, what is the voltage across the 2 Ω resistor?

(A) 1 V

(B) 2 V

(C) 6 V

(D) 12 V

11 Mark for Review

An electron in an atom transitions from an excited state of –16 eV down to a ground state of –20 eV. What is the wavelength of the emitted photon?

(A) 31 nm

(B) 78 nm

(C) 311 nm

(D) 780 nm

GO ON TO THE NEXT PAGE.

12 Mark for Review

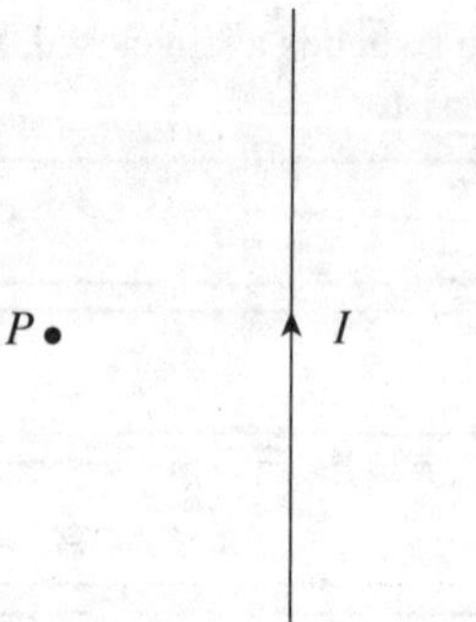

In the diagram above, a straight wire with a current I points vertically upward. At point P, which direction is the magnetic field pointing?

(A) Up

(B) Left

(C) Into the page

(D) Out of the page

13 Mark for Review

Two point charges, q_1 and q_2, are separated by a distance r in a vacuum. What change would cause the magnitude of the electric force between the charges to double?

(A) q_1 is doubled.

(B) r is halved.

(C) Both q_1 and q_2 are doubled.

(D) The sign of the charge on q_1 is flipped.

14 Mark for Review

Two positive point charges with charge $+Q$ and $+2Q$ are placed on a horizontal x-axis separated by a distance r in vacuum. At which of these points could the electric field be zero?

(A) To the left of both charges

(B) To the right of both charges

(C) In between the charges, closer to the $+Q$ charge

(D) In between the charges, closer to the $+2Q$ charge

15 Mark for Review

A student preforms a double-slit experiment and successfully creates an interference pattern on the screen. Which of the following will occur if the distance between the slits, d, is increased?

(A) The fringes will spread out.

(B) The fringes will get closer.

(C) The intensity of the pattern will increase.

(D) The width of the central peak will widen.

GO ON TO THE NEXT PAGE.

16 Mark for Review

An observer stands at rest as a car with a siren drives directly toward them at a constant speed. Which of the following is correct regarding the sound of the siren heard by the observer as the car approaches?

(A) The frequency of the sound is increasing.

(B) The frequency of the sound is constant.

(C) The frequency of the sound is decreasing.

(D) The amplitude of the sound is constant.

17 Mark for Review

Plutonium undergoes alpha decay to form uranium as shown below.

$$^{239}_{94}\text{Pu} \rightarrow \,^{235}_{92}\text{U} + \,^{4}_{2}\text{He}$$

This process causes the daughter nuclei to gain energy. What is the best explanation for this?

(A) The combined mass of the uranium and helium is greater than that of the plutonium.

(B) The combined mass of the uranium and helium is less than that of the plutonium.

(C) Conservation of momentum requires that the velocities of the products cancel out.

(D) Conservation of charge requires that the electrostatic force pushes the products away from each other.

18 Mark for Review

An ideal gas initially has a pressure of 2×10^5 Pa and a volume of 0.5 m^3. It then expands at a constant pressure to a volume of 2 m^3, and after expanding its pressure is increased to 5×10^5 Pa while it's held at a constant volume. What is the total work done by the gas during both transitions?

(A) 3×10^5 J

(B) 4.5×10^5 J

(C) 6×10^5 J

(D) 7.5×10^5 J

19 Mark for Review

If the loudness of a sound wave is increased, which of the following must be true?

(A) The frequency of the wave increased.

(B) The amplitude of the wave increased.

(C) The wavelength of the wave increased.

(D) The period of the wave increased.

GO ON TO THE NEXT PAGE.

Questions 20 through 22 refer to the following.

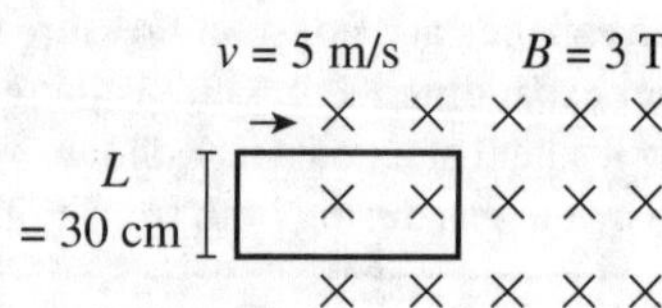

In the above diagram, a rectangular loop of wire is traveling to the right at a constant velocity into a constant magnetic field pointing into the page.

20 Mark for Review

What is the emf induced in the loop as it enters the magnetic field?

(A) 0.5 V

(B) 4.5 V

(C) 50 V

(D) 450 V

21 Mark for Review

What is the direction of the current in and force on the wire on the right side of the loop as it enters the field?

(A) Current is flowing up and force is to the right.

(B) Current is flowing down and force is to the right.

(C) Current is flowing up and force is to the left.

(D) Current is flowing down and force is to the left.

22 Mark for Review

After fully entering the field, the loop comes to a stop and starts to rotate at a constant rate about a horizontal axis. Which choice correctly describes the emf as it spins?

(A) Because it is spinning at a constant rate, the emf will be constant but non-zero.

(B) Because the magnetic field is constant, the emf will be zero.

(C) The emf will be at a maximum when the loop is parallel to the page, because this is where the flux is at a maximum.

(D) The emf will be at a minimum when the loop is parallel to the page, because this is where the change in flux is at a minimum.

23 Mark for Review

A convex lens with a focal length of 20 cm is used to form an image of an object paced 15 cm behind the lens. Which of the following is true about the image formed?

(A) The image is virtual and located 8.6 cm behind the lens.

(B) The image is virtual and located 60 cm behind the lens.

(C) The image is real and located 8.6 cm in front of the lens.

(D) The image is real and located 60 cm in front of the lens.

GO ON TO THE NEXT PAGE.

24 Mark for Review

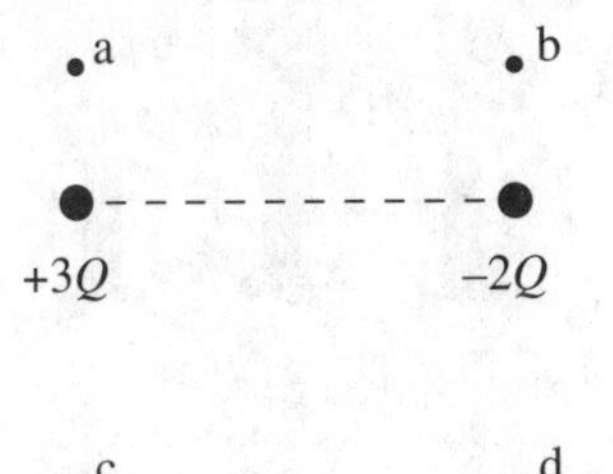

A positive charge of $3Q$ and a negative charge of $2Q$ are arranged as shown above. If the electric potentials are compared between two of the four labeled points, which of the following would have the greatest change in potential?

(A) $|V_a - V_b|$

(B) $|V_a - V_c|$

(C) $|V_c - V_b|$

(D) $|V_c - V_d|$

25 Mark for Review

An image is created by placing an object in front of a concave lens. If the lens is replaced with one of identical shape but with a lower index of refraction, which of the following changes would occur?

(A) The focal length would increase and the image would be formed farther from the lens.

(B) The focal length would increase and the image would be formed closer to the lens.

(C) The focal length would decrease and the image would be formed farther from the lens.

(D) The focal length would decrease and the image would be formed closer to the lens.

26 Mark for Review

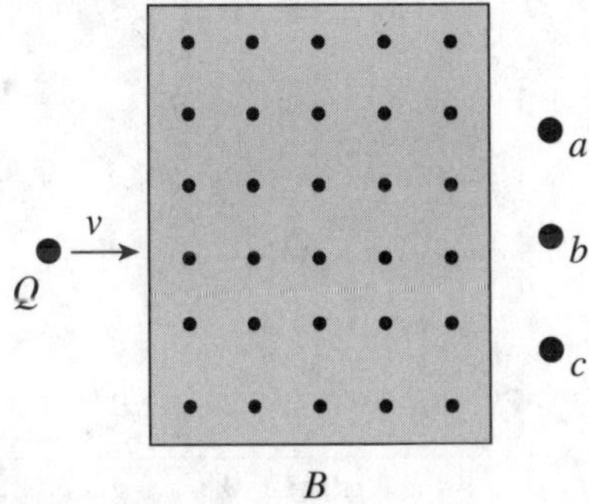

In the above figure, an unknown charge, Q, with negligible mass travels to the right with speed v through a region with uniform magnetic field, B, pointing out of the page. Which of the following is possible?

(A) The charge passes through point a with a speed lower than v.

(B) The charge passes through point b with a speed equal to v.

(C) The charge passes through point c with a speed higher than v.

(D) The charge passes through point a with a speed equal to v.

27 Mark for Review

Three resistors are connected to a battery in parallel. If $R_1 < R_2 < R_3$, what is the relationship between the voltages across each resistor?

(A) $V_1 < V_2 < V_3$

(B) $V_1 > V_2 > V_3$

(C) $V_1 = V_2 = V_3$

(D) Cannot be determined without more information

GO ON TO THE NEXT PAGE.

28 Mark for Review

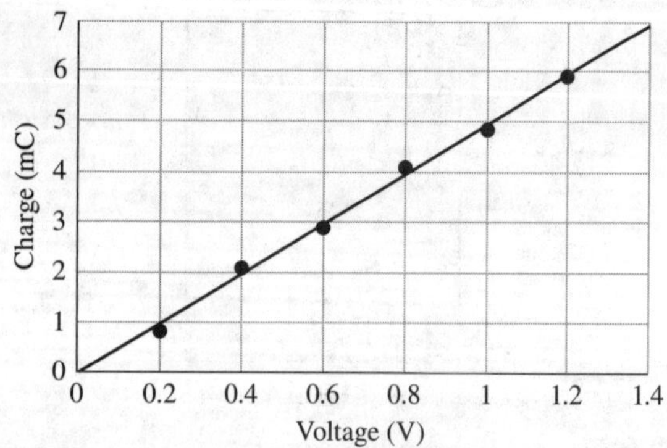

An experiment was performed that measured the charge on a capacitor while connected to a variable battery at several voltages. What is the capacitance of the capacitor that produced the graph above?

(A) 0.5 mF

(B) 1 mF

(C) 5 mF

(D) 10 mF

29 Mark for Review

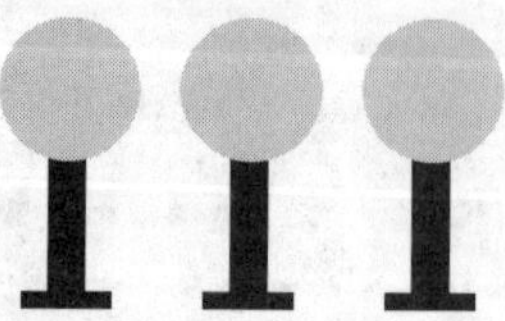

Three neutral conducting spheres sit on insulating bases. The first sphere is given a positive charge. It is then brought in contact with the second sphere and allowed to reach equilibrium. The two spheres are then separated. Finally, the second sphere is brought into contact with the third sphere and allowed to reach equilibrium. If the charge on the first, second, and third spheres after this process is Q_1, Q_2, and Q_3, respectively, what is the correct relationship between these charges?

(A) $Q_1 = Q_2 = Q_3$

(B) $Q_1 > Q_2 > Q_3$

(C) $Q_1 < Q_2 < Q_3$

(D) $Q_1 > Q_2 = Q_3$

30 Mark for Review

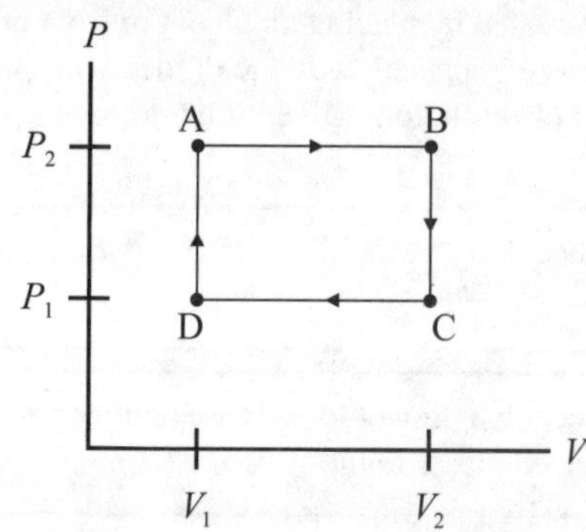

In the above diagram, in ideal gas undergoes the cycle ABCDA. What is the change in internal energy after the cycle?

(A) $(V_2–V_1)(P_2–P_1)$

(B) $(V_2–V_1)P_2$

(C) $(P_2–P_1)V_2$

(D) 0

GO ON TO THE NEXT PAGE.

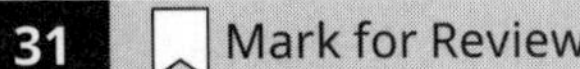

31 Mark for Review

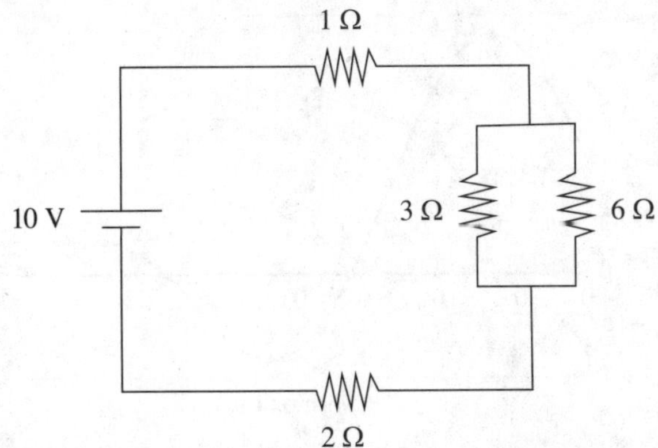

In the circuit shown above, what is the current through the 3 Ω resistor?

Ⓐ 0.83 A

Ⓑ 1.3 A

Ⓒ 2.9 A

Ⓓ 3.3 A

32 Mark for Review

During an isothermal expansion, an ideal gas pushes a piston with a mass of 2 kg up a height of 20 cm. Approximately what is the amount of heat that flows into the gas during this process?

Ⓐ 0 J

Ⓑ 4 J

Ⓒ 10 J

Ⓓ 400 J

33 Mark for Review

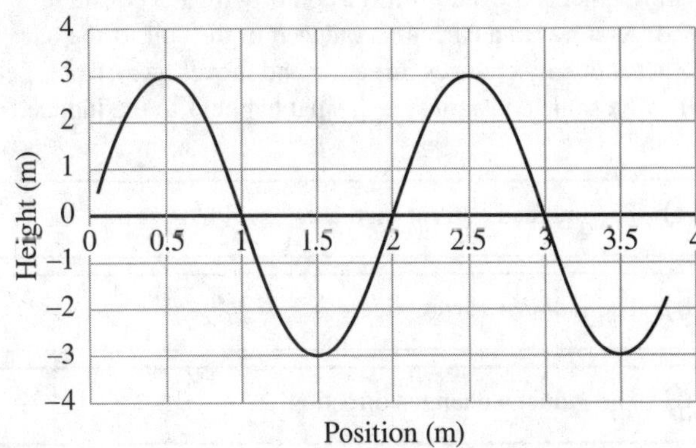

The above graph shows the displacement height of a string against its horizontal position as a wave travels along its length. If the period of the wave is 5 ms, what is the speed of the wave?

Ⓐ 10 m/s

Ⓑ 40 m/s

Ⓒ 200 m/s

Ⓓ 400 m/s

34 Mark for Review

Two point charges, Q_1 and Q_2, are placed 3 m apart. If Q_1 = + 5 μC and Q_2 = -2 μC, what is the magnitude and direction of the electric field at the midpoint between the charges?

Ⓐ 28 kN/C, toward Q_1

Ⓑ 28 kN/C, toward Q_2

Ⓒ 12 kN/C, toward Q_1

Ⓓ 12 kN/C, toward Q_2

GO ON TO THE NEXT PAGE.

35 Mark for Review

A bar magnet is moved toward a coil of wire at a constant speed. As a result, a current is induced in the coil. If the bar magnet suddenly changes direction and moves away from the coil at the same constant speed, what happens to the induced current?

(A) The induced current increases.

(B) The induced current decreases.

(C) The induced changes direction.

(D) The induced current does not change.

36 Mark for Review

A 6 μF capacitor and 15 μF capacitor are connected in series to a 12 V battery. After the circuit reaches equilibrium, what is the approximate charge on the 6 μF capacitor?

(A) 26 μC

(B) 51.5 μC

(C) 126 μC

(D) 252 μC

37 Mark for Review

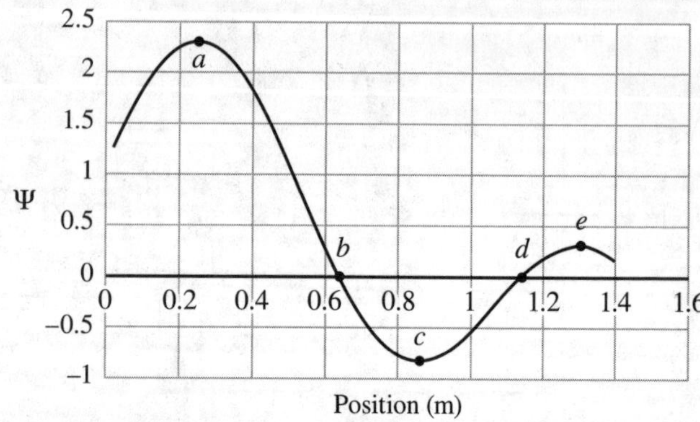

The above graph shows the wave function of a certain particle. Of the labeled points, which one(s) are the least likely to be the particle's position when observed?

(A) *a*

(B) *b* and *d*

(C) *c*

(D) *a*, *c*, and *e*

38 Mark for Review

A cooling fan is used to reduce the temperature of a system by blowing air past it and carrying heat away. Why does this not violate the Second Law of Thermodynamics?

(A) The second law only applies to isolated systems and since the air is leaving the area, this is an open system.

(B) The temperature decrease does not mean an entropy decrease in this scenario.

(C) The entropy of the surroundings will increase less than the entropy of the object decreases.

(D) The heat lost will match the loss of internal energy.

GO ON TO THE NEXT PAGE.

39 Mark for Review

Two charges, one with a mass of 5 kg and a charge of +3 C and the other with a mass of 2 kg and a charge of +2 C, are positioned next to each other. What would the resulting motion be?

(A) The charges move away from each other with the same acceleration.

(B) The charges move away from each other and the +3 C charge has greater acceleration.

(C) The charges move away from each other and the +2 C charge has greater acceleration.

(D) The charges move toward each other with the same acceleration.

40 Mark for Review

A detector is positioned to measure the intensity of an object's blackbody radiation at a distance, *d*, from the object. If the temperature of the object is increased by a factor of 2, what will happen to the intensity measured by the detector?

(A) The intensity will increase by a factor of 2.

(B) The intensity will increase by a factor of 4.

(C) The intensity will increase by a factor of 8.

(D) The intensity will increase by a factor of 16.

END OF SECTION I

GO ON TO THE NEXT PAGE.

AP PHYSICS 2
SECTION II
Time—100 minutes
4 Questions

Directions: Question 1 is a Math Routines question worth 10 points and the recommended time for the question is 20–25 minutes. Question 2 is a Translation Between Representations question worth 12 points and the recommended time for the question is 25–30 minutes. Question 3 is an Experimental Design and Analysis question worth 10 points and the recommended time for the question is 25–30 minutes. Question 4 is a Qualitative/Quantitative Translation question worth 8 points and the recommended time for the question is 15–20 minutes.

1 Mark for Review

Figure 1

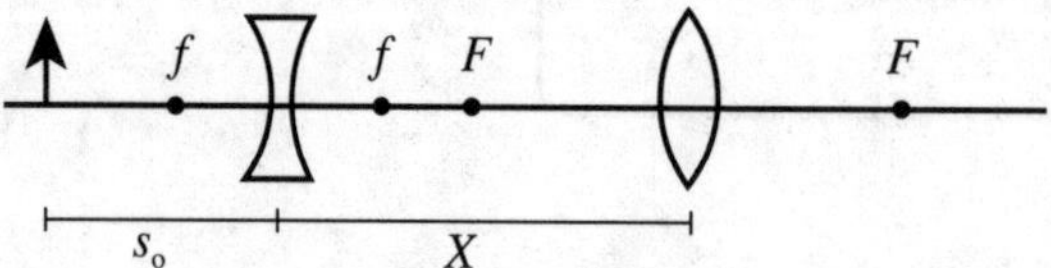

In Figure 1, a concave lens and a convex lens are used together to project an image of an object onto a screen on the right. The focal length of the concave lens is f, the focal length of the convex lens is F, and the object is positioned a distance of s_o to the left of the concave lens. The distance between the lenses is X.

A. Derive an equation for s_i the distance from the concave lens to the image of the object created solely by the concave lens. The equation should be in terms of s_o and f. Is this image real or virtual? Upright or inverted?

B. Add at least two rays to Figure 2 to show how the concave lens creates an image and clearly label s_o and s_i. You do not need to include the convex lens in Figure 2.

Figure 2

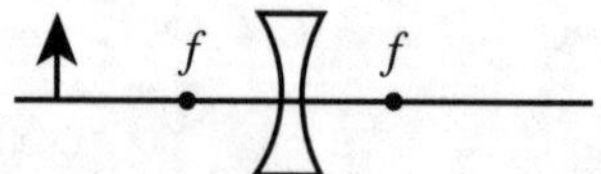

C. The convex lens will use the image created by the concave lens as its object.

i. Derive an equation for S_o, the distance from the convex lens to the image created by the concave lens, in terms of s_o, f, F, and X.

ii. Derive an equation for S_i, the distance from the convex lens to the final image that is to be projected onto the screen, in terms of s_o, f, F, and X.

GO ON TO THE NEXT PAGE.

D. The convex lens is made of a flexible material so that it can be stretched and compressed to change F, the focal length of the convex lens. The screen will always be to the right of the convex lens, but it can be moved to different positions to focus the image onto the screen.

i. Is there a maximum value of F? If so, express it in terms of s_o, f, and X and briefly explain what happens if F becomes too large. If not, briefly explain what happens as F approaches infinity.

ii. Is there a minimum value of F? If so, express it in terms of s_o, f, and X and briefly explain what happens if F becomes too small. If not, briefly explain what happens as F goes to zero.

GO ON TO THE NEXT PAGE.

2 Mark for Review

Figure 1

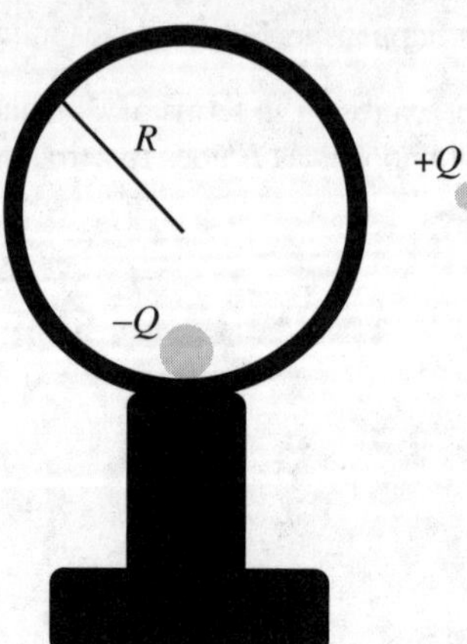

In Figure 1, a negatively charged ball of mass m and charge magnitude Q is placed in a spherical container with inner radius R made of insulating glass; the ball comes to a rest at the bottom of the container. A positive charge of magnitude Q is then positioned next to the container and held in place. This causes the negative charge to move a brief distance before coming to a stop.

A.

i. Below, create two force diagrams. The first showing the forces on the ball immediately after the positive charge was introduced, and the second the forces on the ball after it has moved and come to a stop. Ignore any effects due to polarization of the insulating container throughout this problem.

ii. Briefly explain why the ball moves and why it comes to a stop.

B. When the ball is at the bottom of the container, the distance between the ball and the positive charge is r_1. After the ball has moved and come to a stop, the distance between the ball and the positive charge is r_2.

i. Which is larger, r_1 or r_2?

ii. The positive charge creates an electric potential, V. What is V_i, the electric potential at the ball's initial position?

iii. What is V_f, the electric potential at the ball's final position?

iv. Does the change in potential agree with your description in part A (ii)? Briefly explain why or why not.

GO ON TO THE NEXT PAGE.

C. Assume the energy lost to frictional and drag forces during the ball's motion is negligible.

i. Using Conservation of Energy, derive an equation for the height the ball will rise, h, from the bottom of the container in terms of m, Q, R, r_1, r_2, and physical constants, as appropriate.

Figure 2

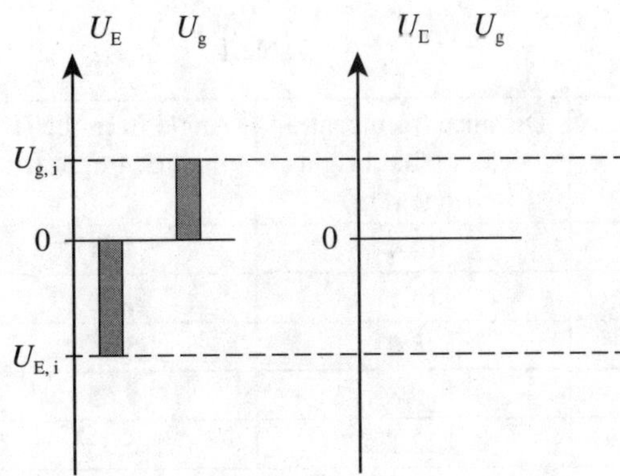

ii. The left bar chart in Figure 2 is complete and represents the initial electric potential energy $U_{E,i}$ and initial gravitational potential energy $U_{g,i}$ of the ball before it starts to move. In the space provided on the right, draw a bar chart to represent a possible final electric potential energy and final gravitational potential energy of the ball after it has moved and come to a stop.

GO ON TO THE NEXT PAGE.

3 Mark for Review

Students perform a double-slit experiment with the goal of precisely measuring the distance between the slits. They set up the wall with a variable monochromatic light source shining through the slit and a projection screen 2 m away from the wall on the opposite side of the light source. They set the wavelength of the light, λ, and measure the distance from the central peak of the diffraction pattern to the first bright fringe, x. They repeat this measurement for several different wavelengths. Additionally, they use the x-value to mathematically derive the angle to the bright fringe, θ. Their data are shown in Table 1 below:

Table 1

Wavelength of the light (nm)	Distance from central peak to first bright fringe (cm)	Angle from the slit to the first bright fringe	
400	14.4	4.118°	
450	16.5	4.745°	
500	18.0	5.143°	
550	19.8	5.654°	
600	22.1	6.334°	

A. Explain one possible way the students derived the angle, θ, from the x-value. Include a diagram if needed.

B. The students want to produce a linear graph of the data so that the distance between the slits, d, can be determined from the line of best fit.

i. Indicate two quantities that could be plotted to produce the desired graph. Use empty columns of the data table in part A to record any values that you need to calculate.

Vertical axis_______________ Horizontal axis_______________

GO ON TO THE NEXT PAGE.

ii. Label the axes in Figure 1 and provide an appropriate scale with units. Plot the data points for the quantities indicated in part B (i) on the axes and draw a best-fit line.

Figure 1

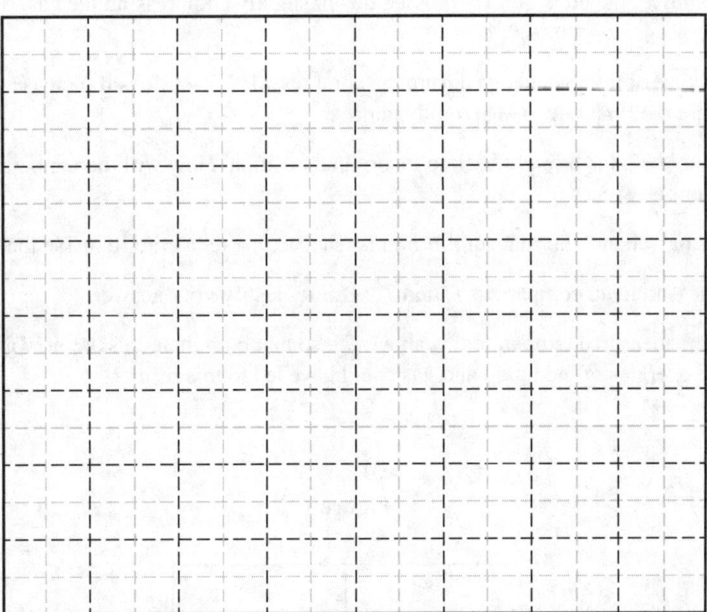

iii. Using your best-fit line, determine the distance between the slits, *d*.

C. A student finds a different wall with one slit. A rough measurement of the slit's width shows that it is 1 cm wide. The student has the idea of setting up an experiment similar to that described in part A to get a more precise measurement of the slit's width, using diffraction instead of interference. Is this feasible? Briefly explain why or why not.

GO ON TO THE NEXT PAGE.

4 Mark for Review

An isolated container is filled halfway with a liquid and the remainder is filled with an ideal gas. The condensation point of the ideal gas is the same as the boiling point of the liquid, T_{bp}. Initially the ideal gas has a temperature of $T_1 > T_{bp}$ and the liquid has a temperature of $T_2 < T_{bp}$.

A. After the container is filled, the lid is quickly pressed down, slightly compressing the gas. Assume the liquid is incompressible.

   i. Will the temperature of the gas change during the compression? If so, does it increase or decrease? Briefly justify your answer using the First Law of Thermodynamics.

   ii. After a short time, the lid is brought back to its original position. How will the temperature of the gas compare to T_1? Briefly justify your answer.

B. After a long time has passed and thermal equilibrium is reached, the temperature inside the container is measured.

   i. How will this measurement compare to T_1 and T_2? Briefly justify your answer.

   ii. It is found that the measured temperature is above T_{bp}, so that both materials are now gasses. What about the respective initial energies of the liquid and gas could have led to this result?

C.

Figure 1

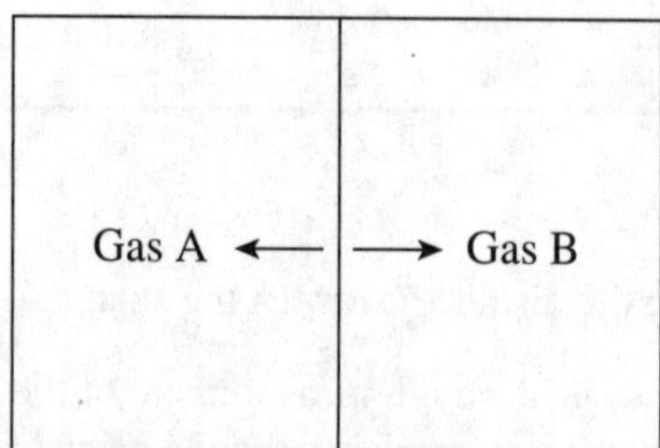

The container is emptied and a partition that is free to move left and right is placed in the middle. Ideal gas *A* is put into the left side of the container and ideal gas B into the right side, as shown in Figure 1. The partition does not allow the gasses to mix, but it does allow heat to flow between the two gasses. Gas A has twice the number of moles and one-fourth the temperature of gas B. If the partition is initially placed in the center of the container, will it feel a net force? If so, which way? Briefly justify your answer.

D. After a long time, thermal equilibrium is reached and the partition is found to be at rest.

   i. What fraction of the container does gas A now occupy?

   ii. Justify your answer.

   iii. Does your answer to the previous question contradict your answer in part C. Briefly explain why or why not.

STOP

END OF EXAM

Practice Test 1: Diagnostic Answer Key and Explanations

PRACTICE TEST 1: DIAGNOSTIC ANSWER KEY

Let's take a look at how you did on Practice Test 1. Follow the three-step process in the diagnostic answer key below and go read the explanations for any questions you got wrong or you struggled with but got correct. Once you finish working through the answer key and the explanations, go to the next chapter to make your study plan.

Check your answers and mark any correct answers with a ✔ in the appropriate column.

Section I: Multiple Choice							
Q #	Ans.	✔	Chapter #, Section Title	Q #	Ans.	✔	Chapter #, Section Title
1	C		**10,** Photons and the Photoelectric Effect	21	C		**7,** Faraday's Law of Electromagnetic Induction
2	C		**3,** The Kinetic Theory of Gases	22	D		**7,** Faraday's Law of Electromagnetic Induction
3	B		**8,** Reflection and Refraction	23	B		**8,** Using Equations to Answer Questions About the Image
4	C		**8,** Reflection and Refraction	24	A		**5,** Electric Potential
5	C		**8,** Ray Tracing for Lenses	25	A		**8,** Thin Lenses
6	A		**9,** Standing Waves	26	D		**7,** The Magnetic Force on a Moving Charge
7	B		**3,** The Laws of Thermodynamics	27	C		**6,** Analysis of Circuits with Resistance
8	A		**6,** Capacitors in Circuits	28	C		**5,** Capacitors and Capacitance
9	B		**6,** Capacitors in Circuits	29	D		**4,** Conductors and Insulators
10	B		**6,** Analysis of Circuits with Resistance	30	D		**3,** The Laws of Thermodynamics
11	C		**10,** The Bohr Model of the Atom	31	B		**6,** Analysis of Circuits with Resistance
12	D		**7,** Magnetic Fields Created by Current-Carrying Wires	32	B		**3,** The Laws of Thermodynamics
13	A		**4,** Coulomb's Law	33	D		**9,** Transverse Traveling Waves
14	C		**4,** The Electric Field	34	B		**4,** The Electric Field
15	B		**9,** Interference and Diffraction	35	C		**7,** Faraday's Law of Electromagnetic Induction
16	B		**9,** The Doppler Effect	36	B		**6,** Capacitors in Circuits
17	B		**10,** Nuclear Reactions	37	B		**10,** The Wave Function
18	A		**3,** The Laws of Thermodynamics	38	A		**3,** The Laws of Thermodynamics
19	C		**9,** Waves, Sound, and Physical Optics	39	C		**4,** Coulomb's Law
20	B		**7,** Faraday's Law of Electromagnetic Induction	40	D		**10,** Blackbody Radiation

Section II: Free Response			
Q #	Ans.	✔	Chapter #, Section Title
1A	See Explanation*		**8,** Using Equations to Answer Questions about the Image
1B	See Explanation*		**8,** Ray Tracing for Lenses
1C	See Explanation*		**8,** Using Equations to Answer Questions about the Image
1D	See Explanation*		**8,** Thin Lenses
2A	See Explanation*		**4,** Coulomb's Law
2B	See Explanation*		**4,** Coulomb's Law **5,** Electric Potential
2C	See Explanation*		**5,** Electric Potential Energy
3A	See Explanation*		**9,** Interference and Diffraction
3B	See Explanation*		**9,** Interference and Diffraction
3C	See Explanation*		**9,** Interference and Diffraction
4A	See Explanation*		**3,** The Laws of Thermodynamics
4B	See Explanation*		**3,** The Laws of Thermodynamics
4C	See Explanation*		**3,** The Laws of Thermodynamics
4D	See Explanation*		**3,** The Laws of Thermodynamics

Explanations begin on page 43.

Once you have checked your answers, remember to return to page 4 and respond to the Reflect questions.

Tally your correct answers from Step 1 by chapter. For each chapter, write the number of correct answers in the appropriate box. Then, divide your correct answers by the number of total questions (which we've provided) to get your percent correct.

CHAPTER 3 TEST SCORE SELF-EVALUATION

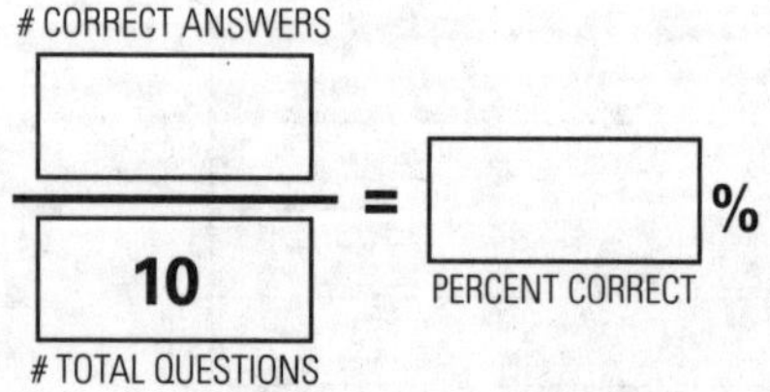

CHAPTER 4 TEST SCORE SELF-EVALUATION

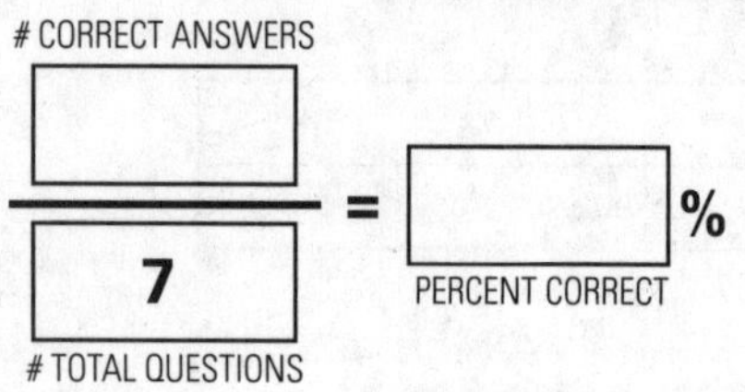

CHAPTER 5 TEST SCORE SELF-EVALUATION

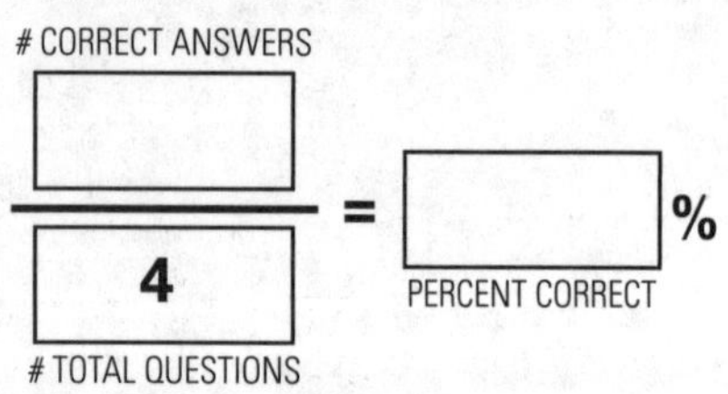

CHAPTER 6 TEST SCORE SELF-EVALUATION

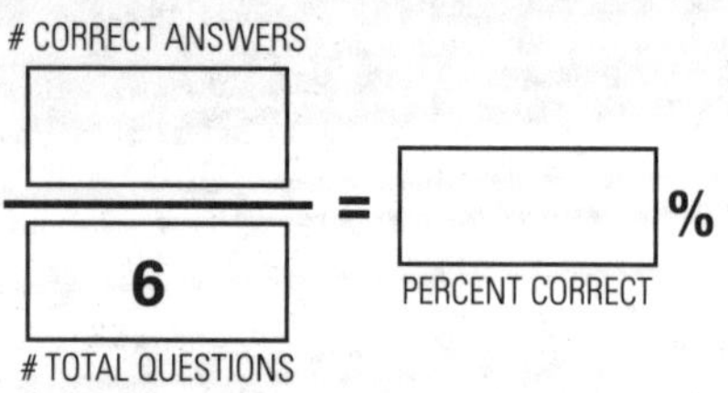

CHAPTER 7 TEST SCORE SELF-EVALUATION

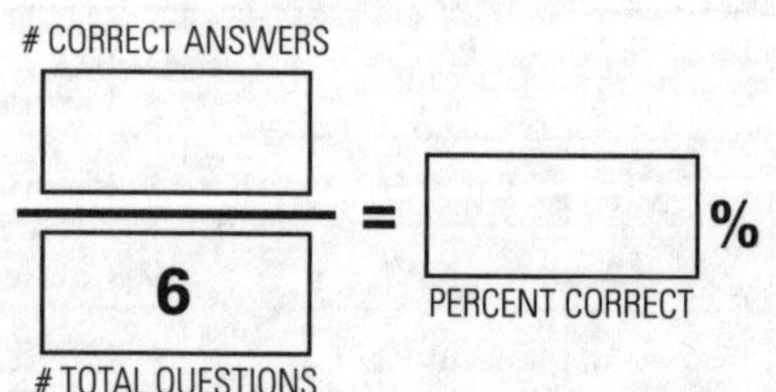

CHAPTER 8 TEST SCORE SELF-EVALUATION

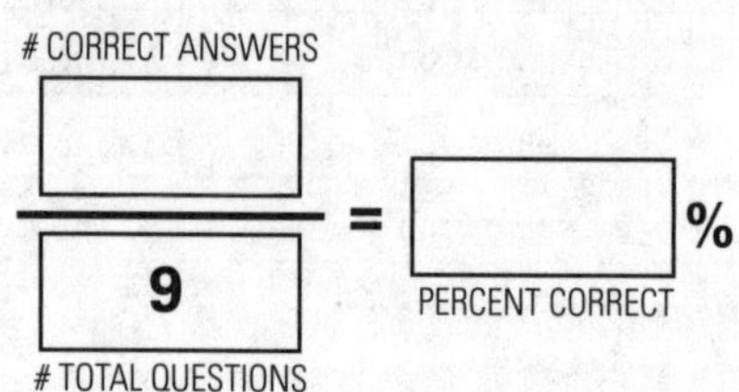

CHAPTER 9 TEST SCORE SELF-EVALUATION

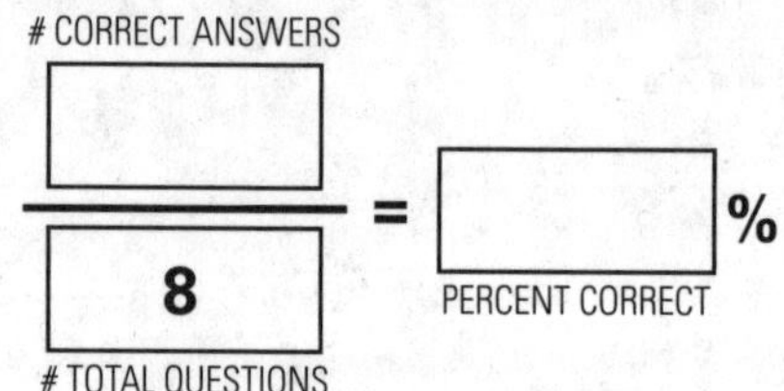

CHAPTER 10 TEST SCORE SELF-EVALUATION

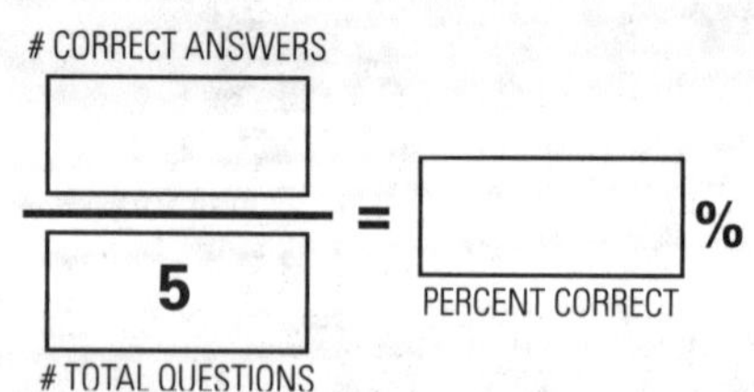

Use the results above to customize your study plan. You may want to start with, or give more attention to, the chapters with the lowest percents correct.

PRACTICE TEST 1: ANSWERS AND EXPLANATIONS

Section I: Multiple Choice

1. **C** To use $K_{max} = hf - \varphi$, find the frequency of the photons by using $\lambda = \frac{v}{f}$. Thus, $f = \frac{3 \times 10^{8}\ \text{m/s}}{550 \times 10^{-9}\ \text{m}} = 5.45 \times 10^{14}\ \text{Hz}$, plugging into the photoelectric effect equation gives you $K_{max} = (4.14 \times 10^{-15}\ \text{eV} \cdot \text{s})(5.45 \times 10^{14}\ \text{Hz}) - 3\ \text{eV} = -0.74\ \text{eV}$. This is a negative value, which means that the energy of the photon is less than the energy of the work function. Therefore, no electrons would be emitted.

2. **C** Because the temperatures are the same, the average kinetic energy of the molecules are the same. Since $K = \frac{1}{2}mv^2$, it must be that $v^2 \propto \frac{1}{m}$ or $v \propto \frac{1}{\sqrt{m}}$. Thus, if m is less by a factor of 4, v increases by a factor of 2.

3. **B**

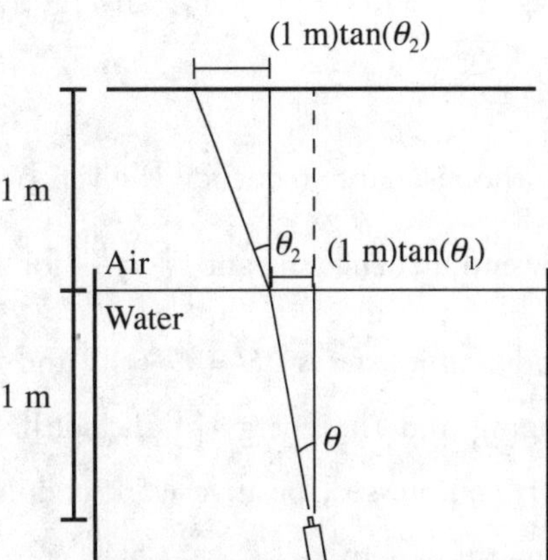

As seen in the above diagram, you can use trigonometry to see that the distance from the center will be $(1\ \text{m})\tan(\theta_1) + (1\ \text{m})\tan(\theta_2)$. θ_1 is the angle given in the problem, 20°, and to find θ_2, apply Snell's Law $n_1 \sin(\theta_1) = n_2 \sin(\theta_2) \rightarrow \theta_2 = \sin^{-1}(\frac{n_1 \sin(\theta_1)}{n_2}) = \sin^{-1}(\frac{1.3\sin(20°)}{1}) = 26.4°$. Plug this in to get $d = (1\ \text{m})\tan(20°) + (1\ \text{m})\tan(26.4°) = 0.86\ \text{m}$.

4. **C**

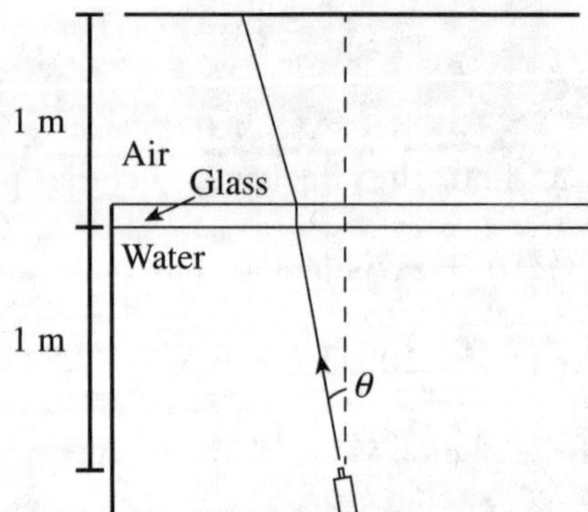

As seen in the image above, adding a glass will introduce a portion where the light travels closer than 20° to the vertical, since the index of glass is higher than that of the index of water and the angle from the vertical will decrease as it transitions from water to glass. This means that the light will not travel as far horizontally while traveling in the glass. After exiting the glass, it will travel at the same angle as when the glass wasn't there because an intermediate medium does not affect the final angle, but it now has a shorter vertical distance to travel. Therefore, the light will not go as far horizontally and will end up closer to the center.

5. **C** A concave lens is diverging. The light will therefore diverge away from the center line after passing through.

6. **A** For an open-ended pipe, the resonant frequency is given by $f_n = n\frac{v}{4L}$. When plugging in the given n and L values, they end up canceling and $f = \frac{v}{n}$ for all three cases.

7. **B** The First Law of Thermodynamics states $\Delta U = Q + W$, and in the case of an isothermal process, temperature is not changing and the energy of the ideal gas would not change. This means $Q = -W$. When the gas is compressed, positive work is done on the system, which means Q is negative and heat is leaving the system.

8. **A** When a capacitor is uncharged, it acts as a wire and current is allowed to flow freely. Therefore, at the start, the current will be at a maximum. As the capacitor is charged, it gains voltage and resists the current more and more until it reaches equilibrium. At equilibrium, charge no longer moves and the current drops to zero.

9. **B** With the switch closed, after a long time, there will be no current in the part of the circuit with the capacitor, and therefore all 12 volts must be across the capacitor. Once the switch is opened, the capacitor will initially act as a 12-volt voltage source for the three resistors, and since no current flows across the open switch, you can ignore that part of the circuit so the three resistors act like they are in series. This means the equivalent resistance is given by 15 Ω + 2 Ω + 10 Ω = 27 Ω and the current is given by I = (12 V)/(27 Ω) = 0.44 A.

10. **B** Because it is in parallel with the battery, the part of the circuit with the 10 Ω and 2 Ω resistors must have a total of 12 V across it. Combining them in series, there will be a total of 12 V across an equivalent resistance of 12 Ω. Therefore, the current through that part of the circuit is I = (12 V)/(12 Ω) = 1 A. The voltage must therefore be V = (1 A)(2 Ω) = 2 V.

11. **C** The change of energy here is –20 eV – (–16 eV) = –4 eV. Since the electron lost 4 eV of energy, that is the energy of the emitted photon. By $E = hf$, $f = \frac{E}{h} = \frac{4 \text{ eV}}{4.14 \times 10^{-15} \text{ eV} \cdot \text{s}} = 9.66 \times 10^{14} \text{ Hz}$. Thus, $\lambda = \frac{v}{f} = \frac{3 \times 10^{8} \text{ m/s}}{9.66 \times 10^{14} \text{ Hz}} = 3.11 \times 10^{-7} \text{ m} = 311 \text{ nm}$.

12. **D** Apply the Right-Hand Rule for a current-generated magnetic field. The thumb of the right hand points upward and the fingers curl so that on the left side they are coming out of the page.

13. **A** By Coulomb's Law, $F_E = k\frac{q_1 q_2}{r^2}$. Thus, doubling q_1 would double the magnitude of the force. Halving r or doubling both charges would increase the force by a factor of 4, and flipping the sign of a charge would change the direction of the force without affecting its magnitude.

14. **C** The electric field points away from positive charge. For the field to be zero, the directions of the field from each of the two charges must be opposite. This will only happen when in between the charges. Additionally, to balance the magnitudes of the fields, the point must be farther away from the more powerful charge and closer to the weaker charge.

15. **B** The fringes in a double-slit experiment are given by $\sin\theta = \pm m\frac{\lambda}{d}$, where d is the distance between the slits. Increasing d will therefore decrease the angle and the fringes will get closer together.

16. **B** The Doppler effect changes the frequency based on the relative speeds of the source and the detector. Because the car is driving at a constant speed, the frequency heard will not be changing, though it would be higher than how it would sound if everything were stationary.

17. **B** Mass can be converted to energy through $E = mc^2$, so a decrease in mass matches an increase in energy. While momentum would be conserved here, it is not a good explanation of why there's a gain of energy, because everything staying stationary would also conserve momentum.

18. **A** The work done on the gas during an isobaric process is given by $W = -P\Delta V = -(2 \times 10^5 \text{ Pa})(2 \text{ m}^3 - 0.5 \text{ m}^3) = -3 \times 10^5 \text{ J}$, and no work is done during an isochoric process. The work done by the gas would then be 3×10^5 J.

19. **C** The intensity of a wave from a point source is given by $I = \frac{P}{4\pi r^2}$, where r is the distance from the source. As a wave travels, its intensity therefore drops as the distance is squared.

20. **B** The magnitude of the emf is the change of the magnetic flux over the change in time. In this case, this is given by $\varepsilon = BLv = (3\text{ T})(0.3\text{ m})(5\,\text{m/s}) = 4.5\text{ V}$.

21. **C** The emf will act to counter the change in magnetic flux. This means the induced current will flow to create a magnetic field pointing out of the page in the middle of the loop. To accomplish this, the current must flow counter-clockwise by the Right-Hand Rule. Thus, the current in the right wire would flow up. Applying the Right-Hand Rule again to find force on the upward-flowing current reveals that the force would be going to the left.

22. **D** The magnitude of emf is the change in magnetic flux over the change in time. As the loop is spinning, the area the magnetic field is passing through would be changing. Eliminate (A) and (B). Choice (C) is eliminated because it does not matter what the flux currently is, just how it is changing. Indeed, the change in flux would be zero at the point the loop is parallel to the page, as that is the point where the magnitude of flux changes from increasing to decreasing.

23. **B** Because convex lenses are converging, $f = +20$ cm. From the lens equation, $\frac{1}{i} = \frac{1}{f} - \frac{1}{o} = \frac{1}{20\text{ cm}} - \frac{1}{15\text{ cm}} = -\frac{1}{60\text{ cm}} \rightarrow i = -60\text{ cm}$. The negative sign indicates that the image is on the same side as the object and is virtual.

24. **A** The potential is more positive the closer the point is to a positive charge and more negative the closer the point is to a negative charge. Therefore, moving from the point closest to the positive charge to the point closest to the negative charge would give the biggest change of potential.

25. **A** The focal length for a concave lens represents where parallel light diverges away from. Having a lower index means the index more closely matches that of air ($n = 1$) and the light would bend less. This means the point it is diverging from is farther from the lens. If the focal length increases, f would become more negative. From the lens equation $\frac{1}{i} = \frac{1}{f} - \frac{1}{o}$, if f is more negative, $1/f$ would be less negative, and therefore the magnitude of $1/i$ would be smaller and i would be larger.

26. **D** The magnetic force never does work and therefore the speed of the charge would not be affected. Eliminate (A) and (C). Additionally, the magnetic force will cause the charge to turn so it cannot go in a straight line through point *b*. Eliminate (B), and the correct answer must be (D).

27. **C** Objects in parallel always have the same voltage regardless of their resistance.

28. **C** $Q = CV \rightarrow C = \frac{Q}{V}$, so use any point on the line to get the charge and matching voltage. Using the point (1 V, 5 mC), for example, gives $C = \frac{5\text{ mC}}{1\text{ V}} = 5\text{ mF}$.

29. **D** If the initial charge given to the first sphere is Q_{tot}, then after it is brought in contact with the second sphere, it will be evenly distributed, so the charge on the first sphere will be $1/2Q_{tot}$ and the charge on the second sphere is also $1/2Q_{tot}$. After the second sphere is brought into contact with the third sphere, it will once again evenly distribute the charge of $1/2Q_{tot}$. This means at the end $Q_1 = 1/2Q_{tot}$, and $Q_2 = Q_3 = 1/4Q_{tot}$.

30. **D** For an ideal gas, internal energy is related to temperature. Completing the loop and coming back to the same pressure and volume means that it is back at the same temperature and internal energy would not change.

31. **B** Start by combining the two resistors in parallel: $\frac{1}{R_{eq}} = \frac{1}{R_1} + \frac{1}{R_2} = \frac{1}{3\ \Omega} + \frac{1}{6\ \Omega} = \frac{1}{2\ \Omega} \rightarrow R_{eq} = 2\ \Omega$. Then combine the result with the other two resistors in series: $R_{eq} = 1\ \Omega + 2\ \Omega + 2\ \Omega = 5\ \Omega$. Use this to find the current through the circuit: $I = \frac{V}{R} = \frac{10\text{ V}}{5\ \Omega} = 2\text{ A}$. This will be the current through the three series resistors, including the 2 Ω equivalent resistor from the two parallel resistors. This means the voltage across the two parallel resistors is $V = IR = (2\text{ A})(2\ \Omega) = 4\text{ V}$. Thus, the current through the 3 Ω resistor is $I = \frac{V}{R} = \frac{4\text{ V}}{3\ \Omega} = 1.3\text{ A}$.

32. **B** During an isothermal process, $\Delta U = 0$ and the first law becomes $Q = -W$. To find the work done, examine the change of potential energy $\Delta U_g = mg\Delta h = (2\text{ kg})(9.8\text{ m/s}^2)(0.2\text{ m}) \approx 4\text{ J}$. This is the work done by the gas, so the work done on the gas is –4 J and the heat is $Q = -(-4\text{ J}) = 4\text{ J}$.

33. **D** From the graph, you can see $\lambda = 2$ m. Because $f = \frac{1}{T}$, you get $v = \lambda f = \frac{\lambda}{T} = \frac{2\text{ m}}{5 \times 10^{-3}\text{ s}} = 400$ m/s.

34. **B** Find the magnitude of each electric field with $|E| = \frac{|q|}{4\pi\varepsilon_0 r^2} = k\frac{|q|}{r^2}$.

$$E_1 = (9 \times 10^9\ \text{N} \cdot \text{m}^2/\text{C}^2)\frac{|5 \times 10^{-6}\ \text{C}|}{(1.5\ \text{m})^2} = 2 \times 10^4\ \text{N/C}$$

$$E_2 = (9 \times 10^9\ \text{N} \cdot \text{m}^2/\text{C}^2)\frac{|-2 \times 10^{-6}\ \text{C}|}{(1.5\ \text{m})^2} = 8 \times 10^3\ \text{N/C}$$

The electric field from a positive charge points away from the positive charge, and the electric field from a negative charge points toward the negative charge. When in between the two charges, these would be the same direction, so add all the magnitudes together: $E_{tot} = E_1 + E_2 = 2 \times 10^4 + 8 \times 10^3 = 28$ kN/C.

35. **C** The magnitude of the emf and induced current is the change of magnetic flux over the change in time. If the magnet moves away at the same speed that it was approaching, the rate of change would be the same and the magnitude of the current would be the same. However, the emf acts to counter the changing flux, so changing from an increase in the flux (moving toward) to a decrease in the flux (moving away) would flip the direction of the induced current.

36. **B** To combine capacitors in series use $\frac{1}{C_{eq}} = \frac{1}{C_1} + \frac{1}{C_2} = \frac{1}{6\ \mu\text{F}} + \frac{1}{15\ \mu\text{F}} = \frac{7}{30\ \mu\text{F}} \rightarrow C_{eq} = 4.29\ \mu\text{F}$. The charge on the equivalent capacitor is given by $Q = CV = (4.29\ \mu\text{F})(12\ \text{V}) = 51.5\ \mu\text{C}$.

Because these capacitors are in series, they would both have this charge.

37. **B** The probability of finding a particle at a position is based on the absolute value of the wave function. The smallest possible probability is therefore where the wave function is zero.

38. **A** The Second Law of Thermodynamics states that the entropy of an isolated system cannot decrease. This drop in temperature would be a drop in entropy, but the second law does not apply to open systems. Choice (C) can be eliminated since it implies an overall decrease of entropy, which does violate the second law.

39. **C** By Newton's Third Law, the two charges would feel equal and opposite forces, and by Newton's Second Law, $F_{net} = ma$. The smaller mass must therefore have greater acceleration.

40. **D** For black-body radiation, $E \propto T^4$, and because intensity is directly related to energy, increasing temperature by a factor of 2 would increase intensity by a factor of $2^4 = 16$.

Section II: Free Response Questions

1. A. From the lens equation, $\frac{1}{i} = \frac{1}{f} - \frac{1}{o}$, so $\frac{1}{s_i} = \frac{1}{f} - \frac{1}{s_o} \rightarrow s_i = \frac{1}{\frac{1}{f} - \frac{1}{s_o}} = \frac{fs_o}{s_o - f}$.

   Because f for a concave lens is negative, s_i will be negative, which indicates that it is a virtual, upright image.

   B.

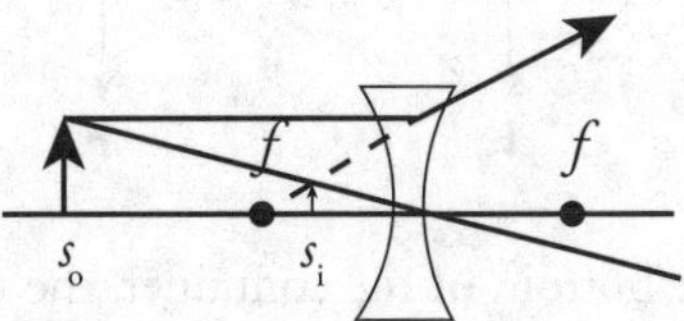

   C.

   i. Because the image from the concave lens is on the left side, the distance from the convex lens to the image created by the concave lens will be $S_o = X + s_i = X + \frac{fs_o}{s_o - f}$.

   ii. From the lens equation $\frac{1}{i} = \frac{1}{f} - \frac{1}{o}$. Applying this to the convex lens gives us

$$\frac{1}{S_i} = \frac{1}{F} - \frac{1}{S_o} \rightarrow S_i = \frac{FS_o}{S_o - F}$$

   Plug in the equation from part C (i) for S_o

$$S_i = \frac{F\left(X + \frac{fs_o}{s_o - f}\right)}{\left(X + \frac{fs_o}{s_o - f}\right) - F} = F\frac{Xs_o - Xf + fs_o}{Xs_o - Xf - Fs_o + Ff + fs_o}$$

   D. i. Yes, there is a maximum value of F. It is $F_{max} = X + s_i = X + \frac{fs}{s_o - f}$. If F is larger than this value, then the effective object for the convex lens, s_i will be closer to the convex lens than the focal point. When an object is closer than the focal length for a convex lens, it creates a virtual image so the image cannot be projected onto the screen on the right.

   ii. No, there is no minimum value of F. As F approaches zero, the real image will form closer and closer to the convex lens, but it will always be to the right of the convex lens.

2. A. i.

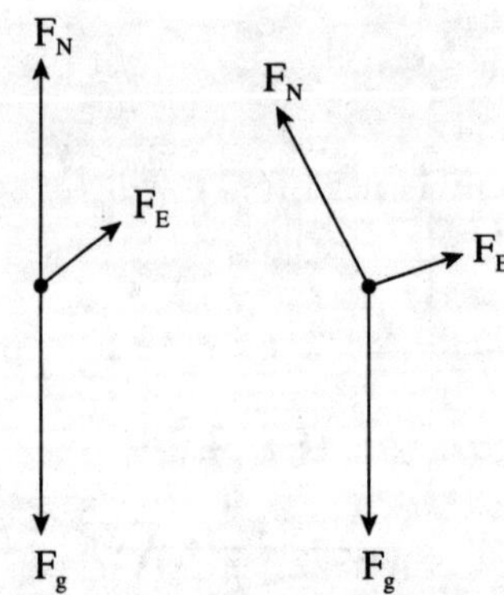

ii. When the ball is at the bottom of the container, the vertical forces cancel, but the horizontal component of the electrostatic force will give the ball an acceleration. The ball will roll up the side of the container, which causes the normal force to gain a horizontal component opposing the electrostatic force. It will eventually reach an equilibrium position where both the vertical and horizontal components cancel out.

B. i. r_1 is larger.

ii. $V = k\frac{|q|}{r} \rightarrow V_i = k\frac{Q}{r_1}$

iii. $V = k\frac{|q|}{r} \rightarrow V_i = k\frac{Q}{r_2}$

iv. This does agree with the description. Because r_1 is larger than r_2, the potential change will be positive. The ball is negatively charged, and it will feel a force toward higher potential. It will then accelerate in that direction until another force cancels the draw toward higher potential, in this case, the normal force.

C. i. Because frictional forces are negligible, $\Delta E = 0 \rightarrow \Delta K + \Delta U_g + \Delta U_E = 0$. The ball starts from rest and ends at rest $\Delta K = 0$ and so

$$\Delta U_g = -\Delta U_E \rightarrow mgh = -(-Q\Delta V) = kQ^2\left(\frac{1}{r_2} - \frac{1}{r_1}\right) \rightarrow h = \frac{kQ^2}{mg}\left(\frac{1}{r_2} - \frac{1}{r_1}\right)$$

ii.

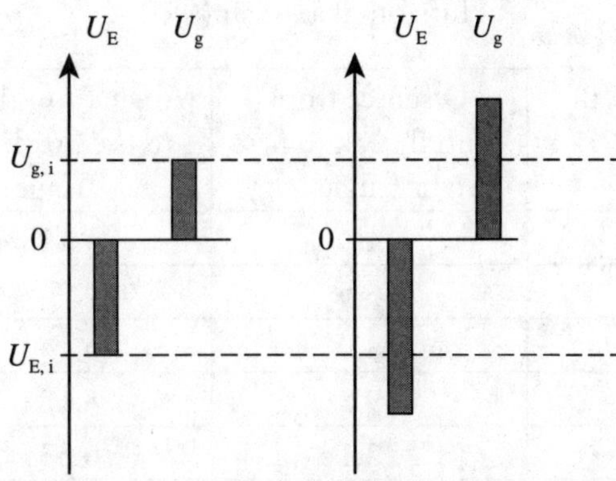

The gravity potential energy increases as the ball moves up the side of the container, and the electric potential decreases as the negative charge moves to a higher electric potential. The two changes in potential energy must cancel out since no energy is lost to friction.

3. A.

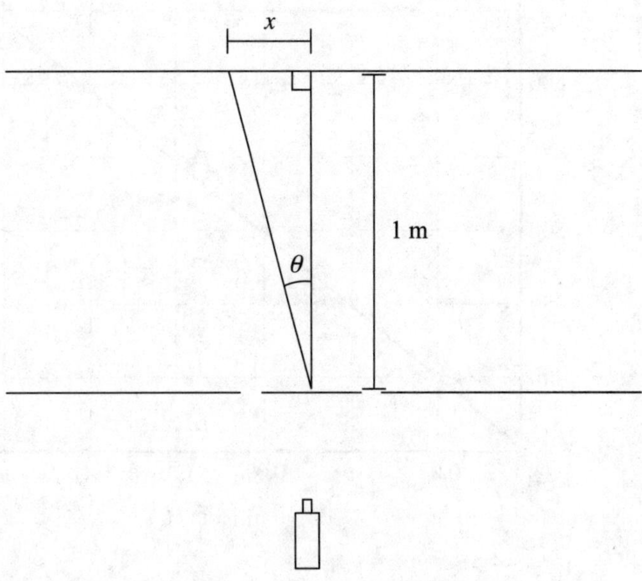

As seen in the image above, the angle can be found by using $\theta = \tan^{-1}(x/1\text{ m})$.

B.

i. Vertical axis: λ Horizontal axis: $\sin(\theta)$

Wavelength of the light (nm)	Distance from central peak to first bright fringe (cm)	Angle from the slit to the first bright fringe	$\sin(\theta)$
400	14.4	4.118°	0.0718
450	16.5	4.745°	0.0827
500	18.0	5.143°	0.0896
550	19.8	5.654°	0.0985
600	22.1	6.334°	0.110

ii.

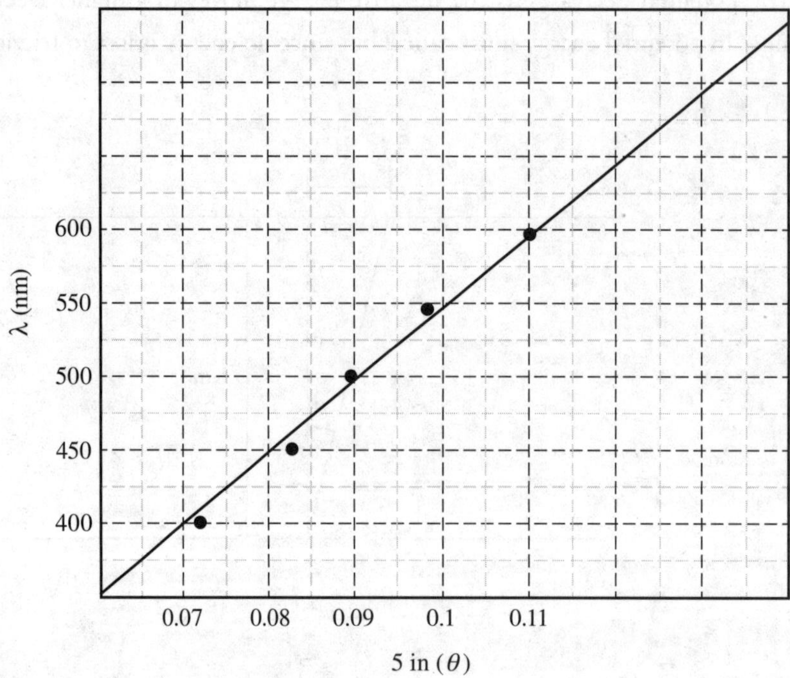

iii. Using the start and end points of the line in the graph above, Slope $= \frac{\text{Rise}}{\text{Run}} =$

$$\frac{650 \times 10^{-9}\text{ m} - 375 \times 10^{-9}\text{ m}}{0.122 - 0.065} = 4.83 \times 10^{-6}\text{ m}$$

C. No, this is not feasible. Diffraction interference only occurs when the size of the hole is roughly the same size as the wavelength of the light. The longest visible wavelength is about 700 nm, which is much smaller than 1 cm.

4. A. i. The temperature will increase. This is because the First Law of Thermodynamics states $\Delta E = Q + W$; since the system is isolated, heat won't flow to the environment. Additionally, since the compression happens quickly, there won't be time for heat to flow between the gas and liquid. This means $Q = 0$ and $\Delta E = W$. When the lid is compressed, positive work is done on the system, the energy increases, and the temperature of the ideal gas increases.

ii. The new temperature will be lower than T_1. During the time it's compressed, heat will flow from the ideal gas to the liquid because the liquid is at a lower temperature. Therefore, when the lid comes back to its original position, there will be less energy in the gas and its temperature will be lower.

B. i. The temperature will be between T_1 and T_2. Heat will flow from the higher-temperature gas to the lower-temperature liquid until they reach the same temperature. This means the gas decreases in temperature and the liquid increases in temperature, making the final temperature between the two initial temperatures.

ii. For the liquid to become a gas, it needs to gain enough energy to raise its temperature to the boiling point and then undergo a phase change. This must happen before the gas lowers its temperature to the boiling point. Therefore, the amount of energy required to drop the gas's temperature to T_{bp} must have been more than the amount of energy required to change the liquid to a gas.

C. The partition will feel a force to the left. If the partition is in the center, then the two volumes of the gases are the same. Therefore, the Ideal Gas Law $PV = nRT$ becomes the proportion $P \propto nT$. Gas A has twice the moles, but it only has one-fourth the temperature. Thus, the pressure of gas A is half of that of gas B and the imbalance in pressure causes a net force to the left.

D. i. Gas A occupies 2/3 of the container.

ii. If the system is in thermal equilibrium, the temperatures of the gases are the same. Additionally, since the partition is a rest, the pressures must be the same. Therefore, the Ideal Gas Law $PV = nRT$ becomes the proportion $V \propto n$. Gas A has twice the number of moles, so gas A has twice the volume of gas B. Gas A occupies 2/3 and gas B occupies 1/3 of the container.

iii. No, it does not contradict the previous answer. While the partition will initially feel a force to the left and move in that direction, heat will flow from the higher-temperature gas B to the lower-temperature gas A. As it does so, gas A will momentarily gain pressure and push the partition to the left to equalize the pressures. This continues to happen until equilibrium is reached and the partition has overall moved to the right.

PRACTICE TEST 1: FREE-RESPONSE QUESTIONS GRADING RUBRICS

Use the following grading rubrics to get a sense of what you should include in your answers to the four Free Response Questions. These align with what the College Board AP Exam graders will be looking for.

Question 1: Mathematical Routines **10 points total**

A. For a derivation that results in the correct value for s_i. This is an example: 1 point

$$\frac{1}{s_i} = \frac{1}{f} - \frac{1}{s_o} \rightarrow s_i = \frac{1}{\frac{1}{f} - \frac{1}{s_o}} = \frac{fs_o}{s_o - f}$$

For identifying the image as virtual and upright 1 point

B. For a ray diagram with the object further than the focal point and the image closer than the focal point 1 point

For a ray diagram showing a virtual, upright image that's smaller than the object. This is an example: 1 point

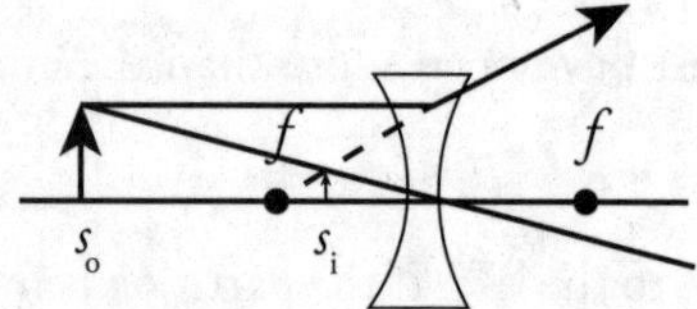

C. i. For identifying the object distance as $S_o = X + s_i$ and correctly plugging in the solution from part A. This is an example: 1 point

$$S_o = X + s_i = X + \frac{fs_o}{s_o - f}$$

ii. For correctly applying the lens equation to the solution from part C (i) and solving for S_i. This is an example: 2 points

$$\frac{1}{S_i} = \frac{1}{F} - \frac{1}{S_o} \rightarrow S_i = \frac{FS_o}{S_o - F} = \frac{F\left(X + \frac{fs_o}{s_o - f}\right)}{\left(X + \frac{fs_o}{s_o - f}\right) - F} = F\frac{Xs_o - Xf + fs_o}{Xs_o - Xf - Fs_o + Ff + fs_o}$$

D. i. For identifying the maximum as the point where the focal length of the convex lens passes the image created by the concave lens and correctly deriving the expression for this maximum using the solution from part A. This is an example: 1 point

$$F_{max} = X + s_i = X + \frac{fs}{s_o - f}$$

For an explanation including mention of the image becoming virtual after F passes the maximum. 1 point

ii. For identifying there is no minimum and an explanation including mention of the image getting closer and closer to the right side of the convex lens as F goes to zero 1 point

Question 2: Translation Between Representations **12 points total**

A. i. For a pair of force diagrams, each on including a normal force, force of gravity, and electrostatic force. The initial diagram should have the normal and gravity in the *y*-directions and the electrostatic force pointing towards the positive charge's position. The second diagram should now have the normal force at an angle to the left, a larger electrostatic force and the forces should roughly cancel. This is an example: 1 point

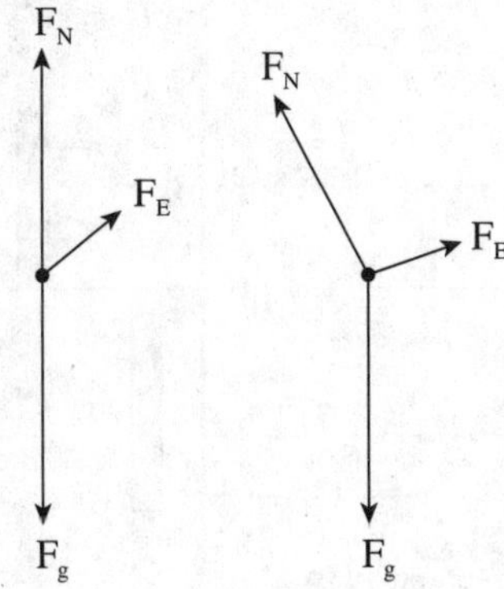

ii. For an explanation including a discussion of the electrostatic force and acceleration 1 point

For an explanation of how the forces don't cancel at the bottom and do cancel after rolling up the side of the container. 1 point

NOTE: While a discussion of friction is not necessary for this portion, points should not be taken off if static friction is included in the second diagram or kinetic friction is discussed during the explanation.

B. i. For stating that r_1 is larger 1 point

ii. $V_i = k\frac{Q}{r_1}$ 1 point

iii. $V_f = k\frac{Q}{r_2}$ 1 point

iv. For identifying whether their answers agree or not 1 point

For an explanation including a discussion of whether the potential is increasing or decreasing and that the negatively charged ball will feel a force towards a higher electric potential. 1 point

C. i. For a derivation starting from the idea that energy does not change 1 point

For a derivation that ends with the correct value for h. This is an example: 1 point

$$\Delta U_g = -\Delta U_E \rightarrow mgh = -(-Q\Delta V) = kQ^2\left(\frac{1}{r_2} - \frac{1}{r_1}\right) \rightarrow h = \frac{kQ^2}{mg}\left(\frac{1}{r_2} - \frac{1}{r_1}\right)$$

ii. For a graph including an increase in gravitational potential energy and a decrease in electric potential energy 1 point

For a graph where the total energy stays constant. This is an example: 1 point

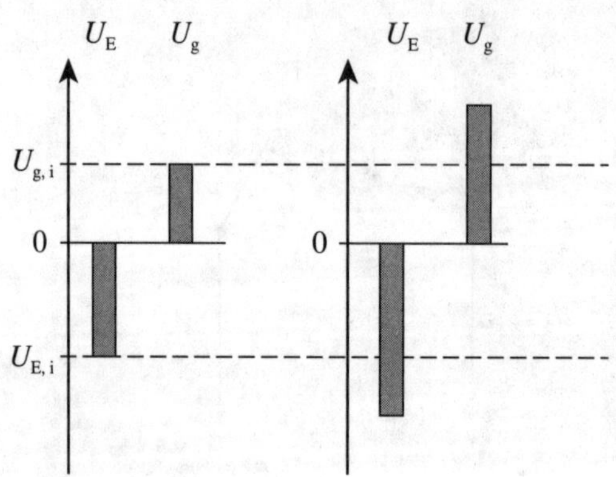

Question 3: Experimental Design and Analysis **10 points total**

A. For a correct description of a method to find θ 1 point

B. i. For indicating quantities that would produce a linear graph that could be used to determine d. 1 point

For correctly calculating the indicated quantities and adding them to the table, including units. This is an example: 2 points

Vertical axis: λ Horizontal axis: $\sin(\theta)$

Wavelength of the light (nm)	Distance from central peak to first bright fringe (cm)	Angle from the slit to the first bright fringe	$\sin(\theta)$
400	14.4	4.118°	0.0718
450	16.5	4.745°	0.0827
500	18.0	5.143°	0.0896
550	19.8	5.654°	0.0985
600	22.1	6.334°	0.110

ii. For a correctly scaled and labeled graph, including units, with the points from the table. This is an example: 2 points

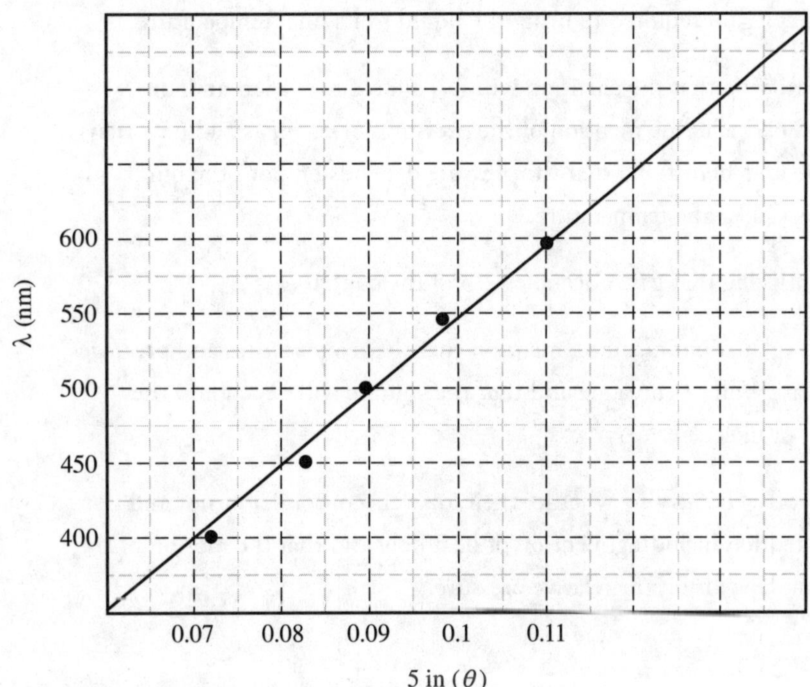

iii. For using a correct method to estimate the slope from their graph 1 point

For a result of roughly 5×10^{-6} m 1 point

C. For indicating that the method will not work 1 point

For an explanation including that the hole is too large for diffraction/interference to occur. 1 point

Question 4: Qualitative/Quantitative Translation **8 points total**

A. i. For identifying the temperature will increase and an explanation including the First Law of Thermodynamics, $\Delta E = Q + W$ and that there is no heat flow in or out of the gas. 1 point

ii. For identifying the final temperature will be less than T_1 and an explanation including mention of heat flowing between the gas and the liquid due to a difference in temperature. 1 point

B. i. For identifying the final temperature as between T_1 and T_2 and an explanation including mention of heat flowing until thermal equilibrium is reached 1 point

ii. For an explanation including mention of heat/energy needed to change the temperature of the substances and mention of heat/energy needed to cause the liquid to change phases. As well as identifying that the heat/energy loss required to condense the gas is higher than the heat/energy gain required to bring the liquid to its final temperature. 1 point

C. For identifying that the partition will feel a force to the left and an explanation including mention of the pressures of each gas and how they compare and identifying that the pressure depends on both the number of moles as well as the temperature. 1 point

D. i. For identifying that gas A occupies 2/3 of the container 1 point

ii. For an explanation including mention that the temperatures and pressures of the gases will be the same and that the volume will depend on the number of moles. 1 point

iii. For correctly identifying whether their answers contradict or not and an explanation including mention of heat flow and that the motion of the partition depends on the relative pressure. 1 point

HOW TO SCORE PRACTICE TEST 1

Section I: Multiple Choice

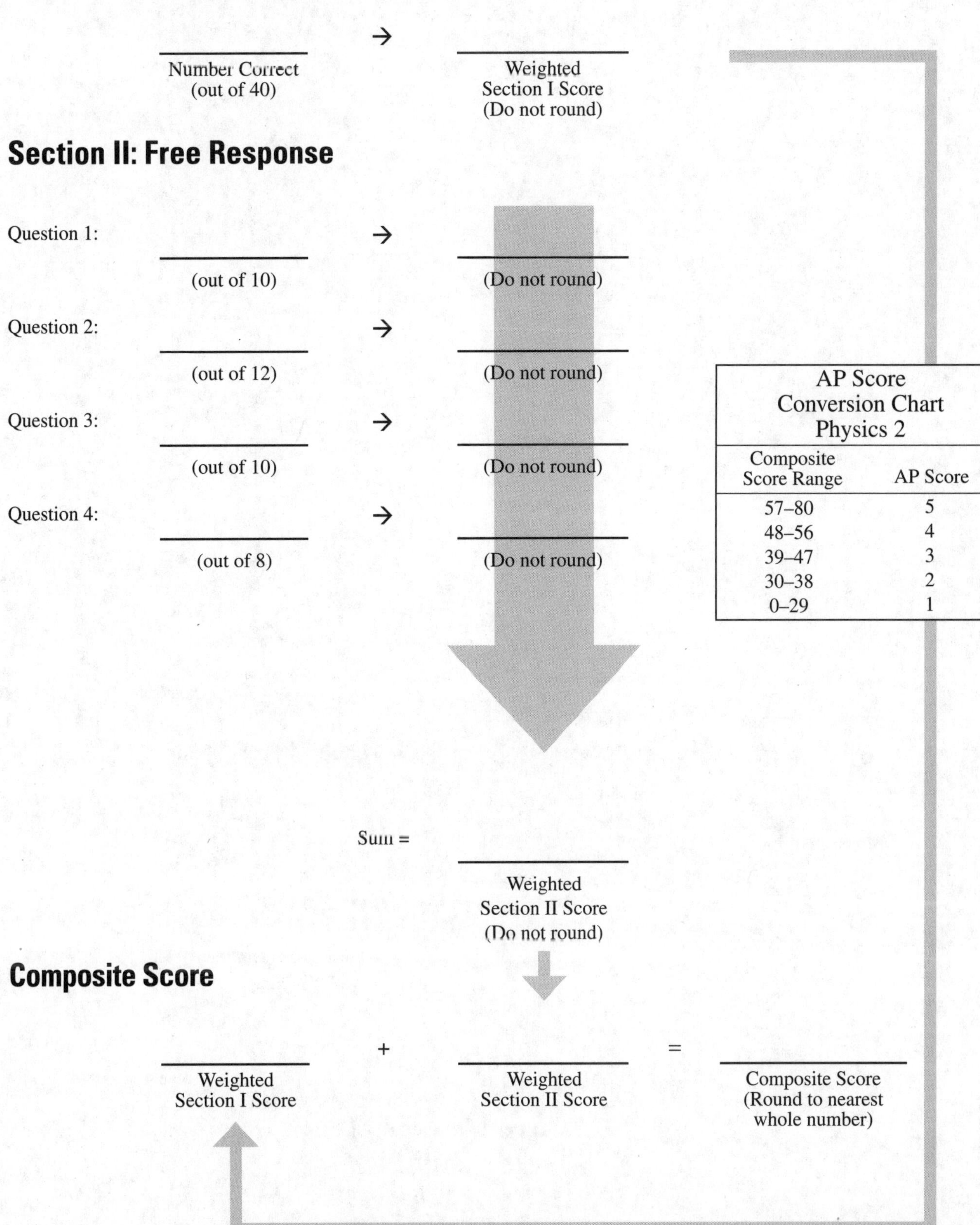

Section II: Free Response

AP Score Conversion Chart Physics 2	
Composite Score Range	AP Score
57–80	5
48–56	4
39–47	3
30–38	2
0–29	1

Composite Score

Part III
About the AP Physics 2 Exam

- The Structure of the AP Physics 2 Exam
- The AP Physics 2 Exam is Hybrid Digital
- How the AP Physics 2 Exam is Scored
- Overview of Content Topics
- How AP Exams Are Used
- Other Resources
- Designing Your Study Plan

THE STRUCTURE OF THE AP PHYSICS 2 EXAM

The AP Physics 2 Exam consists of two sections: a multiple-choice section and a free-response section. Each question in the multiple-choice section is followed by four possible responses. It is your job to choose the right answer(s). Each correct answer is worth one point.

Section	Timing	Scoring	Question Type	Number of Questions
I: Multiple Choice	80 minutes	50% of exam score	Single-select (discrete questions and questions in sets with one correct answer)	40
				Total—40

Section	Timing	Scoring	Question Type	Number of Questions
II: Free Response	100 minutes	50% of exam score	Mathematical Routines (10 points)	1
			Translation Between Representations (12 points)	1
			Experimental Design (10 points)	1
			Qualitative/Quantitative Translation (8 points)	1
				Total—4

Be Calculating

Students may use a four-function, scientific, or graphing calculator on the entire exam. However, be aware that some models have unapproved features (like keyboards, styluses, or wireless capability) and are not permitted, so please check the College Board website.

As you may know, the AP Physics 2 Exam is now a hybrid digital exam, which means that Section I is taken entirely on a computer and then for the free-response section (Section II), you will see the questions in the Bluebook app but the section requires you to write out your solutions by hand in the paper exam booklet. More on that in a moment. The total amount of time for Section II is 100 minutes, so you have an average of 25 minutes per question, although the College Board recommends spending a bit more time on the highest point value question (Translation Between Representations) and less time on the lowest point value question (Qualitative/Quantitative Translation). Unlike the multiple-choice section, which is scored by computer, the free-response section is graded by high school and college teachers. They have guidelines for awarding partial credit, so you don't need to correctly answer every part to get points. You are allowed to use a calculator on the entire exam—both sections. In fact, a version of the Desmos graphing calculator is built into the Bluebook app. That said, four-function, programmable, or graphing calculators are acceptable also, if you wish to use your personal calculator. Note that calculators with a keyboard are not permitted. Also, a table of equations is provided for your use. The two sections—multiple-choice and free-response—are weighted equally, so each is worth 50 percent of your score.

THE AP PHYSICS 2 EXAM IS HYBRID DIGITAL

While the multiple-choice section of AP Physics 2 is administered digitally via the College Board's Bluebook testing app, the free-response questions are viewed digitally, but your answers must be written in the paper exam booklets given to you and returned for scoring.

For the digital portion of the AP Physics 2 Exam, access to the Bluebook app is available via Windows, Mac laptops/desktops, iPads, tablets, and Chromebooks. If your computer or tablet is owned or managed by your school, your school official will likely already have this installed. The test cannot be taken on a smartphone.

The Bluebook app allows you to annotate and highlight texts, eliminate answers, and flag questions for review later. You will also have access to scratch paper to plan responses to essay questions. Unlike other digital exams like the SAT, the digital AP Exam will NOT be adaptive: this means that the difficulty of questions will not change depending on how well you do on earlier sections of the test.

It is important to note that the digital AP Physics 2 Exam includes the same number of sections, types of questions, and timing as the previous paper exam. This means the strategies used for success on paper versions of the exam are just as applicable to the digital version. So don't worry; the practice tests in this book will still prepare you for test day and digital practice tests are available via your student tools.

HOW THE AP PHYSICS 2 EXAM IS SCORED

Grades on the AP Physics 2 Exam are reported as a number: either 1, 2, 3, 4, or 5. The following table provides descriptions of these scores along with the distribution of how students scored for the 2024 exam administrations.

Score	2024 Percentage	Credit Recommendation	College Grade Equivalent
5	19.1%	Extremely Well Qualified	A
4	18.0%	Well qualified	A–, B+, B
3	33.4%	Qualified	B–, C+, C
2	22.9%	Possibly Qualified	–
1	6.6%	No Recommendation	–

Scores from the May 2024 test administration. Data taken from the College Board website.

Colleges are generally looking for a 4 or 5, but some may grant credit for a 3. How well do you have to do to earn such a grade? Each test is curved, and specific cutoffs for each grade vary a little from year to year, but here's a rough idea of how many points you must earn—as a percentage of the maximum possible raw score—to achieve each of the grades 2 through 5:

Check out the Digital AP Exam
View the College Board's walk-through video of what each part of the digital AP Exam will look like here: YouTube.com/watch?v=LfB3tyJjDqc

AP Exam Grade	Percentage Needed
5	≥ 75%
4	≥ 60%
3	≥ 45%
2	≥ 25%

OVERVIEW OF CONTENT TOPICS

You may be using this book as a supplementary text as you take an AP Physics 2 course at your high school, or you may be using it on your own. The College Board is very detailed in what it requires your AP teacher to cover in their AP Physics 2 course. The following summary is the College Board's explanation of what you should know.

> Students explore principles of thermodynamics, electric force, field, and potential, electric circuits, magnetism and electromagnetism, geometric optics, waves, sound, and physical optics, and modern physics. The course is based on seven units and three over-arching science practices.

The course content is broken down into seven units, which all have several topics. In Part V of this book, the course content is organized by topic. This is similar to the structure of the majority of physics textbooks.

The College Board recently revised their AP Physics 1, 2, and C courses and exams. The topic of Fluids moved from AP Physics 2 to AP Physics 1, multi-select questions were removed from all exams entirely, and many more changes were rolled out. Here is exactly what the College Board says about AP Physics 2:

AP Physics 2 is an algebra-based, introductory college-level physics course. Students cultivate their understanding of physics by developing models of physical phenomena through inquiry-based investigations.

Students build their understanding of physical models as they explore and solve problems in these topics:

- Thermodynamics
- Electric Force, Field, and Potential
- Electric Circuits
- Magnetism and Electromagnetism
- Geometric Optics
- Waves, Sound, and Physical Optics
- Modern Physics

To read more about how the AP Physics 2 course and exam has changed, be sure to visit the College Board website and read the complete, new AP course framework found at apcentral.collegeboard.org/media/pdf/ap-physics-2-course-and-exam-description-effective-fall-2024.pdf

The Course Content Units are assigned approximate weightings for the multiple-choice portion of the AP Physics 2 Exam.

Content Unit	Approximate AP Exam Weighting Percentages
Unit 9: Thermodynamics	15–18%
Unit 10: Electric Force, Field, and Potential	15–18%
Unit 11: Electric Circuits	15–18%
Unit 12: Magnetism and Electromagnetism	12–15%
Unit 13: Geometric Optics	12–15%
Unit 14: Waves, Sound, and Physical Optics	12–15%
Unit 15: Modern Physics	12–15%

This book is aligned with the College Board AP Physics 2 Course and Exam Description: Effective Fall 2024.

HOW AP EXAMS ARE USED

Different colleges use AP Exams in different ways, so it is important that you go to a particular college's website to determine how it uses AP Exams. The three items below represent the main ways in which AP Exam scores can be used:

AP Exams Go Digital
As of May 2025, some AP Exams are fully digital and others are hybrid digital. For the latest info regarding specific AP subjects and testing accommodations, visit apcentral.collegeboard.org/exam-administration-ordering-scores/digital-ap-exams

- **College Credit.** Some colleges will give you college credit if you score well on an AP Exam. These credits count toward your graduation requirements, meaning that you can take fewer courses while in college. Given the cost of college, this could be quite a benefit, indeed.
- **Satisfy Requirements.** Some colleges will allow you to "place out" of certain requirements if you do well on an AP Exam, even if they do not give you actual college credits. For example, you might not need to take an introductory-level course, or perhaps you might not need to take a class in a certain discipline at all.
- **Admissions Plus.** Even if your AP Exam will not result in college credit or even allow you to place out of certain courses, most colleges will respect your decision to push yourself by taking an AP course or even an AP Exam outside of a course. A high score on an AP Exam shows mastery of more difficult content than is taught in many high school courses, and colleges may take that into account during the admissions process.

OTHER RESOURCES

There are many resources available to help you improve your score on the AP Physics 2 Exam, not the least of which are your **teachers**. If you are taking an AP class, you may be able to get extra attention from your teacher, such as obtaining feedback on your essays. If you are not in an AP course, reach out to someone who teaches AP Physics 2 and ask if that teacher will review your essays or otherwise help you with content.

Looking for More Help with Your APs?
We now offer specialized AP tutoring and course packages that guarantee a 4 or 5 on the AP. To see which courses are offered and available, visit PrincetonReview.com/college/ap-test-prep

Another wonderful resource is **AP Students**, the official site of the AP Exams. The scope of the information at this site is quite broad and includes:

- the course description, which provides details on what content is covered and sample questions
- free-response prompts from previous years and exam tips
- access to AP Classroom if you are enrolled in a course (teacher assistance required)

The AP Students web page is apstudents.collegeboard.org.

Finally, The Princeton Review offers tutoring for the AP Physics 2 Exam. Our expert instructors can help you refine your strategic approach and add to your content knowledge. For more information, call 1-800-2REVIEW.

DESIGNING YOUR STUDY PLAN

As part of the Introduction, you identified some areas of potential improvement. Now delve further into your performance on Test 1, with the goal of developing a study plan appropriate to your needs and time commitment.

Read the answers and explanations associated with the multiple-choice questions (starting at page 33). After you have done so, respond to the following questions:

- Review the Practice Test 1 Diagnostic Answer Key, and make a list of the chapter topics in Step 1. Next to each topic, indicate your rank of the topic as follows: "1" means "I need a lot of work on this," "2" means "I need to beef up my knowledge," and "3" means "I know this topic well."
- How many days/weeks/months away is your exam?
- What time of day is your best, most focused study time?
- How much time per day/week/month will you devote to preparing for your exam?
- When will you do this preparation? (Be as specific as possible: Mondays and Wednesdays from 3:00 P.M. to 4:00 P.M., for example.)
- Based on the answers above, will you focus on strategy (Part IV) or content (Part V) or both?
- What are your overall goals in using this book?

Prep Like a Pro
Need some help devising a plan of action for your studying? Check out our free AP Physics 2 Exam study guide on your AP Student Tools. See the "Get More (Free) Content" page for details about accessing your online tools.

Part IV
Test-Taking Strategies for the AP Physics 2 Exam

PREVIEW

Review your Practice Test 1 results and then respond to the following questions:

- How many multiple-choice questions did you miss even though you knew the answer?
- On how many multiple-choice questions did you guess randomly?
- How many multiple-choice questions did you miss after eliminating some answers and guessing based on the remaining answers?
- Did you find any of the free-response questions easier or harder than the others—and, if so, why?

HOW TO USE THE CHAPTERS IN THIS PART

For the following Strategy chapters, think about what you are doing now before you read the chapters. As you read and engage in the directed practice, be sure to appreciate the ways you can change your approach.

We've Got You Covered!
For more information on strategy and testing features for the digital AP exam, please see our "How to Approach Digital Testing for AP Students" PDF online via your student tools.

Chapter 1
How to Approach Multiple-Choice Questions

APPROACHING THE MULTIPLE-CHOICE SECTION

All the multiple-choice questions will have a similar format: each will be followed by four answer choices. At times, it may seem that there could be more than one possible correct answer. In fact, answers resulting from common mistakes in calculation are often included in the four answer choices to trap you.

Recent Update
Remember that the AP Physics Exams used to contain 5 multi-select questions (that is, questions with 2 correct answers). They no longer do, so if you have heard about multi-select questions, fear not! You'll be encountering traditional, single-option, multiple-choice questions on test day (40 of them in Part I of the exam).

Let's Get Digital

Remember that the multiple-choice section of the AP Physics 2 Exam will be administered in a digital format, so you will be taking them on the Bluebook app. It's not taken online, but it is taken digitally (Section I, that is). The test will be administered at your school on a desktop or laptop computer that has the Bluebook app installed. To prepare for this, it's best if you practice as much as you can in a digital environment so that you will be used to the look and feel of the test and know how to use the in-app Desmos calculator (if you wish to use that instead of your own calculator). Head to your online Student Tools (found at princetonreview.com/prep) and register your book to access a digital practice test.

The Two-Pass System

The AP Physics 2 Exam covers a broad range of topics. There's no way, even with our extensive review, that you will know everything about every topic in algebra-based physics. So, what should you do?

Use the two-pass system to save time!

Adopt a two-pass system. The two-pass system entails going through the test and answering the easy questions first. Save the more time-consuming questions for later. (Don't worry—you'll have time to do them later!) First, read the question and decide if it is a "now" or "later" question. If you decide this is a "now" question, answer it right away in the test. If it is a "later" question, use the Mark for Review tool to add a little flag next to the question. Then keep on moving.

Once you've finished all the "now" questions, go back to the questions that you Marked for Review and start in on those ones. Start with the easier questions first. These are the ones that require calculations or that require you to eliminate the answer choices (in essence, the correct answer does not jump out at you immediately).

Clock the Time

Have a clock-checking strategy. The Bluebook testing app provides a countdown clock that you can view or hide. We recommend that you keep it hidden as your default, because constantly checking the clock is, in itself, a waste of time and can be distracting. Devise a plan. Try checking the clock after every 15 or 20 questions to see if you are keeping the correct pace or whether you need to speed up; this will ensure that you're cognizant of the time but will not permit you to fall into the trap of dwelling on it.

Know How to Linearize Graphs to Equations

Graphs, data tables, and mathematical relations (equation of fit lines) are all separate ways to represent the outcome of experiments, and the AP Physics 2 Exam will expect you to know how to use all of these representations. For the multiple-choice portion of the AP Physics 2 Exam (as well as the free-response portion), you are supplied with an equation sheet. You should know what graphs of these equations look like (a graphing calculator may be helpful for this), but most often you will be asked which variables need to be graphed to produce a direct proportion. For Coulomb's Law, $F_{el} = \frac{1}{4\pi\varepsilon_0}\frac{Q_1Q_2}{r^2}$, there are several data that would produce a linear graph. If F_{el} and Q_1 are data, then a graph of F_{el} versus Q_1 will produce a linear plot with slope of $\frac{1}{4\pi\varepsilon_0}\frac{Q_2}{r^2}$, but if F_{el} and r are data, then a graph of F_{el} versus $\frac{1}{r^2}$ produces a linear plot with slope of $\frac{1}{4\pi\varepsilon_0}Q_1Q_2$. Producing a linear graph for data or from an equation is called linearization.

Process of Elimination (POE)

On most tests, you need to know your material backward and forward in order to get the right answer. In other words, if you don't know the answer beforehand, you probably won't answer the question correctly. This is particularly true of fill-in-the-blank and essay questions. We're taught to think that the only way to get a question right is by knowing the answer. However, that's not the case on Section I of the AP Physics 2 Exam. You can get a perfect score on this portion of the test without knowing a single right answer—provided you know all the wrong answers!

Proven Technique
Use POE to boost your chances of scoring well.

What are we talking about? This is perhaps the most important technique to use on the multiple-choice section of the exam. Let's take a look at an example.

41 Mark for Review

A gas undergoes an expansion-compression cycle. If, plotted on a *P–V* diagram, the cycle is counterclockwise and the work is 300 J in magnitude, what was the heat transfer during this cycle?

(A) 300 J into the system

(B) 300 J out of the system

(C) 600 J into the system

(D) 600 J out of the system

Apply the Strategy
Which answer choices can you eliminate first?

Now, if this were a fill-in-the-blank-style question, you might be in a heap of trouble. But let's take a look at what we've got. If we can eliminate a few of the answer choices, we can get closer to making an educated guess.

For a cycle, $\Delta U = 0$, we know the magnitude of the work and heat transfer have to be the same. This eliminates (C) and (D). Since the cycle is counterclockwise on a P–V diagram, the work has to be positive, so the heat transfer is negative. A negative heat transfer is heat leaving the system. Choice (B) is correct.

We think we've illustrated our point: Process of Elimination is the best way to approach the multiple-choice questions. Even when you don't know the answer right off the bat, you'll surely know that two or three of the answer choices are not correct. What then?

Aggressive Guessing

You are scored only on the number of questions you get right, so guessing can't hurt you. But can it help you? It sure can. Let's say you randomly guess the same letter on four questions; odds are you'll get one right. So you've already increased your score by one point. Now, let's add POE into the equation. If you can eliminate as many as two answer choices from each question, your chances of getting them right increase, and so does your overall score. Remember, don't leave any bubbles blank on test day!

General Advice

Answering 40 multiple-choice questions in 80 minutes can be challenging. Make sure to pace yourself accordingly, and remember that you do not need to answer every question correctly to do well. Exploit the multiple-choice structure of this section. For all of your multiple-choice questions (40 questions, according to The College Board), there are three wrong answers and only a single correct one, so even if you don't know exactly which one is the right answer, you can eliminate some that you know for sure are wrong. Then you can make an educated guess from among the answers that are left and greatly increase your odds of getting that question correct.

Problems with graphs and diagrams are usually the fastest to solve, and problems with an explanation for each answer usually take the longest to work through. Do not spend too much time on any one problem, or you may not get to easier problems further into the test.

The practice exams in this book are written to give you an idea of the format of the test, to illustrate the difficulty of the questions, and to allow you to practice pacing yourself. Take them in the same circumstances as you will encounter during the real exam.

Chapter 2
How to Approach Free-Response Questions

THE FREE-RESPONSE SECTION: WHAT TO EXPECT

As we have discussed, the AP Physics 2 Exam is now hybrid digital, which means that the Section I (multiple-choice) questions are entirely digital and in Section II (free-response section) you will read the questions in the Bluebook app (digitally) but then answer them in a paper booklet, by hand. Thus, on this section, be sure to show the graders what you're thinking. Write clearly—that is *very* important—and show your steps. If you make a mistake in one part and carry an incorrect result to a later part of the question, you can still earn valuable points if your method is correct. But the graders cannot give you credit for work they can't follow or can't read. And, where appropriate, be sure to include units on your final answers.

Taking time to read the questions—making sure you understand them—will save you time in the long run.

The most important advice we can give you for the free-response section of the AP Physics 2 Exam is to read the questions carefully and answer according to exactly what the questions are asking you to do. Credit for the answers depends not only on the quality of the solutions but also on how they are explained. On the AP Physics 2 Exam, the words "justify," "explain," "calculate," "what is," "determine," and "derive" have specific meanings, and the graders are looking for very precise approaches in your explanations in order to give maximum credit.

Questions that ask you to "justify" or "explain" are looking for you to show an understanding in words of the principles underlying physical phenomena. Equations and mathematical derivations may be part of your justification, but ultimately these questions are looking for qualitative reasoning beyond just listing equations or manipulating variables. In some cases, the text must elucidate physics fundamentals or laws, while in other cases they will serve to analyze the behavior of different parts of an apparatus or variables in an equation.

The word "calculate" or "determine" requires you to show numerical or algebraic work to arrive at the final answer. In contrast, "what is," "estimate," and "indicate" questions signify that full credit may be given without showing mathematical work. Just remember, showing work that leads to the correct answer is always a good idea when possible, especially since showing work may still earn you partial credit even if the answer is not correct.

"Derive" questions are looking for a more specific approach, which entails beginning the solution with one or more fundamental equations and then arriving at the final answer through the proper use of mathematics, usually involving some algebra.

APPROACHING THE FREE-RESPONSE SECTION

Section II is worth 50% of your grade on the AP Physics 2 Exam. This section is composed of four free-response questions worth a total of 40 points. You're given a total of 100 minutes for this section. You will encounter four types of problems on the exam:

Question 1: Mathematical Routines The Mathematical Routines question is worth 10 points and the suggested time to spend on it is 20–25 minutes. This question involves working with variables as well as calculating values. You may also be asked to create representations of a situation, such as drawing a free-body diagram or sketching a graph.

Question 2: Translation Between Representations The Translation Between Representations question is worth 12 points and the suggested time to spend on it is 25–30 minutes. This question involves connecting ideas, either by applying different methods to the same scenario or by applying similar methods to two different scenarios. You may also be asked to justify a claim, create a representation, or argue why the two ideas do/do not agree with each other.

Question 3: Experimental Design The Experimental Design (or LAB) question is worth 10 points and the suggested time to spend on it is 25–30 minutes. This question is quite distinct from the other three and always involves an experiment. There are two main things that the AP Exam expects you to accomplish in this question: develop an experiment and interpret data. Usually, the question gives you a physical scenario and asks you how you would accomplish a goal, such as determining the value of a specific variable. Then, it presents a slightly different scenario with data and asks you to graph the data and use that graph to accomplish a different goal.

Question 4: Qualitative/Quantitative Translation The Qualitative/Quantitative Translation question is worth 8 points and the suggested time to spend on it is 15–20 minutes. This question involves only working with equations and variables. It asks you to make a prediction or claim about a scenario and derive an equation related to that scenario. You may also be asked to justify the claim, make a prediction about another scenario, or predict what would happen if the scenario were altered. Of note is that this is the only question where you definitely will not have to create a diagram or graph, though it's often a good idea to create one anyway to help work out your solution.

You have 100 minutes to complete these four problems, so you have an average of 25 minutes per problem. However, as shown above, these four questions are worth different numbers of points, so you may, for example, want to spend a few more minutes on the Translation Between Representations question than the Qualitative/Quantitative Translation question. Additionally, if a certain question type from this list is your strongest suit, then try to work on that one first and get through it quickly in order to give yourself more time to work on a topic that is more challenging for you.

Room to Write
On exam day, you will view the free-response questions in the Bluebook testing app and handwrite your responses in a paper exam booklet. You can download a sample AP Physics 2 booklet at: apcentral.collegeboard.org/media/pdf/ap-physics-2-hybrid-sample-free-response-booklet.pdf

You can use scrap paper for the FRQs in this book. For the practice tests, we also provide printable free-response booklets, which are available in your Student Tools via PrincetonReview.com/prep

Clearly Explain and Justify Your Answers

Remember that your answers to the free-response questions are graded by readers and not by computers. Communication is a very important part of AP Physics 2. Compose your answers in precise sentences. Just getting the correct numerical answer is not enough. The majority of questions will require a written explanation or justification either in lieu of a numerical answer or in addition to a calculation. You will need to explain your reasoning behind the technique that you selected and communicate your answer in the context of the problem. For many problems, points are given "for indicating" some physical reality of the situation, so either a written justification or a calculated solution is acceptable.

Do not expect the graders to read between the lines. Your grader will be an experienced AP Physics 2 teacher, so you will not need to explain every detail of each step in your thought process, but you need to present your solution in a systematic manner using solid logic and appropriate language. The first step in almost any written response should cite a specific physical principle. And remember: although you won't earn points for neatness, the graders can't give you a grade if they can't read and understand your solution!

Use Only the Space You Need

Do not try to fill up the space provided for each question. The space given is usually more than enough. The people who design the tests realize that some students write in big letters and some students make mistakes and need extra space for corrections. So if you have a complete solution, don't worry about the extra space. Writing more will not earn you extra credit. In fact, many students tend to go overboard and shoot themselves in the foot by making a mistake after they've already written the right answer.

Use the Numbering to Guide Your Answer!

Questions will typically have several subparts. When the subparts are related directly to one another, they will be numbered with letters and lowercase Roman numerals (for example, *a.i* and *a.ii*). Subparts that are not directly related to one another, but are still related to the main stem of the problem, are numbered with different letters (for example, *a* and *b*). It is not uncommon for a question to have a mixture of both types of subparts. (For example, a problem could be numbered *a.i, a.ii, b, c.i, c.ii*. Subparts *a.i* and *a.ii* would be related to each other, while part *b* would be related back to the main stem of the problem but not to either of the *a* parts. Then both of the *c* parts would be related to each other, but not to the *a* or *b* parts.) Try to answer all of the subparts. If one part is giving you trouble, look at the next subpart to see if you can find a way to approach it. Since the answer to *b* does not usually depend on part *a*, you can take a fresh approach to answering when the numbering of the subparts changes.

Read and Respond to the Entire Question

Many times, a single subpart of a question will ask you to explain multiple things. In order to earn full points, you must address every item the question asks. Such answers typically require no more than a single sentence per item to fully explain. For example, a subpart in a circuit question may ask how, if a particular resistor were removed, the total current, the total resistance, and the total voltage of the circuit would change. You could not receive full credit unless you addressed all three of those quantities in your answer.

Use Common Sense

Always use your common sense in answering questions. If your solutions to part A (i) are incorrect, but are substituted correctly into part A (ii), you will likely not be penalized for getting an incorrect answer to part A (ii). However, if your answer to A (ii) is physically impossible, such as getting a speed of light in vacuum greater than 3.00×10^8 m/s or finding a number of electrons of 3.5, you will not receive credit even if your answer accurately relies on the answer to a previous part. To receive the highest possible score, make your errors easily recognizable and make sure all of your answers are physically reasonable.

Think Like a Grader

When answering questions, try to think about what kind of answer the grader is expecting. Look at past free-response questions and grading rubrics on the College Board website. These examples will give you some idea of how the answers should be phrased. The graders are told to keep in mind that there are two aspects to the scoring of free-response answers: showing statistical knowledge and communicating that knowledge. Again, responses should be written as clearly as possible in complete sentences. You don't need to show all the steps of a calculation, but you must explain how you got your answer and why you chose the technique you used.

Think Before You Write

Abraham Lincoln once said that if he had eight hours to chop down a tree, he would spend six of them sharpening his axe. Be sure to spend some time thinking about what the question is, what answers are being asked for, what answers might make sense, and what your intuition is before starting to write. These questions aren't meant to trick you, so all the information you need is given. If you think you don't have the right information, you may have misunderstood the question. In some calculations, it is easy to get confused, so think about whether your answers make sense in terms of what the question is asking. If you have some idea of what the answer should look like before starting to write, then you will avoid getting sidetracked and wasting time on dead-ends.

Reflect

Respond to the following questions:

- How much time will you spend on the short free-response questions? What about the long free-response questions?
- What will you do before you begin writing your free-response answers?
- Will you seek further help, outside of this book (such as a teacher, tutor, or AP Students), on how to approach the questions that you will see on the AP Physics 2 Exam?

Part V
Content Review for the AP Physics 2 Exam

HOW TO USE THE CHAPTERS IN THIS PART

You may need to review the following content chapters more than once. Your goal is to obtain mastery of the content, and a single read of a chapter may not be sufficient. At the end of each chapter, reflect on the following questions to help you determine whether or not you truly have mastered the content of that chapter:

- For which content topics discussed in this chapter do you feel you have achieved sufficient mastery to answer multiple-choice questions correctly?
- For which content topics discussed in this chapter do you feel you have achieved sufficient mastery to confidently answer free-response questions?
- For which content topics discussed in this chapter do you feel you need more work before you can answer multiple-choice questions correctly?
- For which content topics discussed in this chapter do you feel you need more work before you can confidently answer free-response questions?
- What parts of this chapter are you going to re-review?
- Will you seek further help outside of this book (such as a teacher, tutor, or AP Students) on any of the content in this chapter—and, if so, on what content?

One key difference between our book and the official AP Physics 2 Equation Sheet is that our book employs a **bold font** to indicate a vector quantity and a non-bold font for magnitude only, whereas the College Board uses arrow hats to denote vector quantities and the absolute value symbol for magnitudes.

Equation Sheet Sidebars

Throughout the content review chapters in Part V, you'll see the words "Equation Sheet" in the sidebars. This is meant to draw your attention to the physics equations featured on the College Board's equation sheet that will be provided to you during your AP Physics 2 Exam. So these are equations that are extremely important for you to be familiar with.

Chapter 3 Thermodynamics

INTRODUCTION

This chapter looks at heat and temperature, concepts that seem familiar from our everyday experience. Technically, **heat** is defined as thermal energy transmitted from one body to another. While an object can contain **internal energy** (due to the random motion of its molecules), an object doesn't *contain* heat; heat is energy *in transit*. **Temperature**, on the other hand, is a measure of an object's internal energy. To understand how these ideas are related, we need to examine physics at a microscopic level.

THE KINETIC THEORY OF GASES

Unlike the condensed phases of matter—solid and liquid—the atoms or molecules that make up a gas do not move around relatively fixed positions. Rather, the molecules of a gas move freely and rapidly, in a chaotic swarm. The molecules of the gas, which are constantly in motion, bump against one another. To study such a gas, it would need to be put into a container, and the random motion of the molecules would cause the molecules to bump into not just one another but also the walls of the container. The size of the container is the volume (V) of the thermodynamic system, and as the molecules bump against the walls, they create a pressure (P) where pressure is defined the same way as in the study of fluids as $P = F_{\perp}/A$. The **Kinetic Theory of Gases** deals with the relationship between the volume, pressure, and temperature of a gas, as well as the **Ideal Gas Law**.

Equation Sheet

$$\text{Pressure} = F_{\perp}/A$$

One of the difficulties in describing the behavior of gases is that they are comprised of an extremely large number of atoms or molecules. For example, at room temperature, air has a volume of 24 liters per mole. That means an empty one-liter soda bottle will contain 25 sextillion (2.5×10^{22}) air molecules. That is 25,000,000,000 trillions of molecules. Each of these molecules, at any given instant, will have its own position and be moving with its own momentum. A confined gas exerts a force on the walls of its container, and as the molecules are zipping around inside the container, striking the walls and rebounding, they create pressure within the container. It is impossible to apply Newton's laws to each molecule and determine the amount of force that each molecule would apply to the container of the bottle to determine the total pressure. To avoid dealing with individual molecules, the Kinetic Theory of Gases applies the mathematics of statistics to develop the concept of temperature as a measurement relating to the "average" kinetic energy of molecules in a gas sample and the relationship between temperature, pressure, and volume.

When considering the microscopic world, as the molecules move around in the sample and collide with one another, every collision between molecules and every collision between a molecule and the walls of the container is assumed to be a perfectly elastic collision. This means that the sum of the kinetic energy before the collision is equal to the sum of the kinetic energy after the collision. For inelastic collisions—for example, when a lump of clay is dropped and hits the floor—a certain amount of kinetic energy from the clay was used to generate "thermal energy," resulting in deforming the clay and increasing its temperature and the temperature of the floor. When considering collisions between individual molecules or between a molecule and the wall, the kinetic energy of the molecules is no longer distinct from the thermal energy;

it is the motion of the molecules (kinetic energy) that is the thermal energy, and we can no longer use the convenient label of "thermal energy" to account for a decrease in mechanical energy as we did when looking at macroscopic objects. Energy will be contained within the gas, as **internal energy**, which will determine the temperature. Transfers of energy into or out of the gas will be a result of **work**, a volume change in the container, or **heating**, a flow of energy that naturally occurs between systems at unequal temperatures.

The Ideal Gas Law

Three physical properties—pressure (P), volume (V), and temperature (T)—describe a gas. At low densities, all gases approach *ideal* behavior; this means that these three variables are related by the equation

$$PV = nRT = Nk_BT$$

Equation Sheet

where n is the number of moles of gas and R is a constant (8.31 J/mol•K) called the **universal gas constant**. This equation is known as the **Ideal Gas Law**. It can also be written as $PV = Nk_BT$, where N is the number of atoms (or molecules) of gas and k_B is Boltzmann's constant ($k_B = 1.38 \times 10^{-23}$ J/K). The Ideal Gas Law tells us how these variables are related to one another for a gas with no intermolecular forces. This is a bad assumption for a charged gas, where the individual charges would interact electrically (a strong interaction), but it is a good assumption for a neutral gas, where the atoms would only interact gravitationally (a very weak interaction).

An important consequence of this equation is that, for a fixed volume of gas, an increase in P gives a proportional increase in T. The pressure increases when the gas molecules strike the walls of their container with more force, which occurs if they move more rapidly. The macroscopic variables of pressure, volume, and temperature can be related to the average kinetic energy of any particular molecule within the gas using Newton's Second Law. We can find that the pressure exerted by N molecules of gas in a container is related to the average kinetic energy by the equation $PV = \frac{2}{3}NK_{\text{avg}}$. Comparing this with the Ideal Gas Law, we see that $\frac{2}{3}NK_{\text{avg}} = nRT$. We can rewrite this equation in the form $\frac{2}{3}N_AK_{\text{avg}} = RT$, since, by definition, $N = nN_A$. The ratio R/N_A is a fundamental constant of nature called **Boltzmann's constant** ($k_B = 1.38 \times 10^{-23}$ J/K), so our equation becomes

$$\left(K_{\text{avg}} = \frac{3}{2}k_BT\right) = \frac{1}{2}mv^2_{\text{rms}}$$

Equation Sheet

To convert from degrees Celsius to Kelvin, add 273.15.

This tells us that the average translational kinetic energy of the gas molecules is directly proportional to the absolute temperature of the sample. Remember, this means you must use kelvins as your temperature unit.

Since the average kinetic energy of the gas molecules is $K_{avg} = \frac{1}{2}m(v^2)_{avg}$, the equation above becomes $\frac{1}{2}m(v^2)_{avg} = \frac{3}{2}k_B T$, so

$$\sqrt{(v^2)_{avg}} = \sqrt{\frac{3k_B T}{m}}$$

The quantity on the left-hand side of this equation, the square root of the average of the square of v, is called the **root-mean-square velocity**, v_{rms}, so

$$v_{rms} = \sqrt{\frac{3k_B T}{m}}$$

It's important to realize that the molecules in the container have a wide range of speeds; some are much slower and others are much faster than v_{rms}. The root-mean-square speed is important because it gives us a type of average speed that's easy to calculate from the temperature of the gas.

Example 1 In order for the rms velocity of the molecules in a given sample of gas to double, what must happen to the temperature?

Solution. Temperature is a measure of the average kinetic energy. The velocity is determined from the following equation:

$$v_{rms} = \sqrt{\frac{3k_b T}{m}}$$

Since v_{rms} is proportional to the square root of T, the temperature must quadruple, again, assuming the temperature is given in kelvins.

Example 2 A cylindrical container of radius 15 cm and height 30 cm contains 0.6 mole of gas at 433 K. How much force does the confined gas exert on the lid of the container?

Solution. The volume of the cylinder is $\pi r^2 h$, where r is the radius and h is the height. Since we know V and T, we can use the Ideal Gas Law to find P. Because pressure is force per unit area, we can find the force on the lid by multiplying the gas pressure by the area of the lid.

$$P = \frac{nRT}{V} = \frac{(0.6 \text{ mol})(8.31 \text{ J/mol}\cdot\text{K})(433 \text{ K})}{\pi(0.15 \text{ m})^2(0.30 \text{ m})} = 1.018 \times 10^5 \text{ Pa}$$

So, since the area of the lid is πr^2, the force exerted by the confined gas on the lid is

$$F = PA = (1.018 \times 10^5 \text{ Pa}) \cdot \pi(0.15 \text{ m})^2 = 7200 \text{ N}$$

This is about 1600 pounds of force, which seems like a lot. Why doesn't this pressure pop the lid off? Because, while the bottom of the lid is feeling a pressure (due to the confined gas) of 1.018×10^5 Pa that exerts a force upward, the top of the lid feels a pressure of 1.013×10^5 Pa (due to the atmosphere) that exerts a force downward. The net force on the lid is

$$F_{\text{net}} = (\Delta P)A = (0.005 \times 10^5 \text{ Pa}) \cdot \pi(0.15 \text{ m})^2 = 35 \text{ N}$$

which is only about 8 pounds.

THE MAXWELL-BOLTZMANN DISTRIBUTION

The Kinetic Theory of Gases applies to large numbers of particles, which is why the term "average" continually appears in the previous section when describing the energy or speed of any particular molecule. The gas will have a variety of molecules with different energies and therefore different speeds. Some of the molecules will be moving much faster than average and some much slower. A graph of the distribution of the speeds of all of the molecules is called the Maxwell-Boltzmann distribution.

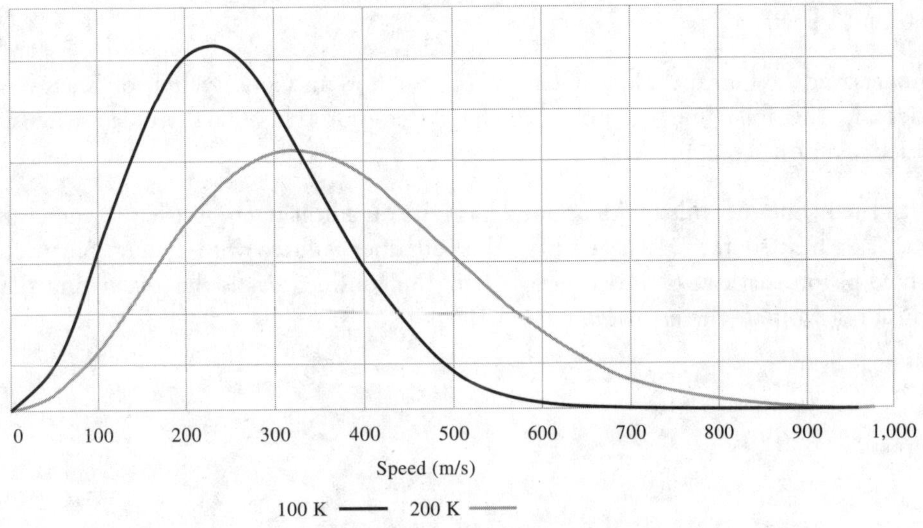

The important features of this type of graph are the position of the peak and the spread of the curve. As a gas sample is heated, the peak (called the most probable speed, which is actually always a bit lower than the v_{rms}) moves to the right, indicating the most probable speed for any particular molecule in the sample is increasing. The graph also spreads out and grows a longer tail on the right-hand side, indicating that there will be more molecules moving at much higher speed than average for hotter gas samples, which intuitively makes sense.

THE LAWS OF THERMODYNAMICS

We've learned about two ways in which energy may be transferred between a system and its environment. One is work, which takes place when a force acts over a distance. The other is heat, which takes place when energy is transferred due to a difference in temperature. The study of the energy transfers involving work and heat, and the resulting changes in internal energy, temperature, volume, and pressure is called **thermodynamics.**

The Zeroth Law of Thermodynamics

When two objects are brought into contact, heat will flow from the warmer object to the cooler one until they reach thermal equilibrium. This property of temperature is expressed by the Zeroth Law of Thermodynamics.

The Zeroth Law of Thermodynamics

If Objects 1 and 2 are each in thermal equilibrium with Object 3, then Objects 1 and 2 are in thermal equilibrium with each other.

The First Law of Thermodynamics

Simply put, the First Law of Thermodynamics is a statement of the Conservation of Energy that includes heat.

To understand how to include a statement of heat flow in Conservation of Energy, start by considering the following example, which is the prototype that's studied extensively in thermodynamics.

An insulated container filled with an ideal gas rests on a heat reservoir (that is, something that can act as a heat source or a heat sink). The container is fitted with a snug, but frictionless, weighted piston that can be raised or lowered. The confined gas is the *system*, and the piston and heat reservoir are the *surroundings*.

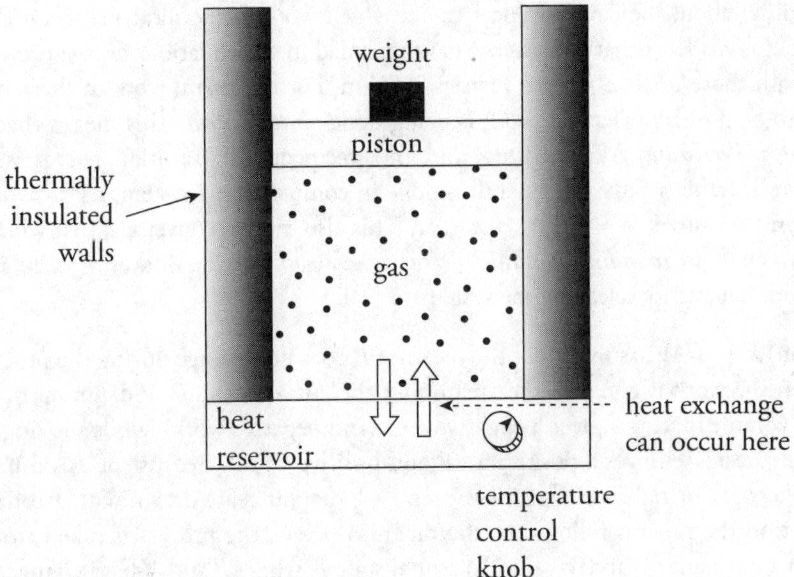

The **state** of the gas is given once its pressure, volume, and temperature are known, and the equation that connects these state variables is the Ideal Gas Law, $PV = nRT$. We'll imagine performing different experiments with the gas, such as heating it or allowing it to cool or increasing or decreasing the weight on the piston, and study the energy transfers (work and heat) and the changes in the state variables. If each process is carried out such that, at each moment, the system and its surroundings are in thermal equilibrium, we can plot the pressure (P) versus the volume (V) on a diagram. By following the path of this ***P–V* diagram**, we can study how the system is affected as it moves from one state to another.

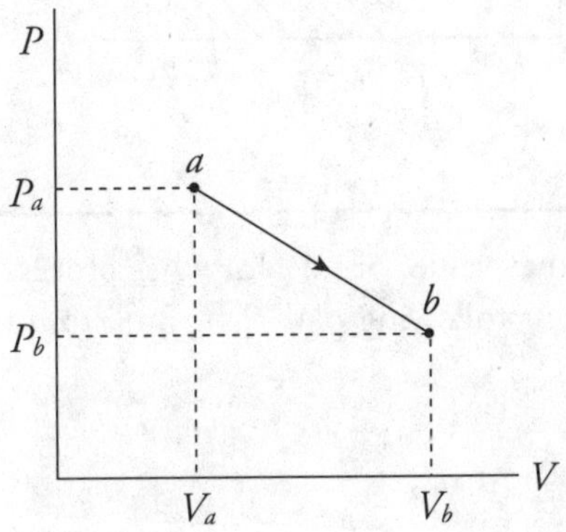

Work is done on or by the system when the piston is moved and the volume of the gas changes. For example, imagine that the weight pushes the piston downward a distance d, causing a decrease in volume. Assume that the pressure stays constant at P. (Heat must be removed via the reservoir to accomplish this.) We can calculate the work done on the gas during this compression as $W = -Fd$, but since $F = PA$, we have $W = -PAd$, and because $Ad = \Delta V$, we have

$$W = -P\Delta V$$

Equation Sheet

Textbooks differ about the circumstances under which work in thermodynamics is defined to be positive or negative. The negative signs we have included in the equations on the previous page are consistent with those used in the AP Physics 2 Exam. For the exam, work in thermodynamics is considered to be positive when the work is being done *on the system.* This means that the volume of the system is *decreasing,* ΔV is *negative,* and, in agreement with intuition, energy is being *added* to the system. In other words, when work is done in compressing a system, ΔV is *negative* and the work done on the system, $W = -P\Delta V$, is *positive.* This also means, conversely, that when the system is doing work *on the surroundings* (volume *increasing,* ΔV *positive*), the work is *negative,* in agreement with intuition that energy is leaving the system.

The equation $W = -P\Delta V$ assumes that the pressure P does not change during the process. If P *does* change, then the work is equal to the area under the curve in the P–V diagram; moving left to right as the volume increases gives a negative area (and negative work), while moving right to left as the volume decreases gives a positive area (and positive work). Because of this difference when moving left to right or right to left, processes in P–V diagrams are drawn with arrows to indicate which direction the process follows. In the diagram below, the process is taken from an initial state a, with a particular P and V value, to a final state b, where P and V have changed.

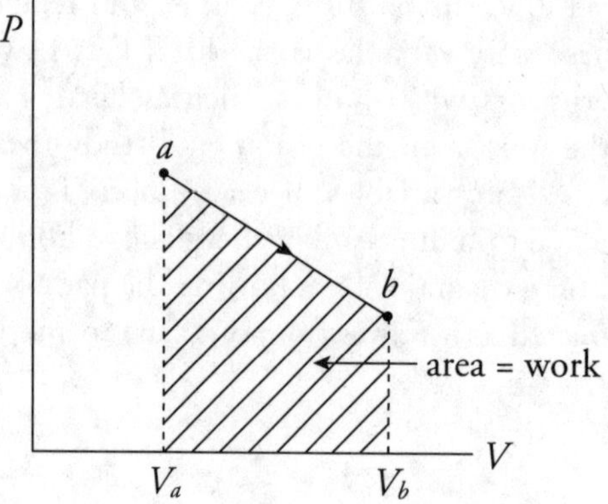

Example 3 What's the value of W for the process ab following path 1 and for the same process following path 2 (from a to d to b), shown in the P–V diagram below?

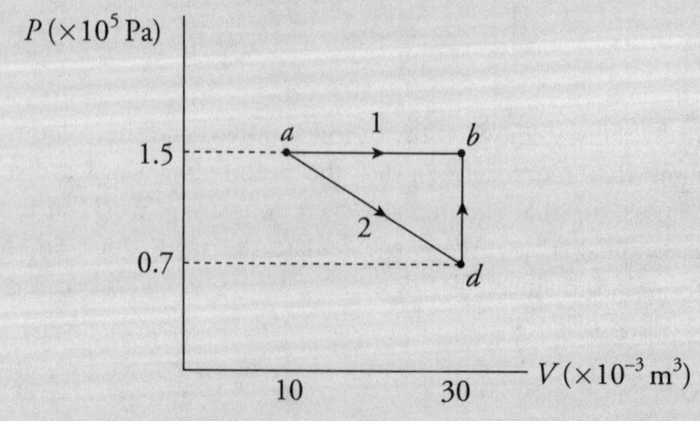

Solution.

Path 1. Since, in path 1, P remains constant, the work done is just $-P\Delta V$:

$$W = -P\Delta V = -(1.5 \times 10^5 \text{ Pa})[(30 \times 10^{-3} \text{ m}^3) - (10 \times 10^{-3} \text{ m}^3)] = -3000 \text{ J}$$

Path 2. If the gas is brought from state a to state b, along path 2, then work is done only along the part from a to d. From d to b, the volume of the gas does not change, so no work can be performed. The area under the graph from a to d is

$$\begin{aligned} W = -\frac{1}{2}h(b_1+b_2) &= -\frac{1}{2}(\Delta V)(P_a + P_d) \\ &= -\frac{1}{2}(20\times10^{-3} \text{ m}^3)[(1.5\times10^5 \text{ Pa})+(0.7\times10^5 \text{ Pa})] \\ &= -2200 \text{ J} \end{aligned}$$

As this example shows, the amount of work done by a gas depends not only on the initial and final states of the system, but also on the path between the two. In general, different paths give different values for W.

Experiments have shown that the value of the sum of the total heat added (or removed), denoted by Q, and the total work done on (or by) the system, denoted by W, does not depend on the path taken by a thermodynamic process; it depends only on the initial and the final state of the system, so it describes a change in a fundamental property. This property is called the system's **internal energy**, denoted U, and the change in the system's internal energy, ΔU, is equal to $Q + W$. This is true regardless of the process that brought the system from its initial to final state. This statement is known as

The First Law of Thermodynamics

$$\Delta U = Q + W$$

Equation Sheet

As stated above, the internal energy, U, of a gas depends only on the state of the gas. As we saw in the section on the Ideal Gas Law, the average kinetic energy of a molecule in a gas sample is directly proportional to the temperature. For ideal gases in particular, this is the only source of internal energy. Because of this, the change in internal energy, ΔU, which arises in the First Law of Thermodynamics, is directly proportional to a change in the temperature of the ideal gas.

This statement of the First Law is consistent with the interpretation of work ($W = -P\Delta V$) explained above. The First Law identifies W and Q as separate physical mechanisms for adding to or removing energy from the system, and the signs of both W and Q are defined consistently: both are positive when they are adding energy to the system and negative when they are removing energy from the system.

Example 4 A 0.5 mol sample of an ideal gas is brought from state a to state b when 7500 J of heat is added along the path shown in the following P–V diagram:

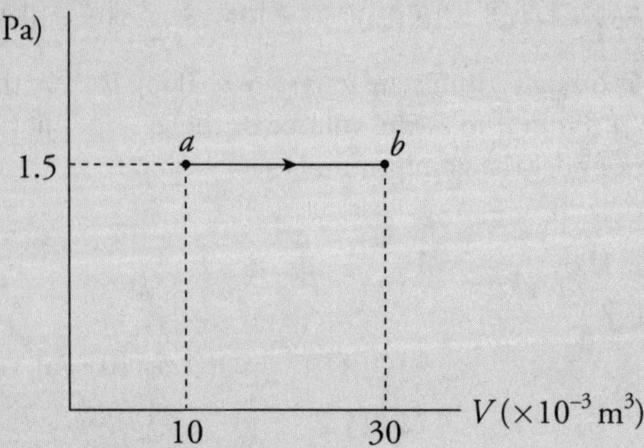

What are the values of each of the following?

(a) The temperature at a
(b) The temperature at b
(c) The work done by the gas during process ab
(d) The change in the internal energy of the gas

Solution.

(a, b) Both of these questions can be answered using the Ideal Gas Law, $T = PV/(nR)$:

$$T_a = \frac{P_a V_a}{nR} = \frac{(1.5\times10^5\text{ Pa})(10\times10^{-3}\text{ m}^3)}{(0.5\text{ mol})(8.31\text{ J/mol}\cdot\text{K})} = 360\text{ K}$$

$$T_b = \frac{P_b V_b}{nR} = \frac{(1.5\times10^5\text{ Pa})(30\times10^{-3}\text{ m}^3)}{(0.5\text{ mol})(8.31\text{ J/mol}\cdot\text{K})} = 1080\text{ K}$$

(c) Since the pressure remains constant during the process, we can use the equation $W = -P\Delta V$. Because $\Delta V = (30 - 10) \times 10^{-3}\text{ m}^3 = 20 \times 10^{-3}\text{ m}^3$, we find that

$$W = -P\Delta V = (1.5 \times 10^5\text{ Pa})(20 \times 10^{-3}\text{ m}^3) = 3000\text{ J}$$

The expanding gas did negative work against its surroundings, pushing the piston upward. Important note: if the pressure remains constant (which is designated by a horizontal line in the P–V diagram), the process is called **isobaric**.

(d) By the First Law of Thermodynamics,

$$\Delta U = Q + W = 7500 - 3000\text{ J} = 4500\text{ J}$$

Example 5 A 0.5 mol sample of an ideal monatomic gas is brought from state *a* to state *b* along the path shown in the following *P*–*V* diagram:

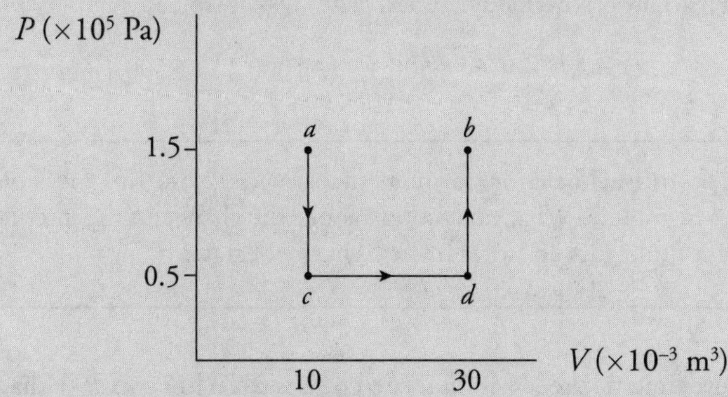

What are the values of each of the following?

(a) The work done by the gas during process *ab*
(b) The change in the internal energy of the gas
(c) The heat added to the gas during process *ab*

Solution. Note that the initial and final states of the gas are the same as in the preceding example, but the path is different.

(a) Let's break the path into 3 pieces:

Over paths 1 and 3, the volume does not change, so no work is done. Work is done only over path 2:

$$W = -P\Delta V = -(0.5 \times 10^5 \text{ Pa})(20 \times 10^{-3} \text{ m}^3) = -1000 \text{ J}$$

Once again, the expanding gas does negative work against its surroundings, pushing the piston upward.

(b) Because the initial and final states of the gas are the same here as they were in the preceding example, the change in internal energy, ΔU, *must* be the same. Therefore, $\Delta U = 4500$ J.

(c) By the First Law of Thermodynamics, $\Delta U = Q + W$, so

$$Q = \Delta U - W = 4500 \text{ J} - (-1000 \text{ J}) = 5500 \text{ J}$$

Example 6 An **isochoric** (or isovolumetric) process is one that takes place with no change in volume. What can you say about the change in the internal energy of a gas if it undergoes an isovolumetric change of state?

Solution. An isovolumetric process is illustrated by a vertical line in a P–V diagram and, since no change in volume occurs, $W = 0$. By the First Law of Thermodynamics, $\Delta U = Q + W = Q$. Therefore, the change in internal energy is entirely due to (and equal to) the heat transferred. If heat is transferred into the system (positive Q), then ΔU is positive; if heat is transferred out of the system (negative Q), then ΔU is negative.

Example 7 A 0.5 mol sample of an ideal gas is brought from state a back to state a along the path shown in the following P–V diagram:

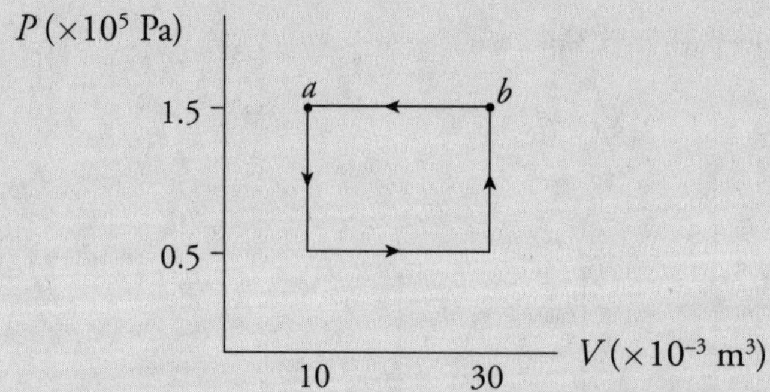

What are the values of each of the following?

(a) The change in the internal energy of the gas
(b) The work done on the gas during the process
(c) The heat added to the gas during the process

Solution. A process such as this, which begins and ends at the same state, is said to be cyclical.

(a) Because the final state is the same as the initial state, the internal energy of the system cannot have changed, so $\Delta U = 0$.

(b) The total work involved in the process is equal to the work done from c to d plus the work done from b to a,

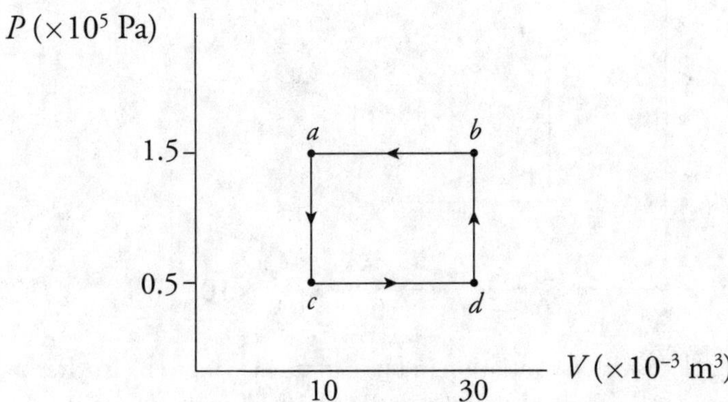

because only along these paths does the volume change. Along these portions, we find that

$$W_{cd} = -P\Delta V_{cd} = (0.5 \times 10^5 \text{ Pa})(+20 \times 10^{-3} \text{ m}^3) = -1000 \text{ J}$$

$$W_{cd} = -P\Delta V_{ba} = (1.5 \times 10^5 \text{ Pa})(-20 \times 10^{-3} \text{ m}^3) = +3000 \text{ J}$$

So, the total work done is $W = +2000$ J. The fact that W is positive means that, overall, work was done *on* the gas by the surroundings. Notice that for a cyclical process, the total work done is equal to the area enclosed by the loop, with clockwise travel taken as negative and counterclockwise travel taken as positive.

(c) The First Law of Thermodynamics states that $\Delta U = Q + W$. Since $\Delta U = 0$, it must be true that $Q = -W$ (which will always be the case for a cyclical process), so $Q = -2000$ J.

Example 8 A 0.5 mol sample of an ideal gas is brought from state a to state d along an **isotherm**, and then isobarically to state c and isochorically back to state a, as shown in the following P–V diagram:

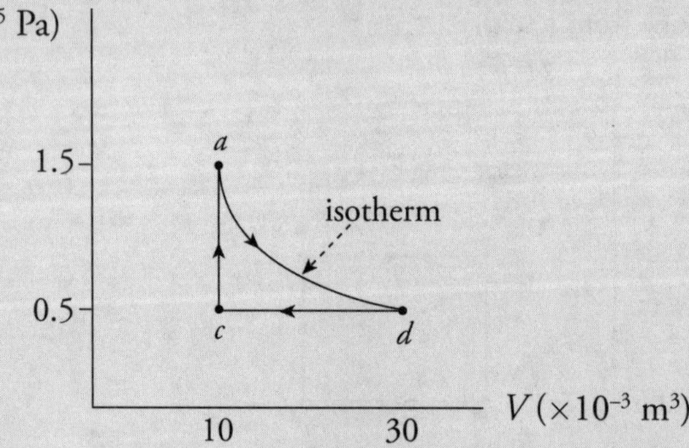

A process that takes place with no variation in temperature is said to be **isothermal**. Given that the work done during the isothermal part of the cycle is –1650 J, how much heat is transferred during the isothermal process, from a to d?

Solution. Be careful that you don't confuse *isothermal* with *adiabatic.* A process is isothermal if the *temperature* remains constant; a process is **adiabatic** if $Q = 0$. You might ask, *How could a process be isothermal without also being adiabatic at the same time?* Remember that the temperature is determined by the internal energy of the gas, which is affected by changes in Q, W, or both. Therefore, it's possible for U to remain unchanged even if Q is not 0 (because there can be an equal but opposite W to cancel it out). In fact, this is the key to this problem. Since T doesn't change from a to d, neither can the internal energy, which depends entirely on T. Because $\Delta U_{ad} = 0$, it must be true that $Q_{ad} = -W_{ad}$. Since W_{ad} equals –1650 J, Q_{ad} must be +1650 J. The gas absorbs heat from the reservoir and uses all this energy to do negative work as it expands, pushing the piston upward.

Here is a summary of special thermal processes:

Process	Condition	Application of 1st Law
Isovolumetric	$\Delta V = 0$	$\Delta U = Q$
Isothermal	$\Delta T = 0$	$\Delta U = 0$
Isobaric	$\Delta P = 0$	$\Delta U = Q - P\Delta V$
Adiabatic	$Q = 0$	$\Delta U = W$

The Second Law of Thermodynamics

The Second Law of Thermodynamics describes how systems evolve over time. Conservation of Energy states that system A and system B will have the same energy without any external work, but it never describes how a system changes from state A into state B. One description of this evolution is about **entropy**, an equivalent description is about **spontaneous heat flow**, and a third equivalent description is about **efficiency** in engines.

Entropy

Consider a box containing two pure gases separated by a partition. What would happen if the partition were removed? The gases would mix, and the positions of the gas molecules would be random.

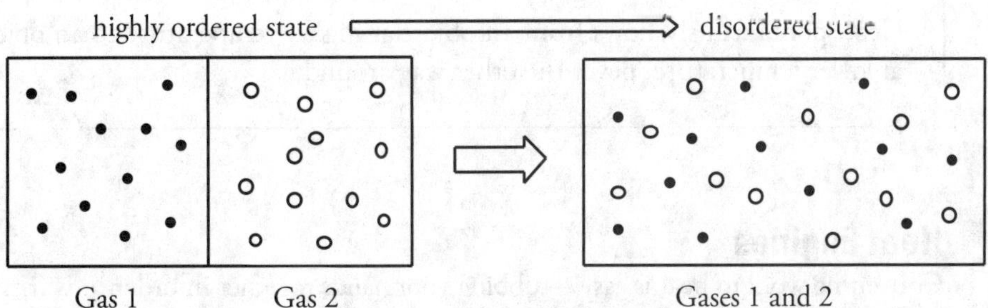

A closed system that shows a high degree of order tends to evolve in such a way that its degree of order decreases. In other words, disorder (or, as it's technically called, entropy) increases. The term "disorder" often makes the concept of entropy sound negative. Rather, entropy is better described as **increasing molecular freedom**. This is the reason why broken glass does not put itself back together and the reason why chemical reactions take place. If we started with the box on the right, containing the mixture of the gases, it would be virtually impossible that at any later time all the molecules of Gas 1 would happen to move to the left side of the box and, at the same time, the molecules of Gas 2 would spontaneously move to the right side of the box. If we were to watch a movie of this process, and saw the mixed-up molecules suddenly separate and move to opposite sides of the box, we'd assume that the film was running backward. In a way, the Second Law of Thermodynamics defines the direction of time. Time flows in such a way that ordered systems become disordered. Disordered states do not spontaneously become ordered without any other changes taking place. The following is the essence of one form of the Second Law of Thermodynamics:

> The total amount of disorder—the total entropy—of a system plus its surroundings will never decrease.

It is possible for the entropy of a system to decrease, but it will always be at the expense of a greater increase in entropy in the surroundings. For example, when water freezes, its entropy decreases. The molecules making up an ice crystal have a more structured order than the random collection of water molecules in the liquid phase, so the entropy of the water decreases when it freezes. But when water freezes, it releases heat energy into its environment, which creates disorder in the surroundings. If we were to figure out the total change in entropy of the water plus its surroundings, we would find that although the entropy of the water itself decreased, it was more than compensated by a greater amount of entropy increase in the surroundings. So, the total entropy of the system and its surroundings increased, in agreement with the Second Law of Thermodynamics.

There are several equivalent statements of the second law. In addition to the entropy form, another form of the Second Law of Thermodynamics says that heat always flows from hot to cold, never cold to hot. Another form, which will be considered more fully in a moment, says it is impossible to convert heat completely into work.

Entropy and Heat
An isolated system never decreases in entropy over time. A non-idealized (real-world) isolated system always increases in entropy over time. This entropy is usually in the form of heat given off. When a system is not isolated, it is possible for the entropy of the system to decrease as long as the surroundings to the system increase in entropy by a greater amount than the decrease in the system. As a result, the entropy of the universe never decreases, and the entropy of the universe will increase over time as real-world systems transfer heat and work.

Heat spontaneously flows from an object at higher temperature to an object at lower temperature, never the other way around.

Heat Engines

Converting work to heat is easy—rubbing your hands together in order to warm them up shows that work can be completely converted to heat. What we'll look at is the reverse process: how efficiently can heat be converted into work? A device that uses heat to produce useful work is called a **heat engine**. The internal-combustion engine in a car is an example. In particular, we're interested only in engines that take their working substance (a mixture of air and fuel in this case) through a cyclic process, so that the cycle can be repeated. The basic components of any cyclic heat engine are simple: energy in the form of heat comes into the engine from a high-temperature source, some of this energy is converted into useful work, the remainder is ejected as exhaust heat into a low-temperature sink, and the system returns to its original state to run through the cycle again.

Since we're looking at cyclic engines only, the system returns to its original state at the end of each cycle, so ΔU must be 0. Therefore, by the First Law of Thermodynamics, $Q_{net} = -W$. That is, the net heat absorbed by the system is equal to the work performed by the system. The heat absorbed from the high-temperature source is denoted Q_H (H for *hot*), and the heat that is discharged into the low-temperature reservoir is denoted Q_C (C for *cold*). Because heat coming *in* is positive and heat going *out* is negative, Q_H is positive and Q_C is negative, and the net heat absorbed is $Q_H + Q_C$. Instead of writing Q_{net} in this way, it's customary to write it as $Q_H - |Q_C|$, to show explicitly that Q_{net} is less than Q_H.

All engines function to convert heat into work by exchanging energy from a reservoir at higher temperature to reservoirs of lower temperature. A good example of this is a refrigerator. In a fridge, a liquid moves through tubes to remove heat from the internal chamber (such as the freezer) to the outer air. Thus, cold is not actually a concept, but the sensation of low heat content. This is why the back of a refrigerator feels hot.

This is one of the forms of

The Second Law of Thermodynamics

For any cyclic heat engine, some exhaust heat is always produced. It's impossible to completely convert heat into useful work.

Example 9 A heat engine draws 800 J of heat from its high-temperature source and discards 450 J of exhaust heat into its cold-temperature reservoir during each cycle. How much work does this engine perform per cycle?

Solution. The absolute value of the work output per cycle is equal to the difference between the heat energy drawn in and the heat energy discarded:

$$|W| = Q_H - |Q_C| = 800\text{ J} - 450\text{ J} = 350\text{ J}$$

Specific Heat

Unlike conduction, there is no easy way to calculate the rate of heat transfer for convection or radiation. However, whenever the temperature of an object changes due to heat transfer, regardless of the method, the **specific heat capacity, *c***, allows us to calculate the total amount of heat required. You can think of the specific heat capacity (or just specific heat) as describing how difficult it is to change the temperature of an object. The amount of heat required to change an object by a temperature, ΔT, is given by

$$Q = mc\Delta T$$

Equation Sheet

The higher the specific heat, the more heat needs to be added to increase the temperature, and vice versa. However, the amount of heat also depends on the mass, m, of the object. A single droplet of water in an empty hot pan will boil very easily, but a full pot of water will take a lot more heat to boil. Note that the equation works the other way, and we can use it to find the amount of heat emitted by an object if it decreases in temperature.

HEAT TRANSFER

There are three principal modes by which energy can be transferred: conduction, convection, and radiation.

Conduction

An iron skillet is sitting on a hot stove, and you accidentally touch the handle. You notice right away that there's been a transfer of thermal energy to your hand. The process by which this happens is known as conduction. The highly agitated atoms in the handle of the hot skillet bump into the atoms of your hand, making them vibrate more rapidly, thus heating up your hand.

Heat conducts from one point to another only if there is a temperature difference between the two objects. The rate at which heat is transferred is given by

Equation Sheet

$$\frac{Q}{\Delta t} = \frac{kA\Delta T}{L}$$

The constant k in the conduction equation is the "thermal conductivity" of an object. This is a property of the object and explains why things like metals get hot faster than plastics. It is also responsible for double-paned windows being "more energy efficient," as the layer of air between the two sheets of glass has a much smaller thermal conductivity than a single sheet of glass.

Convection

As the air around a candle flame warms, it expands, becomes less dense than the surrounding cooler air, and thus rises due to buoyancy. As a result, heat is transferred away from the flame by the large-scale (from the atoms' point of view, anyway) motion of a fluid (in this case, air). This is convection.

Radiation

Sunlight on your face warms your skin. Radiant energy from the Sun's fusion reactions is transferred across millions of kilometers of essentially empty space via electromagnetic waves. Absorption of the energy carried by these light waves defines heat transfer by radiation.

Chapter 3 Review Questions

Answers and explanations can be found in Chapter 11.

Section I: Multiple Choice

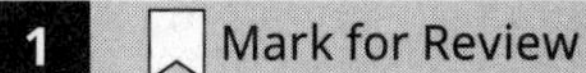

A container holds a mixture of two gases, CO_2 and H_2, in thermal equilibrium. Let K_C and K_H denote the average kinetic energy of a CO_2 molecule and an H_2 molecule, respectively. Given that a molecule of CO_2 has 22 times the mass of a molecule of H_2, the ratio K_C / K_H is equal to

(A) 1/22

(B) 1

(C) $\sqrt{22}$

(D) 22

2 Mark for Review

If the temperature and volume of a sample of an ideal gas are both doubled, then a *P*–*V* diagram will show what sort of shape?

(A) A horizontal line

(B) A vertical line

(C) A curve indicating a positive correlation

(D) A curve indicating a negative correlation

3 Mark for Review

In three separate experiments, a gas is transformed from state P_i, V_i to state P_f, V_f along the paths (1, 2, and 3) illustrated in the figure below:

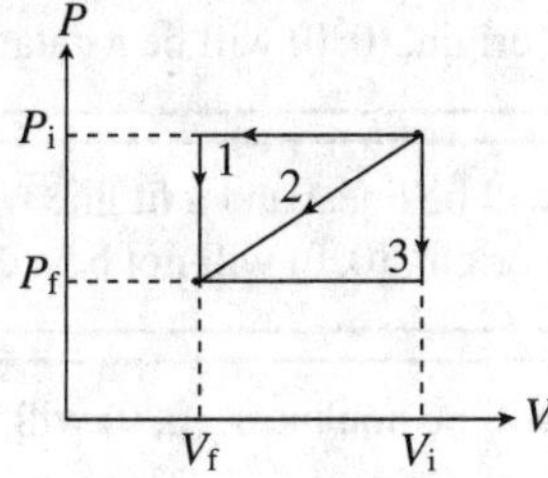

The work done on the gas is

(A) greatest for path 1

(B) least for path 2

(C) greatest for path 2

(D) the same for all three paths

4 Mark for Review

An ideal gas is compressed isothermally from 20 m^3 to 10 m^3. During this process, 5 J of work is done to compress the gas. What is the change of internal energy for this gas?

(A) –10 J

(B) –5 J

(C) 0 J

(D) 5 J

Questions 5 through 6 refer to the following.

An ideal gas is confined in a container with a fixed volume. The amount of gas, *n*, is slowly increased in the container. This experiment is done in such a way that the temperature of the gas remains constant. Pressure data is collected.

5 Mark for Review

Which describes a graph with pressure on the vertical axis and amount of gas on the horizontal axis?

Ⓐ The graph will be linear and a fit line will go through the origin. (0, 0) will be a data point.

Ⓑ The graph will be linear and a fit line will go through the origin. (0, 0) will not be a data point.

Ⓒ The graph will be nonlinear. (0, 0) will be a data point.

Ⓓ The graph will be nonlinear. (0, 0) will not be a data point.

6 Mark for Review

If the temperature of the gas being added is higher than the desired temperature, how could the experiment be done so that, as the amount of gas is increased, the temperature stays at the desired temperature?

Ⓐ Allow the pressure of the gas to decrease as the amount of gas is increased.

Ⓑ Allow work to be done on the gas as the amount of gas is increased.

Ⓒ Allow work to be done by the gas as the amount of gas is increased.

Ⓓ Allow heat to leave the gas as the amount of gas is increased.

7 Mark for Review

Through a series of thermodynamic processes, the internal energy of a sample of confined gas is increased by 560 J. If the net amount of work done on the sample by its surroundings is 320 J, how much heat was transferred between the gas and its environment?

Ⓐ 240 J absorbed

Ⓑ 240 J dissipated

Ⓒ 880 J absorbed

Ⓓ 880 J dissipated

8 Mark for Review

What is the total work performed on the gas as it is transformed from state *a* to state *c*, along the path indicated?

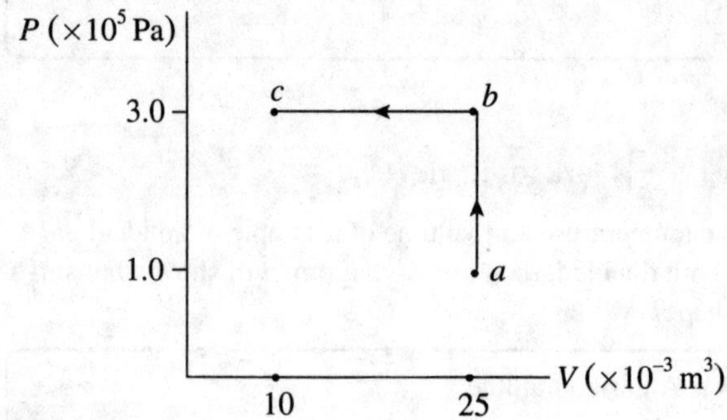

Ⓐ 1500 J

Ⓑ 3000 J

Ⓒ 4500 J

Ⓓ 9500 J

9 Mark for Review

In one of the steps of the Carnot cycle, the gas undergoes an isothermal expansion. Which of the following statements is true concerning this step?

(A) No heat is exchanged between the gas and its surroundings, because the process is isothermal.

(B) The temperature decreases because the gas expands.

(C) The internal energy of the gas remains constant.

(D) The internal energy of the gas decreases due to the expansion.

10 Mark for Review

A cup of hot coffee is sealed inside a perfectly thermally insulating container. A long time is allowed to pass. Which of the following correctly explains the final thermal configuration within the box?

(A) The coffee has not changed temperature because the box is perfectly insulating.

(B) The coffee has gotten warmer, and the air in the box has gotten cooler because of an exchange of thermal energy between the air and the coffee.

(C) The coffee has gotten cooler, and the air in the box has gotten warmer because of an exchange of thermal energy between the air and the coffee.

(D) The coffee has gotten cooler, but the air in the box has not changed its temperature. The energy from the coffee has caused an increase in entropy within the box.

Section II: Free Response

1 Mark for Review

When a system is taken from state *a* to state *b* along the path *acb* shown in Figure 1, 70 J of heat flows into the system, and the system does 30 J of work.

Figure 1

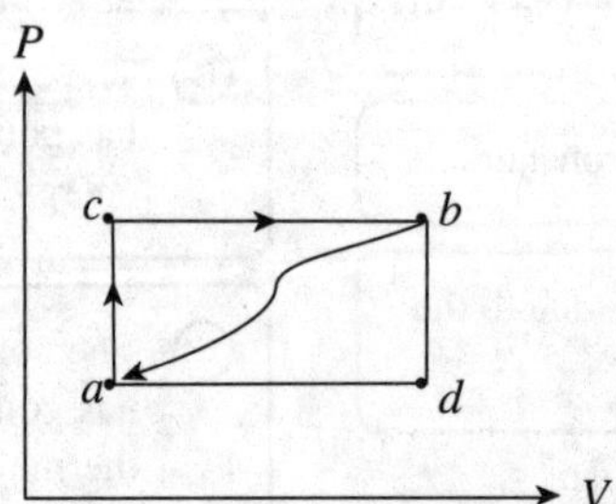

A. When the system is returned from state *b* to state *a* along the curved path shown, 60 J of heat flows out of the system. Does the system perform work on its surroundings, or do the surroundings perform work on the system? How much work is done?

B. If the system does 10 J of work in transforming from state *a* to state *b* along path *adb*, does the system absorb or does it emit heat? How much heat is transferred?

C. If $U_a = 0$ J and $U_d = 30$ J, determine the heat absorbed in the processes *db* and *ad*.

D. For the process *adbca*, identify each of the following quantities as positive, negative, or zero:

$W =$ ____________ $Q =$ ____________ $\Delta U =$ ____________

2 Mark for Review

A 0.4 mol sample of an ideal diatomic gas undergoes slow changes from state *a* to state *b* to state *c* and back to *a* along the cycle shown in the *P*–*V* figure below:

Figure 1

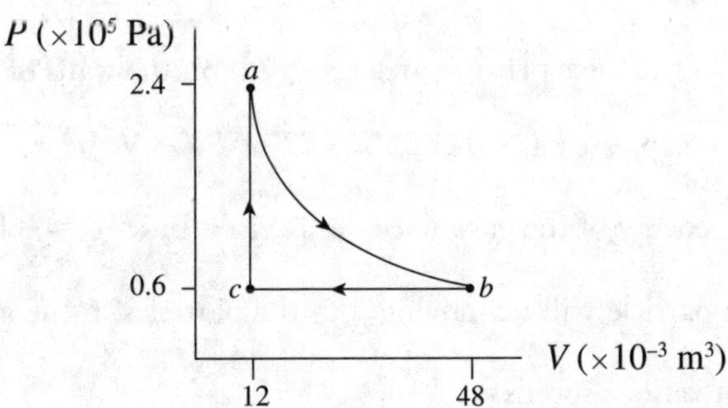

Path *ab* is an isotherm, and it can be shown that the work done by the gas as it changes isothermally from state *a* to state *b* is given by the equation

$$W_{ab} = -nRT \times \ln \frac{V_b}{V_a}$$

A. What's the temperature of

   i. state *a*?

   ii. state *b*?

   iii. state *c*?

B. In order for step *ab* to be isothermal, the gas must be kept in thermal equilibrium with its surroundings at all times. Because heat flows spontaneously from hot to cold systems, does this imply that there is no heat flow during step *ab*? Explain.

C. How much work, W_{ab}, is done by the gas during step *ab*?

D. What is the total work done over cycle *abca*?

Chapter 3 Summary

- For gases, there are a few important ideas to understand:
    - Pressure is the result of the molecules colliding with one another and with the sides of the container. It is defined as the force per unit area ($P = F/A$) on the walls of the container.
    - The Ideal Gas Law is expressed as either $PV = nRT$ or $PV = Nk_BT$.
    - The average kinetic energy of the gas molecules is given by $K_{avg} = \frac{3}{2}k_bT$. This is related to the most likely speed a particle will be moving, but the particles in the gas will be moving with a large distribution of various speeds.
- The work done can be found by the area under a P–V graph. If there is a volume change while the pressure remains constant, then the equation $W = -P\Delta V$ can be used to calculate the work.
- The First Law of Thermodynamics is $\Delta U = Q + W$, where ΔU depends only on the temperature change.
    - W positive means energy is being added to the system, so work is done **on** the system **by** the surroundings. A negative W means that energy is being subtracted from the system, so work is done **by** the system **on** the surroundings.

- Q positive (or negative) means energy is being added to (or subtracted from) the system by means of a flow of heat from the higher temperature surroundings (or system) to the lower temperature system (or surroundings).

  Note that some textbooks define work in thermodynamics in a different way: work is considered to be positive when work is done on the surroundings. This is consistent with the idea that the overall objective of a heat engine is to produce external (positive) work. Under this definition, the First Law of Thermodynamics must be written as $U = Q - W$ (or $U + W = Q$), and W must then be interpreted differently from Q. That is, while Q is still positive when heat is being added to the system, W is now positive when work is being done by the system on the surroundings (thus decreasing the internal energy of the system).

- The rate at which heat is transferred is given by $\Delta Q/\Delta t = \frac{kA\Delta T}{L}$, where k is the thermal conductivity (a property of the material), A is the cross-sectional area, ΔT is the temperature difference between the two sides, and L is the thickness or distance between the two ends of the material.
- Special thermal processes are Isovolumetric, Isothermal, Isobaric, and Adiabatic. Be sure to review the chart on page 90.

Chapter 4
Electric Forces and Fields

INTRODUCTION

The existence of mass in an object results in that object having the ability to experience the gravitational force, and thereby interact with other objects that also have mass. All objects in the universe that have been discovered thus far have mass (although the particle known as the neutrino has such a small mass that it is practically massless). Another fundamental property of objects is charge. A charged object will interact with another charged object through the electromagnetic force, similar to massive objects. The property of charge is as fundamental to physics as the property of mass.

ELECTRIC CHARGE

The basic components of atoms are protons, neutrons, and electrons. Protons and neutrons form the nucleus (and are referred to collectively as *nucleons*), while the electrons keep their distance, swarming around the nucleus. Most of an atom consists of empty space. In fact, if a nucleus were the size of the period at the end of this sentence, then the electrons would be 5 meters away. So what holds such an apparently tenuous structure together? One of the most powerful forces in nature: the *electromagnetic force*. Protons and electrons have a quality called **electric charge** that gives them an attractive force. Electric charge comes in two varieties: positive and negative. A positive particle always attracts a negative particle, and particles of the same charge always repel each other. Protons are positively charged, and electrons are negatively charged.

Protons and electrons are intrinsically charged, but bulk matter is not. This is because the amount of charge on a proton exactly balances the charge on an electron, which is quite remarkable in light of the fact that protons and electrons are very different particles. Since most atoms contain an equal number of protons and electrons, their overall electric charge is 0, because the negative charges cancel out the positive charges. Therefore, in order for matter to be **charged**, an imbalance between the numbers of protons and electrons must exist. This can be accomplished by either the removal or addition of electrons (that is, by the **ionization** of some of the object's atoms). If you remove electrons, then the object becomes positively charged, while if you add electrons, then it becomes negatively charged. Furthermore, charge is **conserved**. For example, if you rub a glass rod with a piece of silk, then the silk will acquire a negative charge and the glass will be left with an *equal* positive charge. *Net charge cannot be created or destroyed.* (*Charge* can be created or destroyed—it happens all the time—but *net* charge cannot.)

The magnitude of charge on an electron (and therefore on a proton) is denoted e. This stands for **elementary charge** because it's the basic unit of electric charge. The charge of an ionized atom must be a whole number times e because charge can be added or subtracted only in lumps of size e. For this reason, we say that charge is **quantized**. To remind us of the quantized nature of electric charge, the charge of a particle (or object) is denoted by the letter q. In the SI system of units, charge is expressed in **coulombs** (abbreviated **C**). One coulomb is a tremendous amount of charge, about 10^{18} electrons. The value of e is about 1.6×10^{-19} C.

COULOMB'S LAW

The electric force between two charged particles obeys a law that is very similar to that describing the gravitational force between two masses: they are both inverse-square laws. The **electric force** between two particles with charges of q_1 and q_2, separated by a distance r, has a magnitude given by the equation

$$\left|\vec{F}_E\right| = \frac{1}{4\pi\varepsilon_0}\frac{|q_1 q_2|}{r^2} = k\frac{|q_1 q_2|}{r^2}$$

Equation Sheet

This is **Coulomb's Law.** If we were to leave off the absolute value bars, we would interpret a negative F_E as an attraction between the charges and a positive F_E as a repulsion. The value of the proportionality constant, k, depends on the material between the charged particles. In empty space (vacuum)—or air, for all practical purposes—it is called **Coulomb's constant** and has the approximate value $k_0 = 9 \times 10^9$ N·m²/C². The expression k_0 is sometimes written in terms of a fundamental constant known as the **permittivity of free space**, denoted ε_0:

$$k_0 = \frac{1}{4\pi\varepsilon_0}$$

You should notice a similarity between the form of Coulomb's Law and Newton's Law of Universal Gravitation you learned in AP Physics 1. Both of these force laws depend on the product of the material property that generates the force (charge or mass), and both are inversely proportional to the square of the separation distance. These two laws are therefore known as **inverse square laws.**

Example 1 Consider the proton and electron in hydrogen. The proton has a mass of 1.6×10^{-27} kg and a charge of $+e$. The electron has a mass of 9.1×10^{-31} kg and a charge of $-e$. In hydrogen, they are separated by a Bohr radius, about 0.5×10^{-10} m. What is the electric force between the proton and electron in hydrogen? What about the gravitational force between the proton and electron?

Solution. The electric force between the proton and the electron is given by Coulomb's Law:

$$F_E = \frac{1}{4\pi\varepsilon_0}\frac{q_1 q_2}{r^2} = (9\times10^9\ \text{N}\cdot\text{m}^2/\text{C}^2)\frac{(1.6\times10^{-19}\ \text{C})(-1.6\times10^{-19}\ \text{C})}{(0.5\times10^{-10}\ \text{m})^2} = -9.24\times10^{-8}\ \text{N}$$

The fact that F_E is negative means that the force is one of *attraction*, which we naturally expect, since one charge is positive and the other is negative. The force between the proton and electron is along the line that joins the charges, as we've illustrated below. The two forces shown form an action/reaction pair.

$$q_1\oplus\xrightarrow{\mathbf{F}_E}\qquad\xleftarrow{\mathbf{F}_E}\ominus q_2$$

Now, the gravitational force between the two charges is given by Newton's Law of Gravitation:

$$\mathbf{F}_G = G \times \frac{(m_1 m_2)}{r^2} = (6.67 \times 10^{-11} \frac{\text{kg} \cdot \text{m}^2}{\text{kg}^2}) \times \frac{(1.67 \times 10^{-27}\ \text{kg})(9.11 \times 10^{-31}\ \text{kg})}{(0.5 \times 10^{-10}\ \text{m})^2} = 4.06 \times 10^{-47}\ \text{N}$$

Now, compare the orders of magnitude of the electric force to the gravitational force. The electric force has an order of magnitude of 10^{-8} N, but the gravitational force has an order of magnitude of 10^{-47} N! This means that the electric force is something like 10^{39} times larger! This is the reason that in problems in which we calculate electrostatic forces, we often neglect gravitational forces.

Addition of Electric Forces

Consider three point charges: q_1, q_2, and q_3. The total electric force acting on, say, q_2 is simply the sum of $\mathbf{F}_{1\text{-on-}2}$, the electric force on q_2 due to q_1, and $\mathbf{F}_{3\text{-on-}2}$, the electric force on q_2 due to q_3:

$$\mathbf{F}_{\text{on } 2} = \mathbf{F}_{1\text{-on-}2} + \mathbf{F}_{3\text{-on-}2}$$

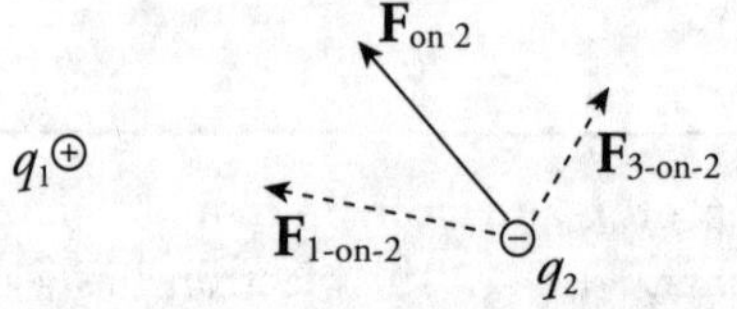

Example 2 Consider four equal, positive point charges that are situated at the vertices of a square. Find the net electric force on a negative point charge placed at the square's center.

Solution. Refer to the diagram below. The attractive forces due to the two charges on each diagonal cancel out: $F_1 + F_3 = 0$, and $F_2 + F_4 = 0$, because the distances between the negative charge and the positive charges are all the same and the positive charges are all equivalent. Therefore, by symmetry, the net force on the center charge is zero.

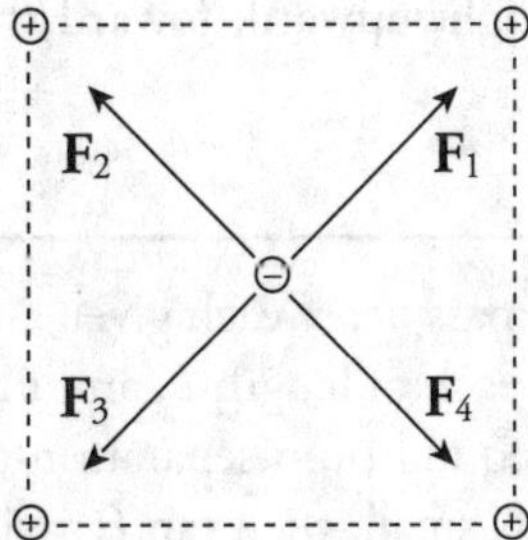

Example 3 If the two positive charges on the bottom side of the square in the previous example were removed, what would be the net electric force on the negative charge? Assume that each side of the square is 4.0 cm, each positive charge is 1.5 ∝C, and the negative charge is –6.2 nC.

Solution. If we break down $\mathbf{F}_1$ and $\mathbf{F}_2$ into horizontal and vertical components, then by symmetry the two horizontal components will cancel each other out, and the two vertical components will add:

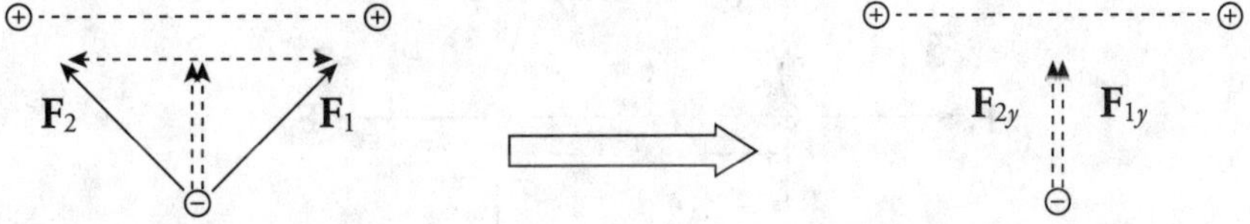

Since the diagram on the left shows the components of $\mathbf{F}_1$ and $\mathbf{F}_2$ making right triangles with legs each of length 2 cm, it must be that $F_{1y} = F_1 \sin 45°$ and $F_{2y} = F_2 \sin 45°$. Also, the magnitude of $\mathbf{F}_1$ equals that of $\mathbf{F}_2$. So the net electric force on the negative charge is $F_{1y} + F_{2y} = 2F \sin 45°$, where F is the strength of the force between the negative charge and each of the positive charges.

If s is the length of each side of the square, then the distance r between each positive charge and the negative charge is $r = \frac{1}{2}s\sqrt{2}$ and

$$F_E = 2F\sin 45° = 2\frac{1}{4\pi\varepsilon_0}\frac{q_1q_2}{r^2}\sin 45°$$

$$= 2(9\times10^9\ \text{N}\cdot\text{m}^2/\text{C}^2)\frac{(1.5\times10^{-6}\ \text{C})(6.2\times10^{-9}\ \text{C})}{(\frac{1}{2}\cdot 4.0\times10^{-2}\cdot\sqrt{2}\ \text{m})^2}\sin 45°$$

$$= 0.15\ \text{N}$$

The direction of the net force is straight upward, toward the center of the line that joins the two positive charges.

Example 4 Two pith balls of mass m are each given a charge of $+q$. They are hung side-by-side from two threads each of length L, and move apart as a result of their electrical repulsion. Find the equilibrium separation distance x in terms of m, q, and L. (Use the fact that if θ is small, then $\tan\theta \approx \sin\theta$.)

Solution. Three forces act on each ball: weight, tension, and electrical repulsion:

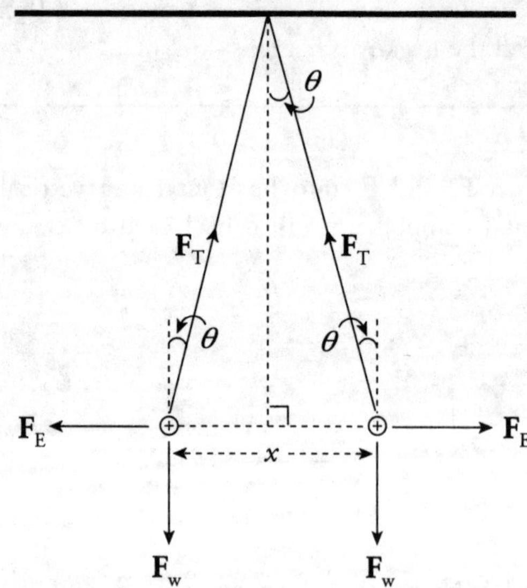

When the balls are in equilibrium, the net force each feels is zero. Therefore, the vertical component of $\mathbf{F}_T$ must cancel out $\mathbf{F}_w$ and the horizontal component of $\mathbf{F}_T$ must cancel out $\mathbf{F}_E$:

$$F_T\cos\theta = F_w \qquad \text{and} \qquad F_T\sin\theta = F_E$$

Dividing the second equation by the first, we get tan $\theta = F_E/F_w$. Therefore,

$$\tan\theta = \frac{k\frac{q^2}{x^2}}{mg} = \frac{kq^2}{mgx^2}$$

Now, to approximate: if θ is small, tan $\theta \approx$ sin θ, and from the diagram, sin $\theta = \frac{1}{2}x/L$. Therefore, the equation above becomes

$$\frac{\frac{1}{2}x}{L} = \frac{kq^2}{mgx^2} \Rightarrow \frac{1}{2}mgx^3 = kq^2L \Rightarrow x = \sqrt[3]{\frac{2kq^2L}{mg}}$$

THE ELECTRIC FIELD

The presence of a massive body such as the Earth causes objects to experience a gravitational force directed toward the Earth's center. For objects located outside the Earth, this force varies inversely with the square of the distance and directly with the mass of the gravitational source. A vector diagram of the gravitational field surrounding the Earth looks like this:

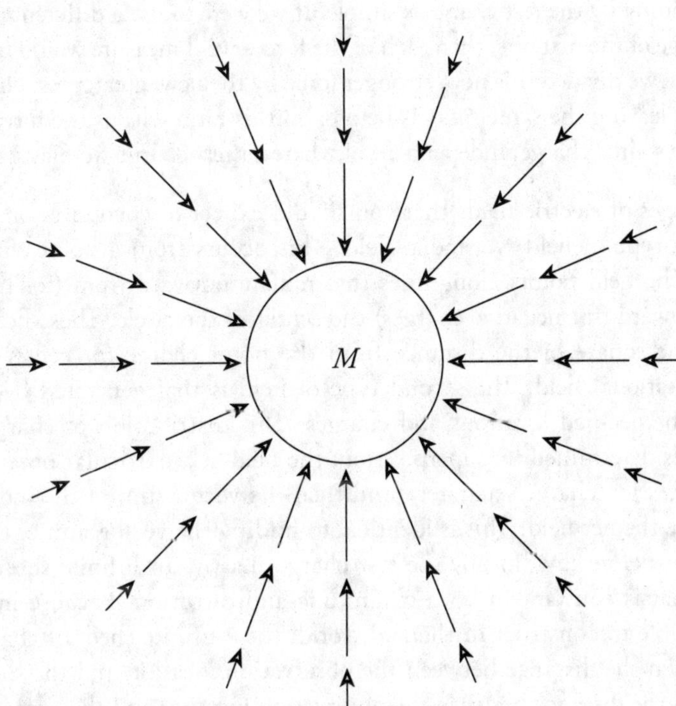

We can think of the space surrounding the Earth as permeated by a **gravitational field** created by the Earth. Any mass placed in this field then experiences a gravitational force due to an interaction with this field.

In general, a field is a physical quantity that has a value at each point in space-time. The values can be scalars (for example, the temperature at each location in a house) or vectors (like the gravitational field in space due to the mass of the Earth). All non-contact forces in physics—such as the gravitational force and the electromagnetic force—are described as the interaction between one object and the field generated by another object.

This is how we describe the electric force. Rather than having two charges reach out across empty space to each other to produce a force, we will instead interpret the interaction in the following way: the presence of a charge creates an **electric field** in the space that surrounds it. Another charge placed in the field created by the first will experience a force due to the field.

Consider a point charge Q in a fixed position and assume that it's positive. Now imagine moving a tiny positive test charge q around to various locations near Q. At each location, measure the force that the test charge experiences, and call it $\mathbf{F}_{\text{on } q}$. Divide this force by the test charge q; the resulting vector is the **electric field vector**, **E**, at that location:

Equation Sheet

$$\vec{E} = \frac{\vec{F}_E}{q}$$

The reason for dividing by the test charge is simple. If we were to use a different test charge with, say, twice the charge of the first one, then each of the forces we'd measure would be twice as much as before. But when we divided this new, stronger force by the new, greater test charge, the factors of 2 would cancel, leaving the same ratio as before. So this ratio tells us the intrinsic strength of the field due to the source charge, independent of whatever test charge we may use to measure it.

There are three types of electric fields that you should expect to encounter on the AP Physics 2 Exam. The first type of field is a radial field. This occurs from a point charge (or from a charged sphere). The field points along lines that radiate outward from (for positive charges) or point inward toward (for negative charges) the center of the circle. These fields are inversely proportional to the square of the distance from the point charge (or center of the sphere), just like the gravitational field. The second type of field is that generated by a collection of point charges with specified locations and charges. This distribution of charge results in an electric field that is determined by superposition; the field at a particular observation location is determined from each charge separately, and then the vector sum of the individual fields is taken to determine the net field. This is identical to finding the vector sum of the forces of the charge distribution before dividing by the test charge. Lastly, an infinite sheet of charge will result in a field that is constant in both magnitude and direction. Because infinite sheets of charge are impossible to construct in the real world, the "infinite sheet of charge" is a useful approximation when the distance between the observation location and the sheet of charge is small compared to the distance between the observation location and the edges of the sheet.

First, let's examine the point charge. Since the test charge used to measure the field is positive, every electric field vector would point radially away from the source charge. *If the source charge is positive, the electric field vectors point away from it; if the source charge is negative, then the field vectors point toward it.* And, since the force decreases as we get farther away from the charge (as $1/r^2$), so does the electric field. This is why the electric field vectors farther from the source charge are shorter than those that are closer.

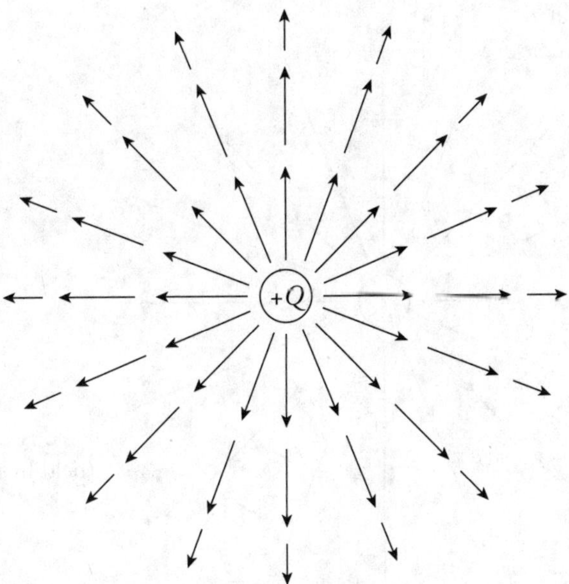

Since the force on any test charge q, due to some source Q, has a strength of $qQ/4\pi\varepsilon_0 r^2$, when we divide this by q, we get the expression for the strength of the electric field created by a point-charge source of magnitude Q:

$$E = \frac{1}{4\pi\varepsilon_0}\frac{Q}{r^2}$$

This is the electric field in the space surrounding a **point charge.**

It is of utmost importance to note that this equation resulted from considering the electric field of a single point charge. Any other distribution of charge, such as placing charge on a sheet to generate a **parallel-plate capacitor** (described later), will have a different electric field and this equation will be invalid. Adding to the confusion, there is one way to distribute a large amount of charge and get the same exact equation for the electric field as having just a single point charge. When the charges were spread out over a sphere (for example, a charged metal ball) instead of a single point charge, at any location outside of the sphere the electric field would be the same as the point charge field.

Sometimes electric fields are drawn, as shown above, with arrows at various locations. The arrows point in the field direction, and their lengths indicate the field strength. However, sometimes they are sketched simply as continuous lines, as shown below, from the source such that the electric field vector is always tangent to the line everywhere it's drawn (a single arrowhead on the line indicates the direction to draw the field line along) and the strength is determined by how close the lines are together.

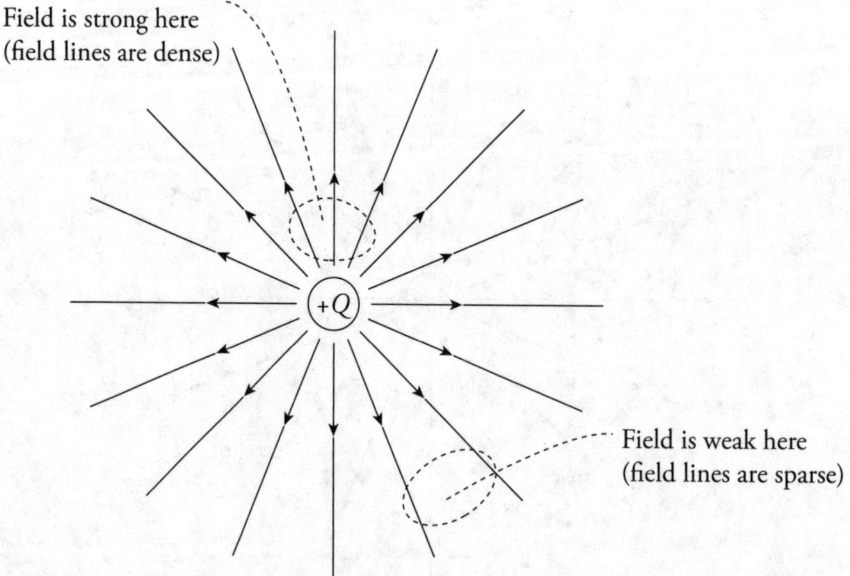

Electric fields obey the same addition properties as the electric force. If we had two source charges, their fields would overlap and effectively add; a third charge would feel the net influence of both charges. At each position in space (referred to as *observation locations*), add the electric field vector due to one of the charges to the electric field vector due to the other charge: $\mathbf{E}_{total} = \mathbf{E}_1 + \mathbf{E}_1$. This extends to any number of source charges. In the diagram below, $\mathbf{E}_1$ is the electric field vector at a particular location due to the charge $+Q$, and $\mathbf{E}_2$ is the electric field vector at that same location due to the other charge, $-Q$. Adding these vectors gives the overall field vector $\mathbf{E}_{total}$ at that location.

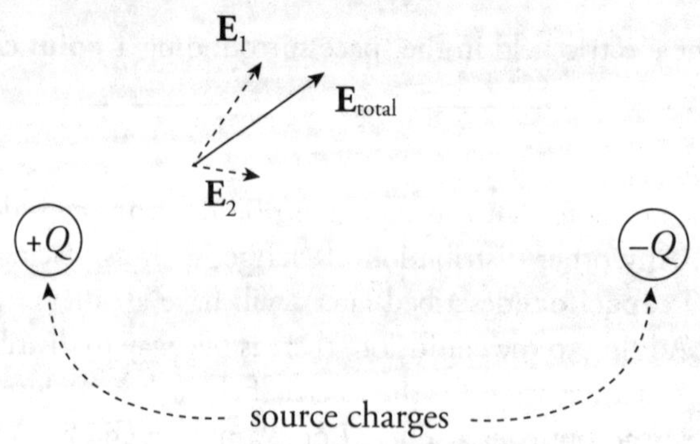

If this is done at enough locations, the electric field lines can be sketched.

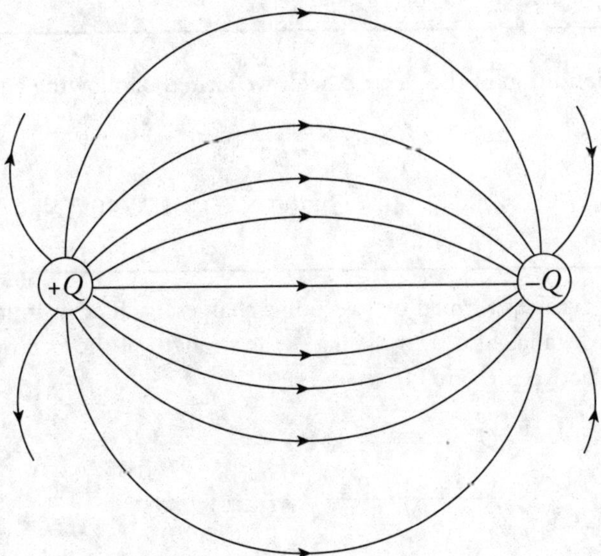

Note that, like electric field vectors, electric field lines always point away from positive source charges and toward negative ones. Two equal but opposite charges, like the ones shown in the diagram above, form a pair called an **electric dipole**.

If a positive charge $+q$ were placed in the electric field above, it would experience a force that is tangent to, and in the same direction as, the field line passing through $+q$'s location. After all, electric field lines indicate the magnitude and direction of the force a positive test charge would experience. On the other hand, if a negative charge $-q$ were placed in the electric field, it would experience a force that is tangent to, but in the direction opposite from, the field line passing through $-q$'s location.

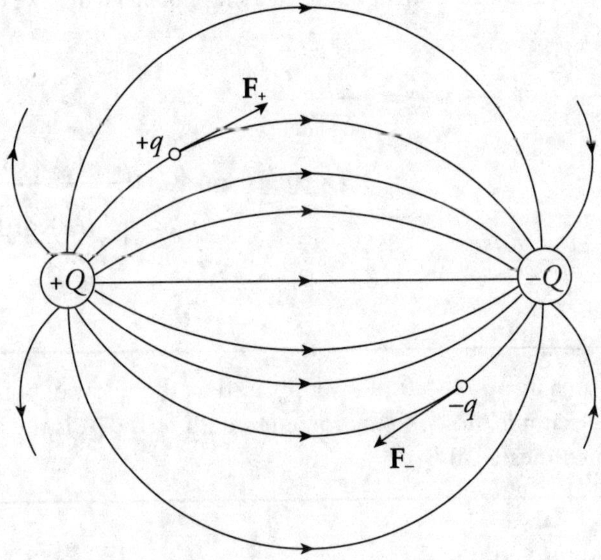

Finally, notice that electric field lines never cross.

Example 5 A charge $q = +3.0$ nC is placed at a location at which the electric field strength is 400 N/C. Find the force felt by the charge q.

Solution. From the definition of the electric field, we have the following equation:

$$\mathbf{F}_{\text{on } q} = q\mathbf{E}$$

Therefore, in this case, $F_{\text{on } q} = qE = (3 \cdot 10^{-9}\text{ C})(400\text{ N/C}) = 1.2 \cdot 10^{-6}\text{ N}$.

Example 6 A dipole is formed by two point charges, each of magnitude 4.0 nC, separated by a distance of 6.0 cm. What is the strength of the electric field at the point midway between them?

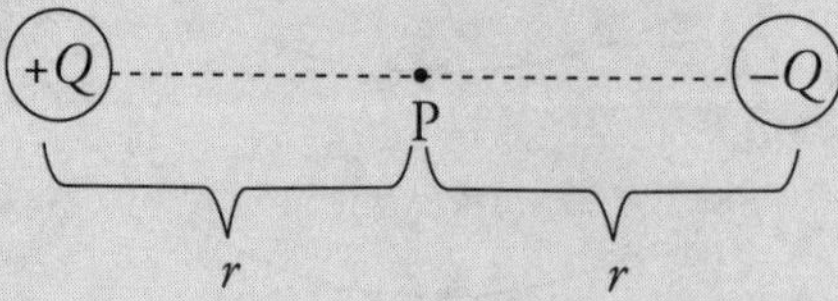

Solution. Let the two source charges be denoted +Q and –Q. At Point P, the electric field vector due to +Q would point directly away from +Q, and the electric field vector due to –Q would point directly toward –Q. Therefore, these two vectors point in the same direction (from +Q to –Q), so their magnitudes would add.

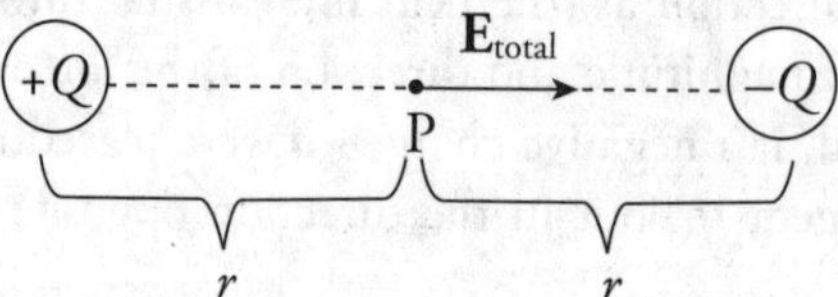

Using the equation for the electric field strength due to a single point charge, we find that

$$E_{\text{total}} = \frac{1}{4\pi\varepsilon_0}\frac{Q}{r^2} + \frac{1}{4\pi\varepsilon_0}\frac{Q}{r^2} = 2\frac{1}{4\pi\varepsilon_0}\frac{Q}{r^2}$$

$$= 2(9\times10^{-9}\text{N}\cdot\text{m}^2/\text{C}^2)\frac{4.0\times10^{-9}\text{ C}}{\left(\frac{1}{2}(6.0\times10^{-2}\text{ m})\right)^2}$$

$$= 8.0\times10^4\text{ N/C}$$

Example 7 If a charge $q = -5.0$ pC were placed at the midway point described in the previous example, describe the force it would feel. ("p" is the abbreviation for "pico-," which means 10^{-12}.)

Solution. Since the field E at this location is known, the force felt by q is easy to calculate:

$\mathbf{F}_{\text{on } q} = q\mathbf{E} = (-5.0 \cdot 10^{-12}\text{ C})(8.0 \cdot 10^4\text{ N/C to the right}) = 4.0 \cdot 10^{-7}\text{ N to the } \textit{left}$

Example 8 What can you say about the electric force that a charge would feel if it were placed at a location at which the electric field was zero?

Solution. Remember that $\mathbf{F}_{\text{on } q} = q\mathbf{E}$. So if $\mathbf{E} = 0$, then $\mathbf{F}_{\text{on } q} = 0$. (Zero field means zero force.)

The Uniform Electric Field

An important subset of problems deals with uniform electric fields. One method of creating this uniform field is to have two large conducting sheets, each storing some charge Q, some distance d apart. Near the edges of each sheet, the field may not be uniform, but near the middle, for all practical purposes, the field is uniform. Having a uniform field means having a constant force (and therefore a constant acceleration), so you can use kinematic equations just as if you had a uniform gravitational field (as you do near the surface of the Earth).

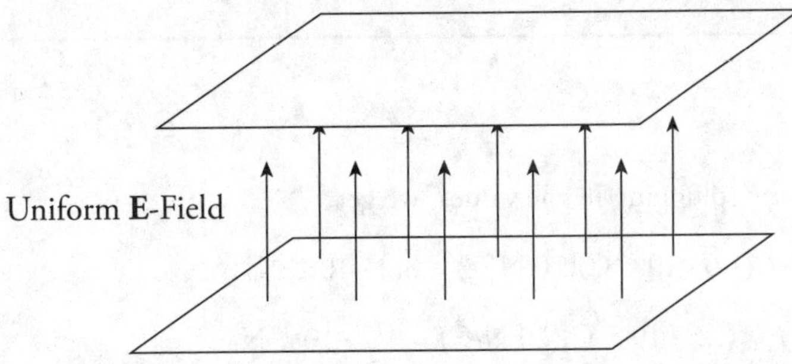

Example 9 Positive charge is distributed uniformly over a large, horizontal plate, which then acts as the source of a vertical electric field. An object of mass 5 g is placed at a distance of 2 cm above the plate. If the strength of the electric field at this location is 10^6 N/C, how much charge would the object need to have in order for the electrical repulsion to balance the gravitational pull?

Solution. Clearly, since the plate is positively charged, the object would also have to carry a positive charge so that the electric force would be repulsive.

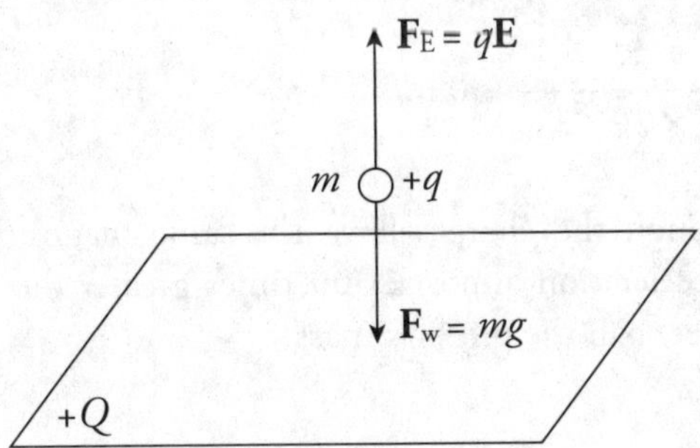

Let q be the charge on the object. Then, in order for F_E to balance mg, we must have

$$qE = mg \quad \Rightarrow \quad q = \frac{mg}{E} = \frac{(5\times10^{-3}\text{ kg})(10\text{ N/kg})}{10^{6}\text{ N/C}} = 5\times10^{-8}\text{ C} = 50\text{ nC}$$

Example 10 A proton, neutron, and electron are in a uniform electric field of 20 N/C that is caused by two large charged plates that are 30 cm apart. The particles are far enough apart that they don't interact with each other. They are released from rest equidistant from each plate.

(a) What is the magnitude of the net force acting on each particle?
(b) What is the magnitude of the acceleration of each particle?
(c) How much work will be done on the particle as it moves to coincide with one of the charged plates?
(d) What is the speed of each particle when it strikes the plate?
(e) How long does it take to reach the plate?

Solution.

(a) Since $F = qE$, plugging in the values, we get

proton: $F = (1.6 \cdot 10^{-19}\text{ C})(20\text{ N/C}) = 3.2 \cdot 10^{-18}\text{ N}$

electron: $F = (1.6 \cdot 10^{-19}\text{ C})(20\text{ N/C}) = 3.2 \cdot 10^{-18}\text{ N}$

neutron: $F = (0\text{ C})(20\text{ N/C}) = 0\text{ N}$

Note: Because the proton and electron have the same magnitude, they will experience the same force. If you're asked for the direction, the proton travels in the same direction as the electric field and the electron travels in the opposite direction as the electric field.

(b) Since $F = ma$, $a = \frac{F}{m}$. Plugging in the values, we get

proton: $a = \frac{3.2\times10^{-18}\text{ N}}{1.67\times10^{-27}\text{ kg}} = 1.9\times10^{9}\text{ m/s}^2$

electron: $a = \frac{3.2\times10^{-18}\text{ N}}{9.11\times10^{-31}\text{ kg}} = 3.5\times10^{12}\text{ m/s}^2$

neutron: $a = \frac{0\text{ N}}{1.67\times10^{-27}\text{ kg}} = 0\text{ m/s}^2$

Notice that although the charges have the same magnitude of force, the electron experiences an acceleration almost 2000 times greater due to its mass being almost 2000 times smaller than the proton's mass.

(c) Since $W = Fd$, we get $W = qEd$. Plugging in the values, recalling that the charges start midway between the plates, we get

proton: $W = (1.6 \cdot 10^{-19}\,\text{C})(20\ \text{N/C})\ (0.15\ \text{m}) = 4.8 \cdot 10^{-19}\,\text{J}$

electron: $W = (1.6 \cdot 10^{-19}\,\text{C})(20\ \text{N/C})\ (0.15\ \text{m}) = 4.8 \cdot 10^{-19}\,\text{J}$

neutron: $W = (0\,\text{C})\ (20\ \text{N/C})(0.15\ \text{m}) = 0\,\text{J}$

(d) From the Work–Energy Theorem,

$W = \Delta K = 1/2mv^2$, so $v = \sqrt{2W/m}$

Using the answer from part (c) for the work, we get

proton: $v_f = \sqrt{2(4.8 \times 10^{-19}\ \text{J})/(1.67 \times 10^{-27}\ \text{kg})} \rightarrow v_f = 24{,}000\ \text{m/s}$

electron: $v_f = \sqrt{2(4.8 \times 10^{-19}\ \text{J})/(9.11 \times 10^{-31}\ \text{kg})} \rightarrow v_f = 1.0 \times 10^6\ \text{m/s}$

neutron: The neutron never hits the plate, so the question about what speed it hits the plate with is not well posed.

Notice that, even though the force is the same and the same work is done on both charges, there is a significant difference in final velocities due to the large mass difference. An alternative solution to this would be using kinematics. You would have obtained the same answers.

(e) Recall the definition of the final velocity of a uniformly accelerated object:

$$v_f = v_i + at \rightarrow t = \frac{v_{fi} - v}{a} \rightarrow t = \frac{v_f}{a}$$

proton: $t = \dfrac{24{,}000\ \text{m/s}}{1.9 \times 10^9\ \text{m/s}^2} \rightarrow 1.3 \cdot 10^{-5}\ \text{s}$

electron: $t = \dfrac{1 \times 10^6\ \text{m/s}}{3.5 \times 10^{12}\ \text{m/s}^2} \rightarrow 2.9 \cdot 10^{-7}\ \text{s}$

neutron: The neutron never accelerates, so it will never hit the plate.

CONDUCTORS AND INSULATORS

Materials can be classified into broad categories based on their ability to permit the flow of charge. If electrons were placed on a metal sphere, they would quickly spread out and cover the outside of the sphere uniformly. These electrons would be free to flow through the metal and redistribute themselves, moving to get as far away from each other as they could. Materials that permit the flow of excess charge are called **conductors**; they conduct electricity. Metals are the best examples of conductors. Aqueous solutions that contain dissolved electrolytes (such as salt water) are also conductors. Metals conduct electricity because they bind all but their outermost electrons very tightly. That outermost electron is free to move about the metal. This creates a sort of sea of mobile (or conduction) electrons.

Insulators, on the other hand, closely guard their electrons—and even extra ones that might be added. Electrons are not free to roam throughout the atomic lattice. Examples of insulators are glass, wood, rubber, and plastic. If excess charge is placed on an insulator, it stays put. This process, called **charging by friction**, involves simply rubbing the insulator against another material, thereby stripping electrons off one material and depositing them on the other material.

Example 11 A solid sphere of copper is given a negative charge. Discuss the electric field inside and outside the sphere.

Solution. In order for the electrons to eventually come to rest, the net force on any electron *that is free* to move must be 0 N. Copper is a conducting material. Therefore, since any excess charges that reside within the body of the sphere must have a net force of 0 N, we conclude that the electric field inside a conductor is zero. Otherwise, charges within the sphere would experience a force and would move, violating the "static" part of electrostatic equilibrium.

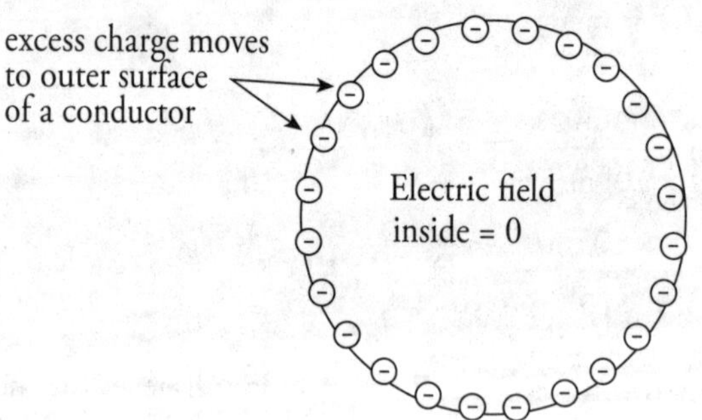

All the excess electrons that are deposited on the sphere arrange themselves on the outer surface, where they are constrained in their motion by the insulating air surrounding the sphere. The electric field within the sphere is 0 N and outside the sphere the field is found to be $E = \frac{1}{4\pi\varepsilon_0}\frac{Q}{r^2}$, which is exactly the same as the field from a single point charge Q located at the center of the conducting sphere.

In fact, you can shield yourself from electric fields simply by surrounding yourself with metal. Charges may move around on the outer surface of your cage, but within the cage, the electric field will be zero. For points outside the sphere, it can be shown that the sphere behaves as if all its excess charge were concentrated at its center. (Remember that this is just like the gravitational field due to a uniform spherical mass.) Also, *the electric field is always perpendicular to the surface, no matter what shape the surface may be.* See the diagram below.

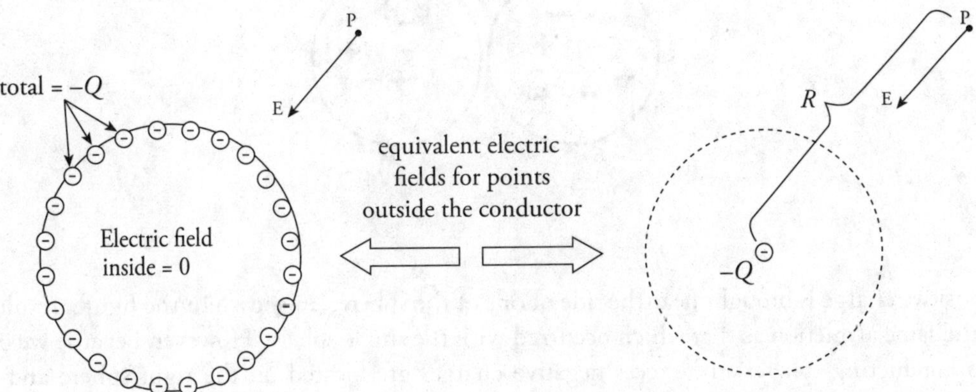

Now, let's take our previous sphere and put it in a different situation. Start with a neutral metal sphere and bring a positive charge Q nearby without touching the original metal sphere. What will happen? The positive charge will attract free electrons in the metal, leaving the far side of the sphere positively charged. Since the negative charge is closer to Q than the positive charge, there will be a net attraction between Q and the sphere. So, even though the sphere as a whole is electrically neutral, the separation of charge induced by the presence of Q will create a force of electrical attraction between them. This process for rearranging charges within a conductor is called **charge conduction**.

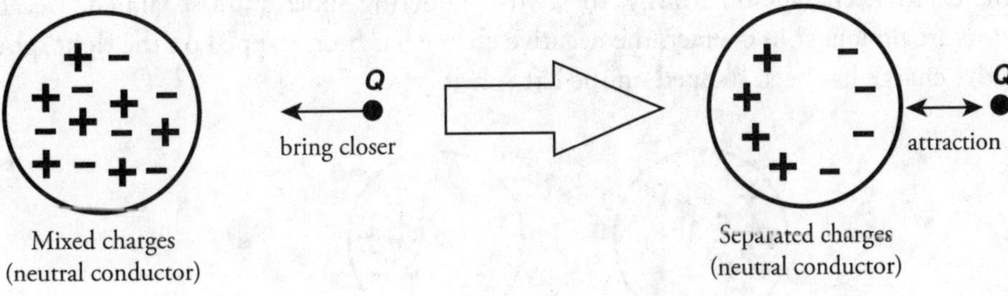

Charging by Induction

The process of charging by induction may be used to redistribute charges among a pair of neutrally charged spheres, so that in the end both spheres are charged. Imagine two neutrally charged spheres that are each set on an insulating stand. The spheres are arranged so that they are in contact with one another.

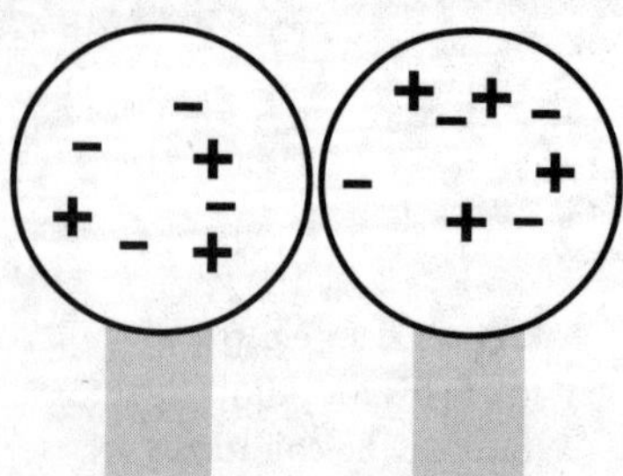

A positive charge is brought near the side of one of the spheres, as shown in the figure, resulting in the same attraction as that which occurred with the single sphere. However, because we have two conducting spheres, the excess negative charges are located on the right sphere and the positive excess charges are on the left sphere.

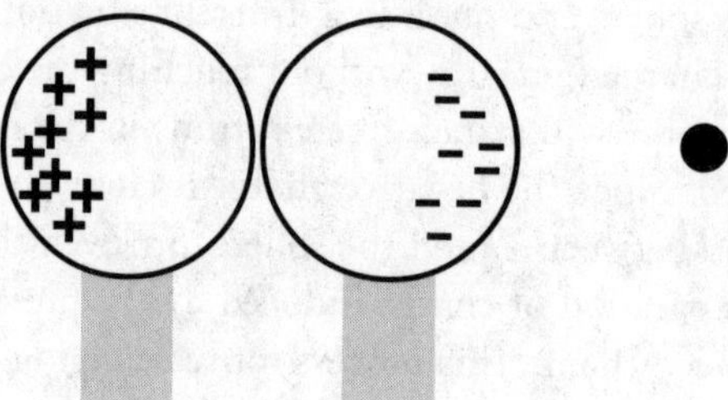

With the external charge still nearby, the two conducting spheres are separated. Because the conductors are no longer in contact, the negative charge has been trapped on the right sphere and the positive charge has been trapped on the left sphere.

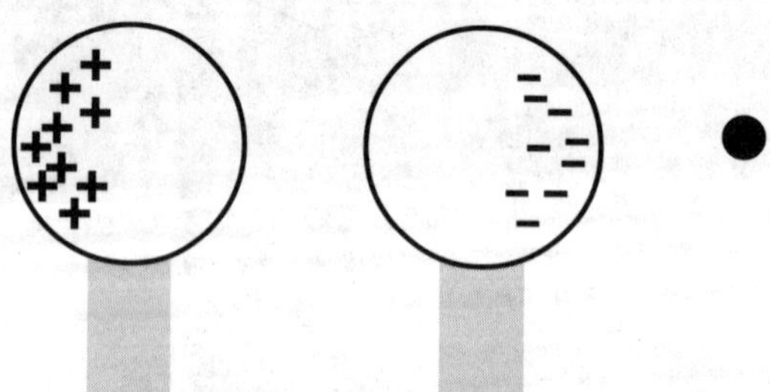

Note that the net charge is unchanged between the two spheres—the distribution of the charges has simply been changed.

Now, what if the sphere was made of glass (an insulator)? Although there aren't free electrons that can move to the near side of the sphere, the atoms that make up the sphere will become polarized. That is, their electrons will feel a tug toward Q, and so will spend more time on the side of the atom closer to Q than on the side opposite Q, causing the atoms to develop a distribution of charge that is more negative on the side nearby Q (and a partial positive charge on the other side from Q). The effect isn't as dramatic as the mass movement of free electrons in the case of a metal sphere, but the polarization is still enough to cause an electrical attraction between the sphere and Q. For example, if you comb your hair, the comb will pick up extra electrons, making it negatively charged. If you place this electric field source near little bits of paper, the paper will become polarized and will then be attracted to the comb.

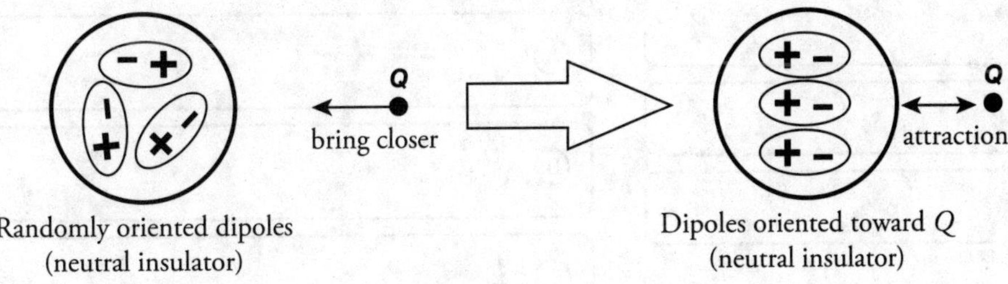

The same phenomenon, in which the presence of a charge tends to cause polarization in a nearby collection of charges, is responsible for a kind of intermolecular force. Dipole-induced forces are caused by a shifting of the electron cloud of a neutral molecule toward positively charged ions or away from negatively charged ions; in either case, the resulting force between the ion and the atom is attractive.

Chapter 4 Review Questions

Answers and explanations can be found in Chapter 11.

Section I: Multiple Choice

1 Mark for Review

An experiment is conducted measuring the electrostatic force, F, on a test object at various distances, r, from a point source. In order to create a plot with a straight line, what should be graphed?

(A) F versus r^2

(B) F versus r

(C) F versus r^{-1}

(D) F versus r^{-2}

2 Mark for Review

Two 1 kg spheres each carry a charge of magnitude 1 C. How does F_E, the strength of the electric force between the spheres, compare to F_G, the strength of their gravitational attraction?

(A) $F_E < F_G$

(B) $F_E = F_G$

(C) $F_E > F_G$

(D) If the charges on the spheres are of the same sign, then $F_E > F_G$; but if the charges on the spheres are of the opposite sign, then $F_E < F_G$.

3 Mark for Review

The figure below shows three point charges, all positive. If the net electric force on the center charge is zero, what is the value of y/x?

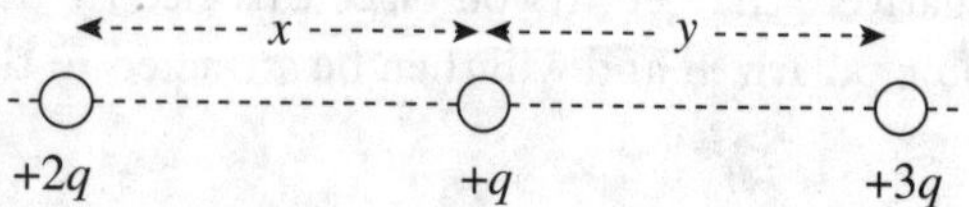

(A) $\frac{4}{9}$

(B) $\sqrt{\frac{2}{3}}$

(C) $\sqrt{\frac{3}{2}}$

(D) $\frac{3}{2}$

4 Mark for Review

+Q　　P　　−Q

The figure above shows two point charges, $+Q$ and $-Q$. If the negative charge were absent, the electric field at Point P due to $+Q$ would have strength E. With $-Q$ in place, what is the strength of the total electric field at P, which lies at the midpoint of the line segment joining the charges?

(A) 0

(B) $\frac{E}{2}$

(C) E

(D) $2E$

5 Mark for Review

A sphere of charge $+Q$ is fixed in position. A smaller sphere of charge $+q$ is placed near the larger sphere and released from rest. The small sphere will move away from the large sphere with

(A) decreasing velocity and decreasing acceleration

(B) decreasing velocity and increasing acceleration

(C) increasing velocity and decreasing acceleration

(D) increasing velocity and increasing acceleration

6 Mark for Review

An object of charge $+q$ feels an electric force $\mathbf{F}_E$ when placed at a particular location in an electric field, **E**. Therefore, if an object of charge $-2q$ were placed at the same location where the first charge was, it would feel an electric force of

(A) $\frac{-\mathbf{F}_E}{2}$

(B) $-2\mathbf{F}_E$

(C) $-2q\mathbf{F}_E$

(D) $\frac{-2\mathbf{F}_E}{q}$

7 Mark for Review

A charge of $-3Q$ is transferred to a solid metal sphere of radius r. How will this excess charge be distributed?

(A) $-Q$ at the center, and $-2Q$ on the outer surface

(B) $-3Q$ at the center

(C) $-3Q$ on the outer surface

(D) $-Q$ at the center, $-Q$ in a ring of radius $\frac{1}{2}r$, and $-Q$ on the outer surface

Section II: Free Response

1 Mark for Review

In Figure 1, four charges are situated at the corners of a square of side length s. The charges on opposite corners are equal to one another and are labeled Q for corners A and C and labeled q for corners B and D. The charge on Q is positive for all experiments.

Figure 1

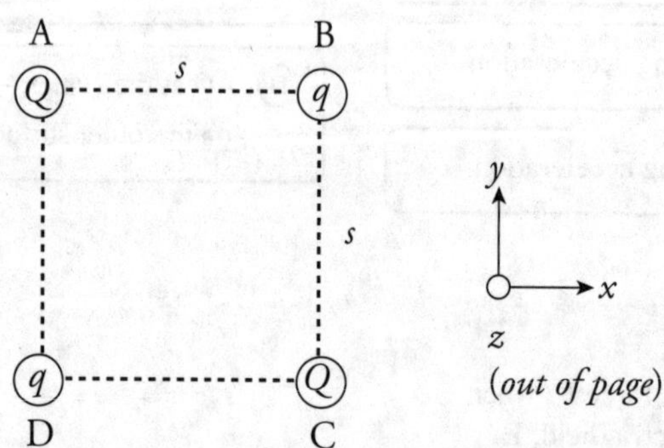

A. In a first experiment, it is found that the force on the charge at position C is 0 N.

  i. Justify the assertion that the charges q cannot have different magnitudes in this experiment.

  ii. Derive an expression for the magnitude and sign of charge on q in this experiment.

B. In a second experiment, the charge at point C is removed. If the charges q are positive, is there anywhere within the boundary square where the electric field could be 0 N/C?

C. The charges are reassembled into their original positions. In a clear, coherent paragraph-length response, explain why the electric field at the center of the square must be 0 N/C regardless of the magnitudes of the charges q or Q and regardless of their signs.

2 Mark for Review

Two charges, $+Q$ and $+2Q$, are fixed in place along the y-axis of an xy-coordinate system as shown in Figure 1. Charge 1 is at the point $(0, a)$, and Charge 2 is at the point $(0, -2a)$.

Figure 1

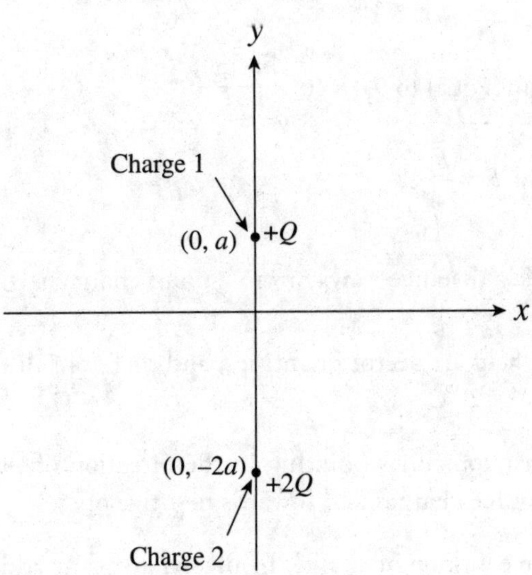

A. Find the electric force (magnitude and direction) felt by Charge 1 due to Charge 2.

B. Find the electric field (magnitude and direction) at the origin created by both Charges 1 and 2.

C. Is there a point on the x-axis where the total electric field is zero? If so, where? If not, explain briefly.

D. Is there a point on the y-axis where the total electric field is zero? If so, where? If not, explain briefly.

Chapter 4 Summary

- Coulomb's Law describes magnitude of the force acting on two point charges and is given by

$$\left|\vec{F}_E\right| = \frac{1}{4\pi\varepsilon_0}\frac{|q_1q_2|}{r^2} = k\frac{|q_1q_2|}{r^2}$$

  where $\frac{1}{4\pi\varepsilon_0} = k$ is a constant equal to $9.0 \times 10^9\ \frac{\text{N}\cdot\text{m}^2}{\text{C}^2}$.

- The electric field is given by $\vec{E} = \frac{\vec{F}_E}{q}$.

- The electric field magnitude a distance r away from a point charge is $\left|\vec{E}\right| = \frac{1}{4\pi\varepsilon_0}\frac{|q|}{r^2} = k\frac{|q|}{r^2}$.

- Both the electric force and field are vector quantities, and therefore all the rules for vector addition apply.

- Electric field lines are continuous lines pointing in the direction of the electric field. They always point away from positive source charges and towards negative ones.

- Conductors allow for the free motion of charge. Insulators keep any added charge in place. The electric field inside any conductor is always zero.

Chapter 5
Electric Potential and Capacitance

INTRODUCTION

When an object moves in a gravitational field, it usually experiences a change in kinetic energy and in gravitational potential energy due to the work done on the object by gravity. Similarly, when a charge moves in an electric field, it generally experiences a change in kinetic energy and in electrical potential energy due to the work done on it by the electric field. By exploring the idea of electric potential, we can simplify our calculations of work and energy changes within electric fields.

ELECTRICAL POTENTIAL ENERGY

When a charge moves in an electric field, unless its displacement is always perpendicular to the field, the electric force does work on the charge. If W_E is the work done by the electric force, then the change in the charge's **electrical potential energy** is defined by

$$\Delta U_E = -W_E$$

Notice that this is the same equation that defined the change in the gravitational potential energy of an object of mass m undergoing a displacement in a gravitational field ($\Delta U_G = -W_G$).

Example 1 A positive charge $+q$ moves from position A to position B in a uniform electric field **E**:

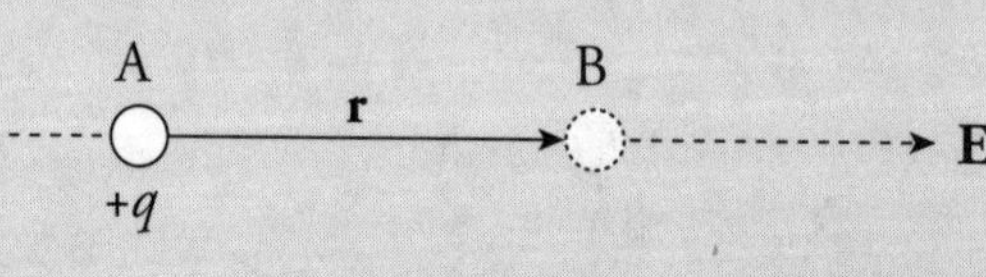

What is its change in electrical potential energy?

Solution. Since the field is uniform, the electric force that the charge feels, $\mathbf{F}_E = q\mathbf{E}$, is constant. Since q is positive, $\mathbf{F}_E$ points in the same direction as **E**, and, as the figure shows, they point in the same direction as the displacement, r. This makes the work done by the electric field equal to $W_E = F_E r = qEr$, so the change in the electrical potential energy is

$$\Delta U_E = -qEr$$

Note that the change in potential energy is negative, which means that potential energy has decreased; this always happens when the field does positive work. It's just like dropping a rock to the ground: gravity does positive work, and the rock loses gravitational potential energy.

Example 2 Do the previous problem, but consider the case of a negative charge, $-q$.

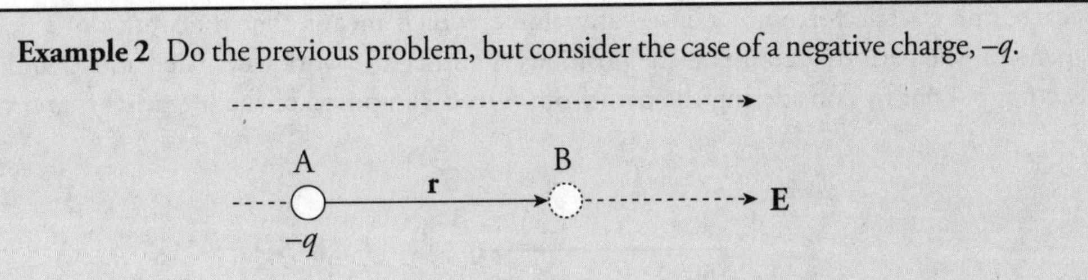

Solution. In this case, an outside agent must be pushing the charge to make it move, because the electric force *naturally* pushes negative charges against field lines. Therefore, we expect that the work done by the electric field is negative. The electric force, $\mathbf{F}_E = (-q)\mathbf{E}$, points in the direction opposite to the displacement, so the work it does is $W_E = -F_E r = -qEr = -qEr$. Thus, the change in electrical potential energy is positive: $\Delta U_E = -W_E = -(-qEr) = qEr$. Because the change in potential energy is positive, the potential energy increases; this always happens when the field does negative work. It's like lifting a rock off the ground: gravity does negative work, and the rock gains gravitational potential energy.

Example 3 A positive charge $+q$ moves from position A to position B in a uniform electric field **E**:

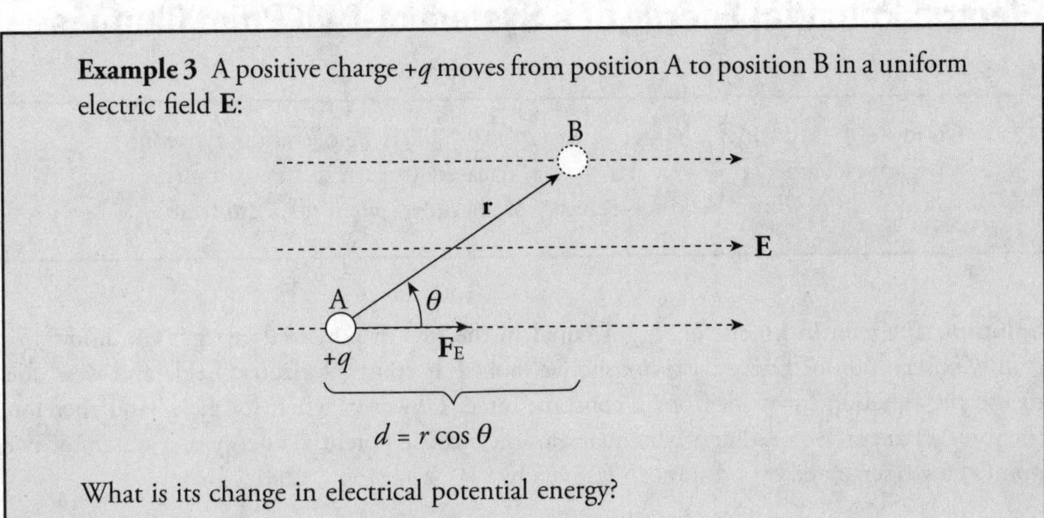

What is its change in electrical potential energy?

Solution. The electric force felt by the charge q is $\mathbf{F}_E = q\mathbf{E}$, and this force is parallel to **E** because q is positive. In this case, because $\mathbf{F}_E$ is not parallel to r (as it was in Example 1), we will use the more general definition of work for a constant force:

$$W_E = \mathbf{F}_E \cdot \mathbf{r} = F_E\, r \cos\theta = qEr\cos\theta$$

But $r\cos\theta = d$, so

$$W_E = qEd \text{ and } \Delta U_E = -W_E = -qEd$$

Because the electric force is a conservative force, which means that the work done does not depend on the path that connects the positions A and B, the work calculated above could have been figured out by considering the path from A to B composed of the segments $\mathbf{r}_1$ and $\mathbf{r}_2$:

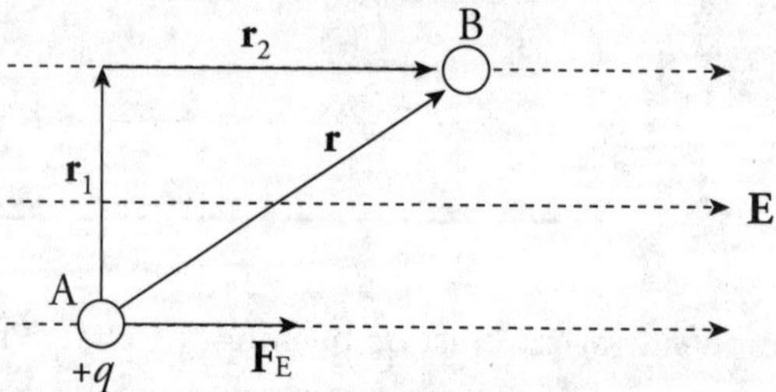

Along $\mathbf{r}_1$, the electric force does no work since this displacement is perpendicular to the force. Thus, the work done by the electric field as q moves from A to B is simply equal to the work it does along $\mathbf{r}_2$. And since the length of $\mathbf{r}_2$ is $d = r \cos \theta$, we have $W_E = F_E d = qEd$, just as before.

Electric Potential Energy of a System of Two Point Charges

Example 4 A positive charge, $q_1 = +2 \cdot 10^{-6}$ C, is held stationary, while a negative charge, $q_2 = -1 \cdot 10^{-8}$ C, is released from rest at a distance of 10 cm from q_1. Find the kinetic energy of charge q_2 when it's 1 cm from q_1.

Solution. The gain in kinetic energy is equal to the loss in potential energy; you know this from Conservation of Energy. Previously, we looked at constant electric fields and were able to use the equation for work from a constant force. However, when the field (and therefore the force) changes, we need another equation. The electric potential energy of a system of two point charges separated by a distance r is given by

Equation Sheet

$$U_E = \frac{1}{4\pi\varepsilon_0}\frac{q_1 q_2}{r} = k\frac{q_1 q_2}{r}$$

Therefore, if q_1 is fixed and q_2 moves from r_A to r_B, the change in potential energy is

$$\begin{aligned}\Delta U_E &= U_B - U_A \\ &= \frac{q_1}{4\pi\varepsilon_0}\left(\frac{q_2}{r_B} - \frac{q_2}{r_A}\right) \\ &= \frac{q_1 q_2}{4\pi\varepsilon_0}\left(\frac{1}{r_B} - \frac{1}{r_A}\right) \\ &= (9\times10^9\ \text{N}\cdot\text{m}^2/\text{C}^2)(+2\times10^{-6}\ \text{C})(-1\times10^{-8}\ \text{C})\left(\frac{1}{0.01\ \text{m}} - \frac{1}{0.10\ \text{m}}\right) \\ &= -0.016\ \text{J}\end{aligned}$$

Since q_2 lost 0.016 J of potential energy, the gain in kinetic energy is 0.016 J. Since q_2 started from rest (with no kinetic energy), this is the kinetic energy of q_2 when it's 1 cm from q_1.

Example 5 Two positive charges, q_1 and q_2, are held in the positions shown below. How much work would be required to bring (from infinity) a third positive charge, q_3, and place it as shown so that the three charges form the corners of an equilateral triangle of side length *s*?

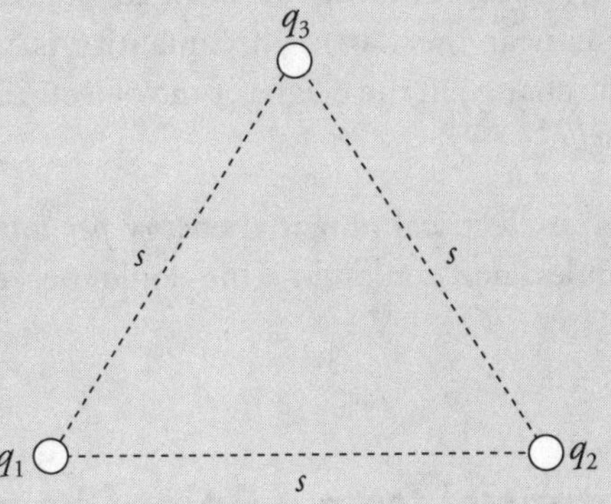

Solution. An external agent would need to do positive work, equal in magnitude to the negative work done by the electric force on q_3 as it is brought into place, so let's first compute this quantity. Let's first compute the work done *by the electric force* as q_3 is brought in. Since q_3 is fighting against both q_1's and q_2's electric field, the total work done on q_3 by the electric force, W_E, is equal to the work done on q_3 by q_1 ($W_{1\text{-}3}$) plus the work done on q_3 by q_2 ($W_{2\text{-}3}$). Using the equation $W_E = -\Delta U_E$ and the one we gave above for ΔU_E, we have

$$\begin{aligned}W_{1\text{-}3} + W_{2\text{-}3} &= -\Delta U_{1\text{-}3} + -\Delta U_{2\text{-}3} \\ &= \left(-\frac{q_1 q_3}{4\pi\varepsilon_0}\frac{1}{s} - 0\right) + \left(-\frac{q_2 q_3}{4\pi\varepsilon_0}\frac{1}{s} - 0\right) \\ &= -\frac{1}{4\pi\varepsilon_0}\frac{q_1 q_3}{s} + \frac{1}{4\pi\varepsilon_0}\frac{q_2 q_3}{s}\end{aligned}$$

Therefore, the work that an external agent must do to bring q_3 into position is

$$-W_E = \frac{1}{4\pi\varepsilon_0}\frac{q_1 q_3}{s} + \frac{1}{4\pi\varepsilon_0}\frac{q_2 q_3}{s}$$

NOT the Same
Electric potential and electric potential energy, although very closely related and having similar names, are not the same thing. Even the units are measured differently—electric potential is measured in joules per coulomb and electric potential energy in joules.

Equation Sheet

ELECTRIC POTENTIAL

Let W_E be the work done by the electric field on a charge q as it undergoes a displacement. If another charge, say $2q$, were to undergo the same displacement, the electric force would be twice as great on this second charge, and the work done by the electric field would be twice as much, $2W_E$. Since the work would be twice as much in the second case, the change in electrical potential energy would be twice as great as well, but the ratio of the change in potential energy to the charge would be the same: $W_E/q = (2W_E)/2q$. This ratio says something about the work done by the field and the *displacement* but not the charge that made the move. The change in **electric potential**, ΔV, is defined as this ratio:

$$\Delta V = \frac{\Delta U_E}{q}$$

There is a similar concept with gravitational fields. The work done by gravity (near the Earth) on an object as it changes its position is $W = mgh$. If twice the mass were moved, the work would have to be $W = (2m)gh$. The ratio of the work to the mass moved is technically the "gravitational potential," but near the Earth, this quantity is simply the gravitational field strength, g, multiplied by the change in the height of the object. *Electric potential is the electrical equivalent of a change in height.*

Potential Difference
Finding the potential at a certain point in space is meaningless. With potential, what is more important is an initial position and final position. Just like in gravity, finding the potential energy at a certain point is meaningless if there is no relation to another point. Once we establish this potential difference, when we move a charge across it, we can generate energy. This will be key in the next chapter, when potential difference is needed in order to have an electrical circuit.

Electric potential is electrical potential energy *per unit charge*; the units of electric potential are joules per coulomb. One joule per coulomb is called one **volt** (abbreviated V), so 1 J/C = 1 V.

Electric Potential from a Point Charge

Consider the electric field created by a point source charge Q. If a charge q moves from a distance r_A to a distance r_B from Q, then the change in the potential energy is

$$U_B - U_A = \frac{Qq}{4\pi\varepsilon_0}\left(\frac{1}{r_B} - \frac{1}{r_A}\right)$$

The difference in electric potential between positions A and B in the field created by Q is

$$V_B - V_A = \frac{U_B - U_A}{q} = \frac{Q}{4\pi\varepsilon_0}\left(\frac{1}{r_B} - \frac{1}{r_A}\right)$$

If we designate $V_A \to 0$ as $r_A \to \infty$ (an assumption that's stated on the AP Physics 2 Exam), then for a point charge Q, the electric potential at a distance r from Q is

$$V = \frac{1}{4\pi\varepsilon_0}\frac{Q}{r}$$

Note that the potential depends on the strength of the source charge making the field and the distance from the source charge.

Example 6 Let $Q = 2 \cdot 10^{-9}$ C. What is the potential at a Point P that is 2 cm from Q?

Solution. Relative to $V = 0$ at infinity, we have

$$V = \frac{1}{4\pi\varepsilon_0}\frac{Q}{r} = \left(9\times10^9\ \text{N}\cdot\text{m}^2/\text{C}^2\right)\frac{2\times10^{-9}\ \text{C}}{0.02\ \text{m}} = 900\ \text{V}$$

This means that the work done by the electric field on a charge of q coulombs brought to a point 2 cm from Q would be $-900q$ joules.

Note that, like potential energy, electric potential is a *scalar.* In the preceding example, we didn't have to specify the direction of the vector from the position of Q to the point P, because it didn't matter. Imagine a sphere with a surface of 2 cm from Q; at any point on that sphere, the potential will be 900 V. These spheres around Q are called **equipotential surfaces**, and they're surfaces of constant potential. The equipotentials are always perpendicular to the electric field lines.

Remember
Equipotential surfaces are often imaginary. Whether we put a metal sphere in that location or whether we visualize a sphere that isn't really there doesn't change the electric potential in that region.

Example 7 How much work is done by the electric field as a charge moves along an equipotential surface?

Solution. If the charge always remains on a single equipotential, then, by definition, the potential, V, never changes. Therefore, $\Delta V = 0$, so $\Delta U_E = 0$. Since $W_E = -\Delta U_E$, the work done by the electric field is zero.

Addition of Electric Potential

The formula $V = kq/r$ tells us how to find the potential due to a single point charge, q. Potential is scalar (we will not be concerned with direction, just the sign of charge). When we add up individual potentials, we're simply adding numbers; we're not adding vectors.

Equation Sheet

$$V = \frac{1}{4\pi\varepsilon_0}\sum_i \frac{q_i}{r_i}$$

Just like Gravitational Potential Energy
The electric force is conservative. All we care about is the change in position when calculating the change in potential energy.

Example 8 How much work would it take to move a charge $q = +1 \cdot 10^{-2}$ C from Point A to Point B (the point midway between $q_1 = 4$ nC and $q_2 = -6$ nC)?

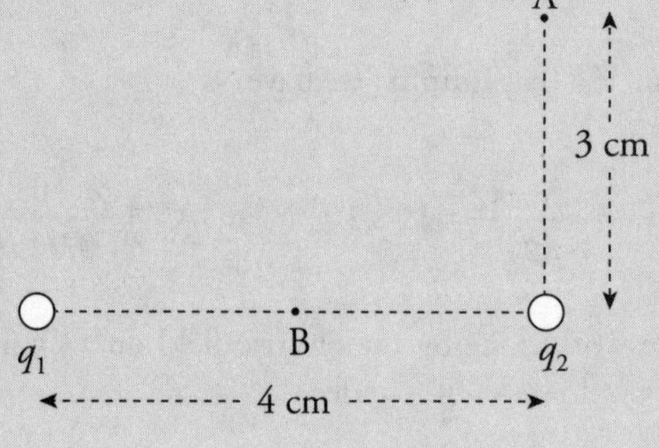

Solution. $\Delta U_E = q\Delta V$, so if we calculate the potential difference between Points A and B and multiply by q, we will have found the change in the electrical potential energy: $\Delta U_{A\to B} = q\Delta V_{A\to B}$. Then, since the work by the electric field is $-\Delta U$, the work required by an external agent is ΔU.

First, we need the potential at point A, V_A. Since there are two charges, q_1 and q_2, contributing to the potential at point A, we calculate the contribution of each using $V = kQ/r$ and sum the results. Remember that the potential is a scalar quantity, so no vector addition is required here. It is very important to keep track of the positive and negative signs, however.

$$V_A = \frac{1}{4\pi\varepsilon_0}\sum_i \frac{q_i}{r_i} = k\left(\frac{q_1}{r_1} + \frac{q_2}{r_2}\right) = \left(9.0\times 10^9\ \tfrac{\text{N}\cdot\text{m}^2}{\text{C}^2}\right)\left(\frac{4\times 10^{-9}\,\text{C}}{0.05\ \text{m}} + \frac{-6\times 10^{-9}\,\text{C}}{0.03\ \text{m}}\right)$$

$$V_A = -1080\ \text{V}$$

A similar calculation can be carried out for point B, although the distance from the charges will be different for point B than they were for point A.

$$V_B = \frac{1}{4\pi\varepsilon_0}\sum_i \frac{q_i}{r_i} = k\left(\frac{q_1}{r_1} + \frac{q_2}{r_2}\right) = \left(9.0\times 10^9\ \tfrac{\text{N}\cdot\text{m}^2}{\text{C}^2}\right)\left(\frac{4\times 10^{-9}\,\text{C}}{0.02\ \text{m}} + \frac{-6\times 10^{-9}\,\text{C}}{0.02\ \text{m}}\right)$$

$$V_B = -900\ \text{V}$$

$\Delta V_{A \rightarrow B} = V_B - V_A = (-900 \text{ V}) - (-1080 \text{ V}) = +180 \text{ V}$. This means that the change in electrical potential energy as q moves from A to B is

$$\Delta U_{A \rightarrow B} = q\Delta V_{A \rightarrow B} = (+1 \cdot 10^{-2} \text{ C})(+180 \text{ V}) = 1.8 \text{ J}$$

This is the work required by an external agent to move q from A to B.

EQUIPOTENTIAL CURVES AND EQUIPOTENTIAL MAPS

Electric field diagrams are a simple way to represent a sometimes very complex result of the influence of a charge distribution created by several charges. A similar image may be formed using isolines of electric potential instead of electric field vectors. In a downward gravitational field, if an object were to move from one position on a line at a constant height to another position on the line, no work would be done because the displacement would be perpendicular to the gravitational field. Electric equipotential curves are similar. A charge may be moved from one position on an equipotential curve to another without any work being required. A drawing of several equipotential curves at various values of the potential for a charge distribution (which may or may not be specified) is called an **equipotential map**. These images may look somewhat familiar as they are very similar to topographical maps, which plot lines of constant height in a uniformly downward directed gravitational field.

Here are the electric fields (which should look very familiar from Chapter 5) and some equipotential lines for isolated positive or negative point charges. The field lines are shown in solid and the equipotentials are shown as dashed lines.

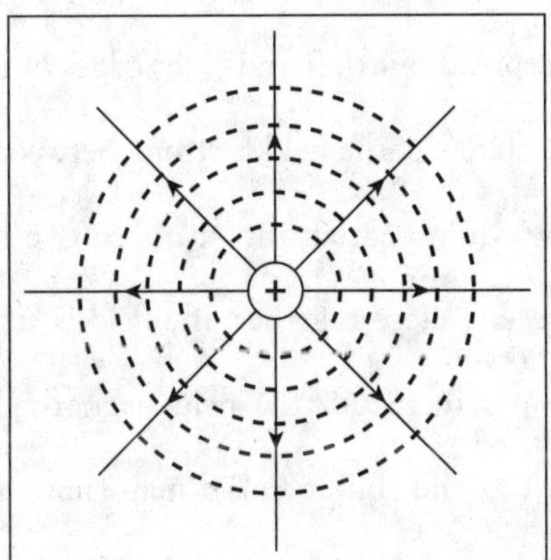

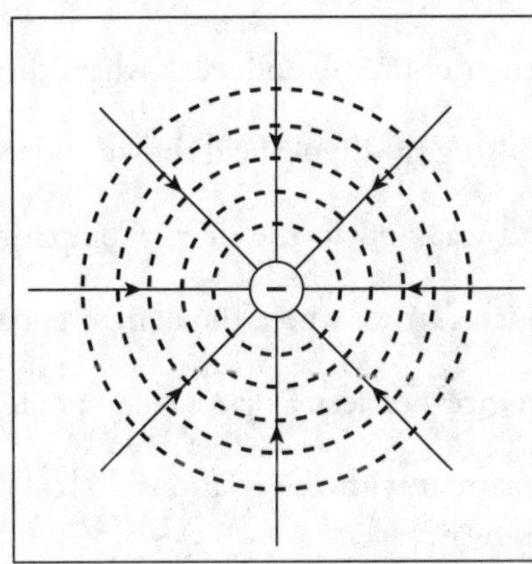

The equipotential map for an isolated positive charge is exactly the same shape as that for an isolated negative charge. Both equipotential maps are made of concentric circles. Notice that everywhere the equipotential curves meet the field lines, the two types of curves meet at a right angle. This must be the case for an equipotential curve to require no work to move along. Imagine if a field component were directed opposite the motion along an equipotential curve. Then an external agent would have to push harder to maintain the same speed along the curve, and that would mean some outside work was being done. Despite the fields for positive and negative charges being identical, it is possible to determine whether the charge at the center is positive or negative whenever two or more of the equipotential lines are labeled with values. By definition, the potential will be higher (or more positive) close to a positive charge, and lower (or more negative) close to a negative charge.

Below is the equipotential map and some electric field lines for an electric dipole.

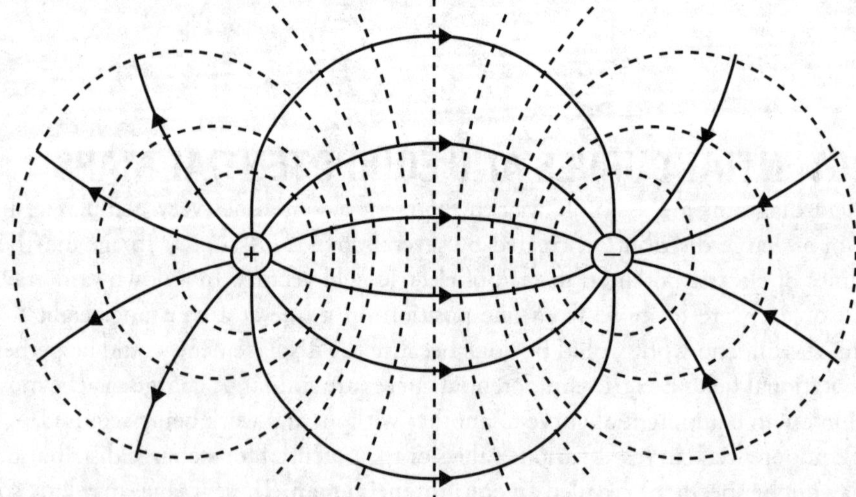

Example 9 Describe the equipotential lines in areas where the electric field is relatively strong compared to the areas where the electric field is relatively weak.

Solution. The electric field is strong where the field lines (solid in drawing above) are densely packed (center of image) and weak where they are spread out (left and right edges of drawing). Therefore, using the dipole field shown above as a reference, the field is strong between the two charges and weak off to the sides. Comparing the equipotential lines at those two locations, we can conclude that where the equipotential lines are close together, the field is strong, and as the distance between adjacent equipotential lines increases, the field strength decreases. This is consistent with the relationship $|\vec{E}| = \left|\frac{\Delta V}{r}\right|$, as the equipotential map shows lines at a constant electric potential difference ΔV, so the field strength, $|\vec{E}|$, is strong when the distance, r, between the lines is small.

THE ELECTRIC POTENTIAL OF A UNIFORM FIELD

Example 10 Consider a very large, flat plate that contains a uniform surface charge density σ. At points that are not too far from the plate, the electric field is uniform and given by the equation

$$E = \frac{\sigma}{2\varepsilon_0}$$

What is the potential at a point which is a distance d from the sheet (close to the plates) relative to the potential of the sheet itself?

Solution. Let A be a point on the plate and let B be a point a distance d from the sheet. Then

$$V_B - V_A = \frac{-W_{E, A \to B} \text{ on } q}{q}$$

Since the field is constant, the force that a charge q would feel is also constant, and is equal to

$$F_E = qE = q\frac{\sigma}{2\varepsilon_0}$$

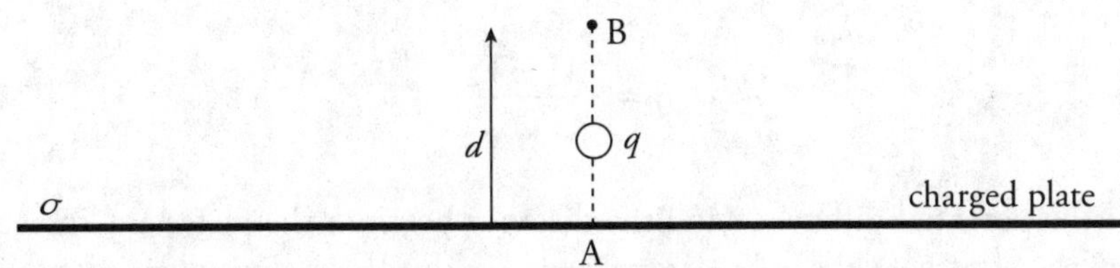

Therefore,

$$\begin{aligned} W_{E, A \to B} &= F_E d \\ &= \frac{q\sigma}{2\varepsilon_0} d \end{aligned}$$

so applying the definition gives us

$$V_B - V_A = \frac{-W_{E, A \to B}}{q} = -\frac{\sigma}{2\varepsilon_0} d$$

This says that the potential decreases linearly as we move away from the plate.

Example 11 Two large flat plates—one carrying a charge of $+Q$, the other $-Q$—are separated by a distance d. The electric field between the plates, **E**, is uniform. Determine the potential difference between the plates.

Solution. Imagine a positive charge q moving from the positive plate to the negative plate:

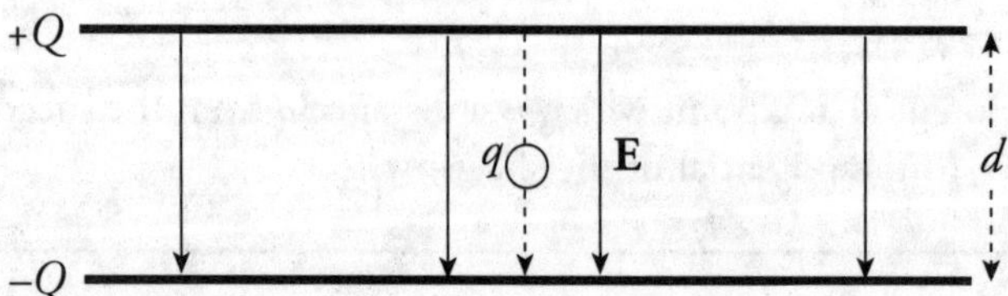

Since the work done by the electric field is

$$W_{E,+\to-} = F_E d = qEd$$

the potential difference between the plates is

$$V_- - V_+ = \frac{-W_{E,+\to-}}{q} = \frac{-qEd}{q} = -Ed$$

This tells us that the potential of the positive plate is greater than the potential of the negative plate, by the amount Ed. This equation can also be written as

$$E = -\frac{V_- - V_+}{d}$$

Therefore, if the potential difference and the distance between the plates are known, then the magnitude of the electric field can be determined quickly. The magnitude is simply

Equation Sheet

$$E = \left|\frac{\Delta V}{\Delta r}\right| = \left|\frac{\Delta V}{\Delta}\right|$$

CAPACITORS AND CAPACITANCE

Consider two conductors, separated by some distance, that carry equal but opposite charges, $+Q$ and $-Q$. Such a pair of conductors comprises a system called a **capacitor**. Work must be done to create this separation of charge, and, as a result, potential energy is stored. Capacitors are basically storage devices for electrical potential energy.

The conductors may have any shape, but the most common conductors are parallel metal plates or sheets. These types of capacitors are called **parallel-plate capacitors**. We'll assume that the distance d between the plates is small compared to the dimensions of the plates since, in this case, the electric field between the plates is uniform. When this is not the case, you must account for the **fringing fields**, which are discussed later in this chapter. The electric field due to *one* such plate, if its surface charge density is $\sigma = Q/A$, is given by the equation $E = \sigma/(2\varepsilon_0)$, with **E** pointing away from the sheet if σ is positive and toward the plate if σ is negative.

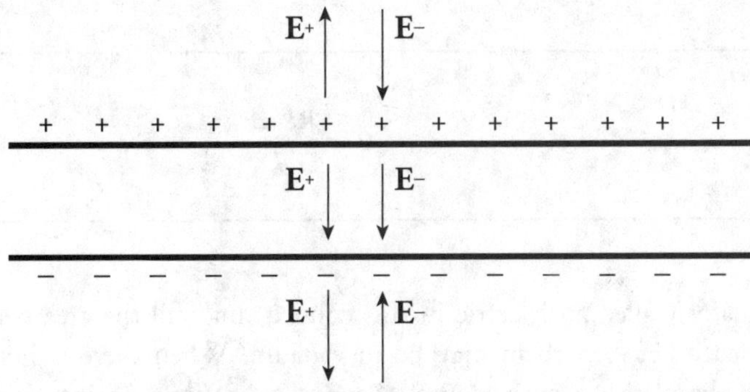

Therefore, with two plates, one with surface charge density $+\sigma$ and the other $-\sigma$, the electric fields combine to give a field that's zero outside the plates and that has the magnitude

$$E_{\text{total}} = \frac{\sigma}{2\varepsilon_0} + \frac{\sigma}{2\varepsilon_0} = \frac{\sigma}{\varepsilon_0}$$

in between.

$$E_{\text{total}} = \frac{Q}{\varepsilon_0 A}$$

In Example 11, we learned that the magnitude of the potential difference, ΔV, between the plates satisfies the relationship $\Delta V = Ed$, so combining this with the previous equation, we get

$$E = \frac{\sigma}{\varepsilon_0} \Rightarrow \frac{\Delta V}{d} = \frac{\sigma}{\varepsilon_0} \Rightarrow \frac{\Delta V}{d} = \frac{Q/A}{\varepsilon_0} \Rightarrow \frac{Q}{\Delta V} = \frac{\varepsilon_0 A}{d}$$

Q is the total charge stored on either plate of a capacitor, and ΔV is the potential difference between the plates. The ratio of Q to ΔV, for *any* capacitor, is defined as its **capacitance** (C).

Equation Sheet

$$C = Q/\Delta V$$

The capacitance measures the capacity for holding charge. The greater the capacitance, the more charge can be stored on the plates at a given potential difference. The capacitance of any capacitor depends only on the size, shape, and separation of the conductors, and the "dielectric constant," κ (the Greek letter kappa), which depends on what material is between the plates of the capacitor. For a parallel-plate capacitor, we get

Equation Sheet

$$C = \frac{\kappa \varepsilon_0 A}{d}$$

What Determines Capacitance?
Capacitance does NOT determine the charge on the capacitor, Q, or the potential difference, V, across it. It only shows a relationship between Q and V. The greater the potential difference applied to a capacitor, the greater the amount of charge the capacitor can hold. Think of capacitance as a property of a physical object. Physically, capacitance is determined by three things: the area of the plates, A; the plate separation, d; and the dielectric constant, κ.

An insulator (called a dielectric in this context) may fill the area between the plates, or the space between them may be in vacuum. When there is nothing in the gap between the plates except vacuum, then $\kappa = 1$. When an insulating material is in the gap between the plates, then $\kappa > 1$. From the definition, $C = Q/\Delta V$, the units of C are coulombs per volt. One coulomb per volt is renamed one **farad** (abbreviated F): 1 C/V = 1 F.

Example 12 A 10-nanofarad parallel-plate capacitor holds a charge of magnitude 50 ∝C on each plate.

(a) What is the potential difference between the plates?

(b) If the plates are separated by vacuum with a distance of 0.2 mm, what is the area of each plate?

Solution.

(a) From the definition, $C = Q/\Delta V$, we find that

$$\Delta V = \frac{Q}{C} = \frac{50 \times 10^{-6}\ \text{C}}{10 \times 10^{-9}\ \text{F}} = 5000\ \text{V}$$

(b) Because $\kappa = 1$, we have the equation $C = \varepsilon_0 \frac{A}{d}$, and can calculate the area, A, of each plate:

$$A = \frac{Cd}{\varepsilon_0} = \frac{(10 \times 10^{-9}\ \text{F})(0.2 \times 10^{-3}\ \text{m})}{8.85 \times 10^{-12}\ \text{C}^2/\text{N} \cdot \text{m}^2} = 0.23\ \text{m}^2$$

ELECTRIC FIELD AND CAPACITORS

For point charges, the electric field created by one or more point source charges varies, depending on the location. For example, as we move farther away from the source charge, the electric field gets weaker. Even if we stay at the same distance from, say, a single source charge, the direction of the field changes as we move around. Therefore, we could never obtain an electric field that was constant in both magnitude and direction throughout some region of space from point-source charges. However, the electric field that is created between the plates of a charged parallel-plate capacitor is constant in both magnitude and direction throughout the region between the plates; in other words, a charged parallel-plate capacitor can create a uniform electric field. The electric field, **E**, always points from the positive plate toward the negative plate, and its magnitude remains approximately the same at every point between the plates, whether we choose a point close to the positive plate, closer to the negative plate, or between them.

"Uniform electric field" is only approximately true. The electric field is only constant everywhere in space for an infinitely large sheet of charge, and such a thing is impossible to construct. However, whenever the observation location is sufficiently close to the plates, or the edges of the plates are sufficiently far away, the constant electric field approximation holds true.

Example 13 The charge on a parallel-plate capacitor is $4 \cdot 10^{-6}$ C. If the distance between the plates is 2 mm and the capacitance is 1 μF, what's the strength of the electric field between the plates?

Solution. Since $C = Q/\Delta V$, we have $\Delta V = Q/C = (4 \cdot 10^{-6}\text{ C})/(10^{-6}\text{ F}) = 4$ V. Now, using the equation $\Delta V = Ed$,

$$E = \Delta V/d = (4\text{ V})/(2 \cdot 10^{-3}\text{ m}) = 2000\text{ V/m}$$

Example 14 The plates of a parallel-plate capacitor are separated by a distance of 2 mm. The device's capacitance is 1 μF. How much charge needs to be transferred from one plate to the other in order to create a uniform electric field whose strength is 10^4 V/m?

Solution. Because $Q = C\Delta V$ and $\Delta V = Ed$, we find that

$$Q = CEd = (1 \cdot 10^{-6}\text{ F})(1 \cdot 10^{4}\text{ V/m})(2 \cdot 10^{-3}\text{ m}) = 2 \cdot 10^{-5}\text{ C} = 20\ \mu\text{C}$$

Fringing Fields

The mathematical analysis of capacitors up until now assumes that the plates of the capacitor are infinitely large. Because of this approximation, the analysis is very good in the region near the centers of the plates, but is not as precise near the edges of the plates. In the areas beyond the edges of the plates, we can approximate the field as the sum of the capacitor field and a dipole formed by the charges at the edges of the plates.

THE ENERGY STORED IN A CAPACITOR

To figure out the electrical potential energy stored in a capacitor, imagine taking a small amount of negative charge off the positive plate and transferring it to the negative plate. This requires that positive work be done by an external agent, and this is the reason that the capacitor stores energy. If the final charge on the capacitor is Q, then we transferred an amount of charge equal to Q, doing work to move the charge through a potential difference at each stage. If the final potential difference is ΔV, then the average potential difference during the charging process is $\frac{1}{2}\Delta V$; so, using the definition of potential difference, $\Delta V = \Delta U_E/Q$, we can write that for a capacitor $\Delta U_C = Q \cdot \frac{1}{2}\Delta V = \frac{1}{2}Q\Delta V$. At the beginning of the charging process, when there was no charge on the capacitor, we had $U_i = 0$, so $\Delta U_C = U_f - U_i = U_f - 0 = U_f$; therefore, we have

Equation Sheet

$$U_C = \frac{1}{2}Q\Delta V$$

This is the electrical potential energy stored in a capacitor. Because of the definition $C = Q/\Delta V$, the equation for the stored potential energy can be written as

$$U_C = \frac{1}{2}(C\Delta V)\cdot \Delta V = \frac{1}{2}C(\Delta V)^2$$

or

$$U_C = \frac{1}{2}Q \cdot \frac{Q}{C} = \frac{Q^2}{2C}$$

Interestingly, this work is typically done by a battery as the external agent. The battery supplies a total work of $Q\Delta V$, but the capacitor ends up storing only half of this energy. The other 50% of the energy supplied by the battery is dissipated as heat during this charging process.

CAPACITORS AND DIELECTRICS

One method of keeping the plates of a capacitor apart, which is necessary to maintain charge separation and store potential energy, is to insert an insulator (called a **dielectric**) between the plates.

Insulator?
Why not a conductor? A capacitor stores energy by holding charges apart from one another. If a conductor is placed between the plates, the charges are no longer separated and the device is no longer a capacitor.

dielectric

A dielectric always increases the capacitance of a capacitor.

Let's see why this is true. Imagine charging a capacitor to a potential difference of ΔV with charge $+Q$ on one plate and $-Q$ on the other. Now disconnect the capacitor from the charging source and insert a dielectric. What happens? Although the dielectric is not a conductor, the electric field that exists between the plates causes the molecules within the dielectric material to polarize; there is more electron density on the side of the molecule near the positive plate.

The effect of this is to form a layer of negative charge along the top surface of the dielectric and a layer of positive charge along the bottom surface; this separation of charge induces its own electric field ($\mathbf{E}_i$), within the dielectric, which opposes the original electric field, $\mathbf{E}$, within the capacitor.

So the overall electric field has been reduced from its previous value: $\mathbf{E}_{total} = \mathbf{E} + \mathbf{E}_i$, and $E_{total} = E - E_i$. Let's say that the electric field has been reduced by a factor of κ from its original value as follows:

$$E_C = \frac{Q}{\kappa \varepsilon_0 A}$$

Equation Sheet

Since $\Delta V = Ed$ for a parallel-plate capacitor, we see that ΔV must have decreased by a factor of κ. But $C = \frac{Q}{\Delta V}$, so if ΔV decreases by a factor of κ, then C increases by a factor of κ:

$$C_{\text{with dielectric}} = \kappa C_{\text{without dielectric}}$$

The value of κ, called the **dielectric constant**, varies from material to material, but it's always greater than 1. In general, the capacitance of parallel-plate capacitors is

Equation Sheet

$$C = \kappa \varepsilon_0 \frac{A}{d}$$

Chapter 5 Review Questions

Answers and explanations can be found in Chapter 11.

Section I: Multiple Choice

1 Mark for Review

An experiment is conducted and data is gathered for the electric potential V at various positions r away from a uniformly charged sphere. All measurements are taken outside of the sphere. Which of the following graphs yields a straight line?

(A) V as a function of $\frac{1}{r^2}$

(B) V as a function of $\frac{1}{r}$

(C) V as a function of r

(D) V as a function of r^2

2 Mark for Review

Negative charges are accelerated by electric fields toward

(A) points of lower electric potential

(B) points of higher electric potential

(C) points where the electric field is weaker

(D) points where the electric field is stronger

3 Mark for Review

A section near the center of a parallel-plate capacitor is shown below. There are 4 labeled positions between the plates shown as A, B, C, and D. Relative to A, which point has the largest potential difference and why?

+Q

D●

C●

A● B●

–Q

(A) Point A because the potential difference is infinite when the position between points is 0 m

(B) Point B because it is the same distance from the $-Q$ plate as A

(C) Point C because it is farther in distance from Point A

(D) Point D because it is closest to the $+Q$ plate

4 Mark for Review

A charge q experiences a displacement within an electric field from Position A to Position B. The change in the electrical potential energy is ΔU_E, and the work done by the electric field during this displacement is W_E. Then

(A) $V_A - V_B = qW_E$

(B) $V_B - V_A = qW_E$

(C) $V_A - V_B = \Delta U_E/q$

(D) $V_B - V_A = \Delta U_E/q$

5 Mark for Review

+Q

1 •

4 • •2

•3

–Q

Which points in this uniform electric field (between the plates of the capacitor) shown above lie on the same equipotential?

(A) 1 and 3 only

(B) 2 and 4 only

(C) None lie on the same equipotential.

(D) 1, 2, 3, and 4 all lie on the same equipotential since the electric field is uniform.

6 Mark for Review

A charge Q creates an electric field through which a second charge q moves, as shown below. q is initially at point B and is moved to point A, farther away. The potential from Q at position A is $V_A = 100\ V$ and at B is $V_B = 200\ V$. The charge on q is negative. What is the sign of Q and the sign of the work done by the electric field of Q as q is moved from B to A?

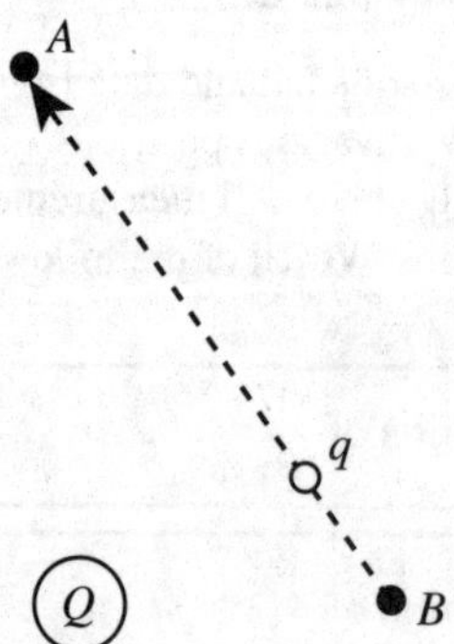

(A) Q is positive and the work done by the electric field is positive.

(B) Q is positive and the work done by the electric field is negative.

(C) Q is negative and the work done by the electric field is positive.

(D) Q is negative and the work done by the electric field is negative.

Section II: Free Response

1 Mark for Review

In Figure 1, four charges, each of magnitude Q, are situated at the corners of a square with side lengths s. The two charges on the top of the square are positively charged, while the two on the bottom of the square are negatively charged.

Figure 1

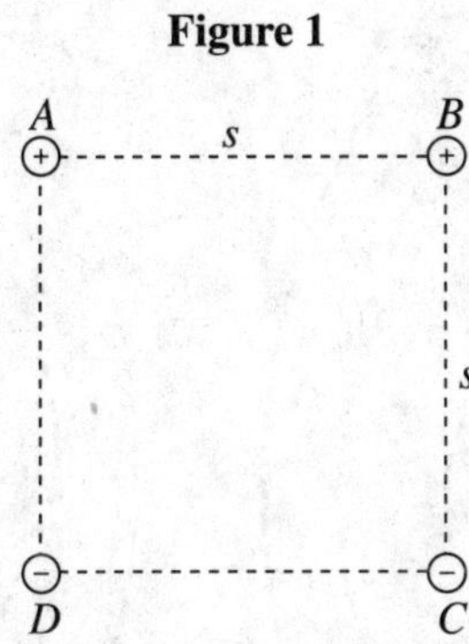

A. These charges were assembled in order by first bringing in charge A, then bringing in charge B, then bringing in C, and finally bringing in charge D. Rank the amount of energy in the charge distribution in the presence of only charge A, only charges A and B, only charges A, B, and C, and in the presence of all four charges. Negative numbers should be taken as smaller than positive numbers. Justify your answer.

Greatest (most positive) ____ ____ ____ ____ Least (most negative)

B. Show that the potential at the exact center of the square is 0 V by calculating the potential from each charge at that location.

C. Sketch (on the diagram) the portion of the equipotential surface that lies in the plane of the figure and passes through the center of the square.

D. As shown in Figure 2, solid conducting bars are placed to connect points A to B and also to connect points C to D. The charge is allowed to distribute over these conductors. Sketch the electric field at each of the dots. Explain why the field is constant at the dots in the center but not at the dots on the edges.

Figure 2

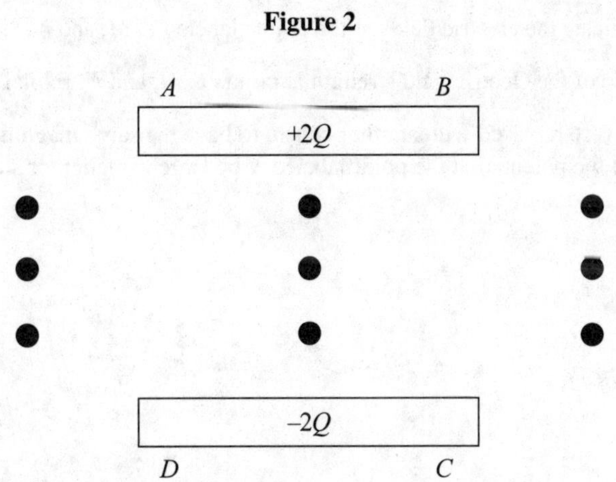

2 Mark for Review

Figure 1 shows the isolines of electric potential surrounding two charged spheres, labeled L and R. The spheres carry opposite charges and the potential difference between adjacent pairs of lines is $\Delta V = 15$ V. The isolines with electric potentials of –25 V and –10 V are indicated.

Figure 1

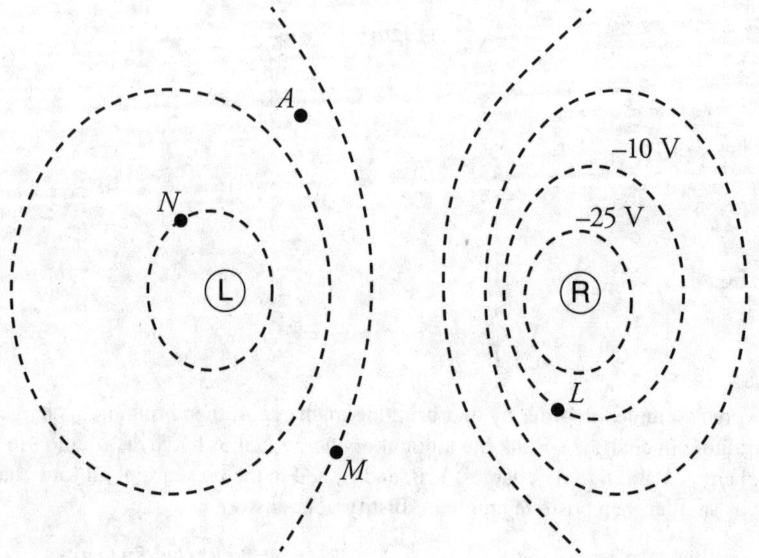

A. Sketch in the line of potential 0 V on the drawing.

B. Approximately what is the value of the potential at the point labeled *A*?

C.

i. Which sphere, L or R, carries a negative charge? Explain your answer.

ii. Which sphere, L or R, carries a greater magnitude of charge? Explain how you know.

D.

i. Draw arrows to indicate the electric fields at the points labeled *L*, *M*, and *N*.

ii. Rank the magnitude of the electric field strength at points *L*, *M*, and *N*. Explain your answer.

E. If the sphere labeled L were replaced with another sphere to have the same magnitude of charge but of the opposite sign, would the value of the potential at the point labeled *N* be larger, smaller, or stay the same? Justify your answer.

3 Mark for Review

A solid conducting sphere of radius a carries an excess charge of Q.

A. Determine the electric field magnitude, $E(r)$, as a function of r, the distance from the sphere's center.

B. Determine the potential, $V(r)$, as a function of r. Take the zero of potential at $r = \infty$.

C. On Figure 1, sketch $E(r)$ and $V(r)$. (Cover at least the range $0 < r < 2a$.)

Figure 1

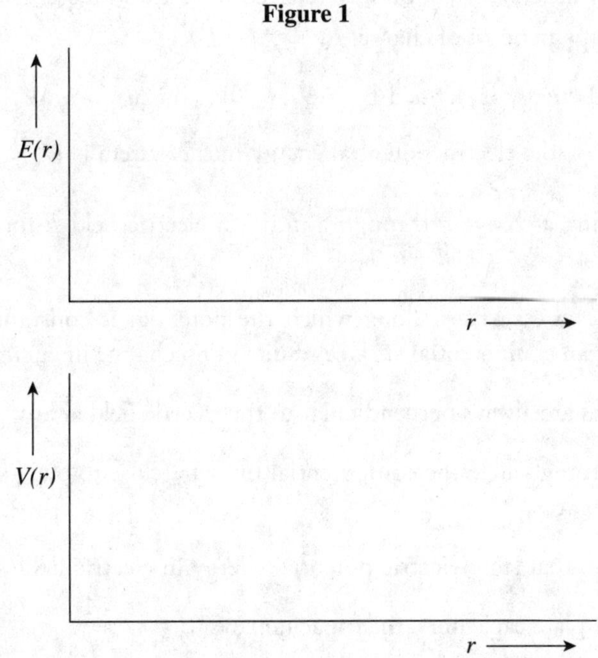

Chapter 5 Summary

- Electric potential energy is a type of energy arising from the interaction of a charge and an external electric field. As with all types of energy, it may be converted into other types of energy or may be added to the system or removed from the system through work.
- The electric potential difference (commonly referred to as the voltage) is defined as the change in electric potential energy per unit of charge: $\Delta V = \Delta U_E / q$.
- The electric potential energy is defined by $\Delta U_E = -W_E$ or $\Delta U_E = q\Delta V$. For a pair of interacting point charges q_1 and q_2, the electric potential energy in the system is $U_E = \frac{1}{4\pi\varepsilon_0}\frac{q_1 q_2}{r}$.
- The work done moving a charge q through a uniform electric field E for a distance d is given by $W = qEd$.
- Equipotential surfaces are surfaces along which the potential is constant. Moving a charge at a constant speed along an equipotential surface results in no change in energy of the charge.
- Equipotential surfaces are always perpendicular to the electric field at any point on the surface.
- The electric field is strong where the equipotential lines are close to one another and weaker where the lines are farther apart.
- Capacitors are devices that store electric potential energy in electric fields. Capacitance is given by $C = \frac{Q}{\Delta V}$. For parallel-plate capacitors, the capacitance is $C = \kappa\varepsilon_0\frac{A}{d}$.
- A parallel-plate capacitor has a uniform electric field between the plates for regions close to the plates and far from the outer edges of the capacitor.
- The electrical energy stored in a capacitor is given by $U_C = \frac{1}{2}Q\Delta V = \frac{1}{2}C(\Delta V)^2$.
- When a capacitor is filled with a dielectric, its capacitance increases from the capacitance it had when it had a vacuum between its plates.

Chapter 6
Electric Circuits

INTRODUCTION

Electric circuits are comprised of an energy source (typically a battery or a wall outlet), one or more conducting materials (such as wires), and circuit components such as resistors and capacitors. The broad topic of electric circuits is vast, but for the AP Physics 2 Exam, we need to focus on only a small set of components in relatively simple arrangements. We will focus on direct current (DC) voltage supplies in resistor and capacitor (RC) circuits in steady state.

The most basic electric circuit is a collection of conductors, such as wire, and a source of electrical energy, such as a battery. Connecting the components together so that there is no insulating material (such as an air gap) between the components completes the circuit and allows for a continuous flow of charge, known as **current**, in the circuit. In this chapter, we will analyze circuits constructed of one or more batteries, wires, switches, resistors, and capacitors. Other circuit components, such as diodes or inductors, are beyond the scope of AP Physics 2.

BATTERIES AND VOLTAGE

A battery is a device that maintains an electric potential difference between the two terminals. When a wire is connected between the two terminals of the battery, there is a potential difference between the ends of the wire. As we saw previously, a potential difference in a conducting material causes an electric field. Previously, when we were looking at electrostatics, the potential difference was always 0 V because we require, as a condition of electrostatics, that the electric field within a wire be 0 N/C. In circuits, when there is a battery present to *maintain* a potential difference, there will be a nonzero electric field within the conductor. As a result, there will be a flow of charge. We call this flow of charge electric current.

A Note on Convention for this Chapter

In the previous chapter, we referred to a change in potential as a voltage, ΔV. In this chapter, we will also use the standard convention, V. Make note of this as you move forward so you won't be confused.

A common misconception is that the battery is the *supply* of the charges which create the current. As we have already seen, conducting materials such as metals contain vast numbers of electrons. The battery supplies the energy which is needed to arrange the electrons in the wire into a configuration where the current continuously flows. However, the electrons present within the wire constitute the current. The electrons present in the wire also create the electric field that is the source of the current.

Voltage Is like Electric Pressure

Voltage is a power source that allows for the movement of charges. It is like a region in the atmosphere of high pressure compared to a region of low pressure, which will result in a wind.

Any particular conduction electron will move through the circuit with a **drift velocity** of some millimeters per second, but the influence of the "electricity" moves through the circuit at the speed of light. As we will see, were the battery the source of the electric charge, not only would it take several hours for light bulbs to turn on, because of the slow drift velocity of electrons, but circuits with capacitors would not function at all.

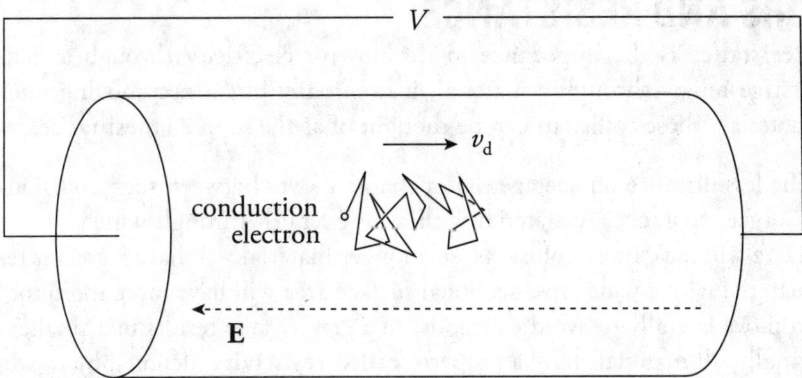

In a circuit diagram, a battery will be represented by one of these two diagrams:

The diagram to the left shows a *single-cell* battery and the one on the right shows a battery with multiple cells. The longer line in the battery diagram represents the positive (higher potential) terminal, and the shorter line is the negative (lower potential) terminal.

The voltage source is a power source. Whether it is a wall plug, a battery, a capacitor, or an electric generator, it supplies energy to the circuit. When two points in a circuit have different voltages, charges will flow from one point to the other. As long as the voltage difference is maintained, the flow of charge will continue.

Voltage Has Many Names but One Meaning
When scientists were discovering the properties of electricity, it was not clear that all the concepts they were analyzing were the same phenomenon. In different contexts, voltage is referred to as potential difference, voltage drop, voltage, or electromotive force.

> Voltage between two points in a conducting material causes charge to flow.

The voltage supplied by an ideal battery is often, for historical reasons, referred to as EMF (electromotive force, ε). Whether referred to as the potential difference, voltage drop, voltage, or EMF, the voltage is measured in units of **volts** (abbreviated V).

A battery never changes which pole is at high potential and which pole is at low potential. As a result, the battery always causes electricity to flow in the same direction through the circuit. This is called a **direct current** circuit.

RESISTORS AND RESISTANCE

Resistance vs. Resistivity
Resistance and resistivity are NOT the same. Resistance is the extent to which a resistor impedes the passage of electric current. Resistivity is a property of what molecules a material consists of and how the molecules that constitute a material are bound together. The units of resistivity are different as well.

Resistance is the impedance to the flow of electricity through a material. As a charge moves through a material, it eventually hits a non-moving nucleus in the material. These collisions can be thought of as the source of resistance.

The less distance on average that a charge moves between such collisions results in a larger resistance. A material with a long length through which the charge must move will have more collisions, so a longer material will have a greater resistance. A material with a wide cross-sectional surface area will have more room for the charge to move laterally to avoid collisions, so a greater area results in a smaller resistance. Finally, all materials have a property called **resistivity**, denoted by ρ, which can be thought of in this context as the density of the nuclei the electrons may strike. A greater density will cause more collisions. The resistance, R, of a material depends on its resistivity and its geometrical configuration, in particular, its length, l, and its cross-sectional area, A.

Equation Sheet

$$R = \frac{\rho \ell}{A}$$

This equation applies only to shapes with constant cross-sectional area, such as cylinders or rectangular prisms, and will not apply to those with a varying cross-sectional area, such as spheres or cones.

Resistors Are like Drinking Straws
A very narrow straw will be harder to drink through than a wide straw. A very long straw will be harder to drink through than a short one.

Resistance is measured in **ohms** (Ω, *omega*) and resistivity is measured in **ohm-meters** (Ω·m).

In a circuit diagram, a resistor will be represented by

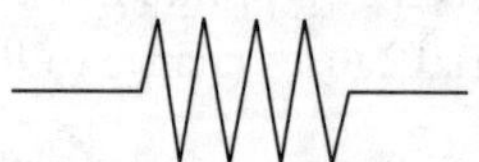

A resistor is awkwardly named because all wires are technically "resistors." Materials with a very low resistivity are referred to as conductors. These include aluminum, silver, and copper, which all have resistivity on the order of 1×10^{-8} Ω·m. Electrical components called "resistors" are made of materials like carbon graphite, which have resistivity on the order of 10×10^{-5} Ω·m, approximately 10,000 times higher than the materials which comprise wire. Materials with a high resistivity, such as glass with resistivity of 10×10^{10} Ω·m, are referred to as insulators.

Example 1 The resistivity of copper is about 1.7×10^{-8} Ω·m. What is the resistance of a 5 m long 16-gauge light-duty extension cord whose diameter is 1.3 mm?

Solution.

$$R = \frac{\rho \ell}{A}$$

To find the cross-sectional area, we use $A = \pi r^2 = (3.14)(1.3 \times 10^{-3}\ \text{m}/2)^2$.

$$R = \frac{(1.7 \times 10^{-8}\ \Omega\cdot\text{m})(5\text{m})}{((3.14)(1.3 \times 10^{-3}\ \text{m}/2)^2)} = 0.064\ \Omega$$

Example 2 A wire with cross-sectional area 1.33 mm^2 has an unknown resistivity.

(a) To find the resistivity, you have a spool of this wire, a meter stick, wire cutters, and a device to measure resistance (such a device is called an Ohmmeter). Briefly describe what data you will collect and how you will determine the resistivity of the wire.

(b) Another student gathers data and makes the following plot. What is the resistivity of the wire in her experiment?

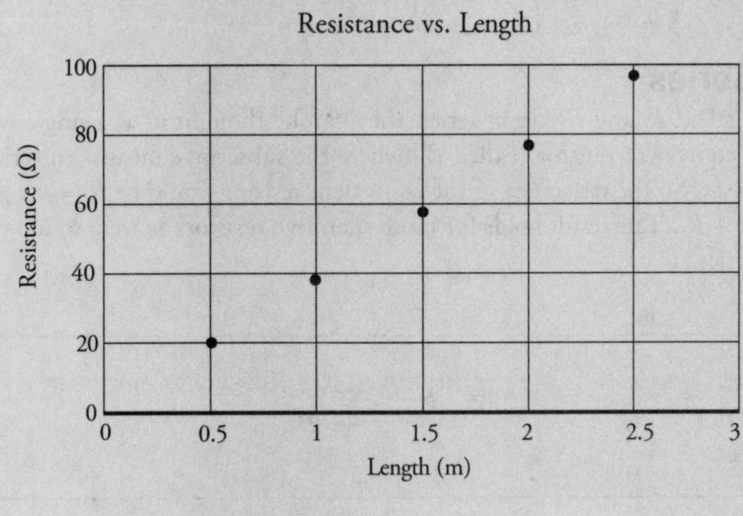

Solution.

(a) Cut a length of the wire with wire cutters. Measure the length of the wire, L, with the meterstick. Measure the resistance of the wire with the Ohmmeter. The resistivity can be found using $\rho = AR/L$.

(b) From the graph, the slope is (98 Ω – 20 Ω)/(2.5 m – 0.5 m) = 39 Ω·m, which is R/L. Since we determined in (a) that $\rho = AR/L$, we can find the resistivity as $\rho = (1.33 \times 10^{-6}\ \text{m}^2)(39\ \Omega\cdot\text{m}) = 5.19 \times 10^{-5}\ \Omega\cdot\text{m}$.

Multiple Resistors Are like Multiple Straws

If one very narrow straw will be hard to drink through, two very narrow straws side by side will act like a wider straw, decreasing the overall resistance.

COMBINING RESISTORS AND EQUIVALENT RESISTANCE

To deal with arrangements that have multiple resistors, combinations of resistors can be considered. Resistors may be arranged in two ways: one after the other or side by side. In an arrangement in which the resistors are one after the other, we say the resistors are "in series." In an arrangement in which the resistors are side by side, we say the resistors are "in parallel." When two or more resistors are combined mathematically, the resulting resistance is called *equivalent* resistance.

Two resistors R_1 and R_2 arranged in series

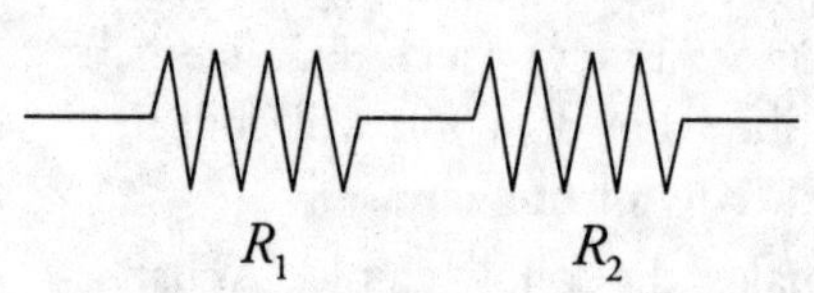

Two resistors R_1 and R_2 arranged in parallel

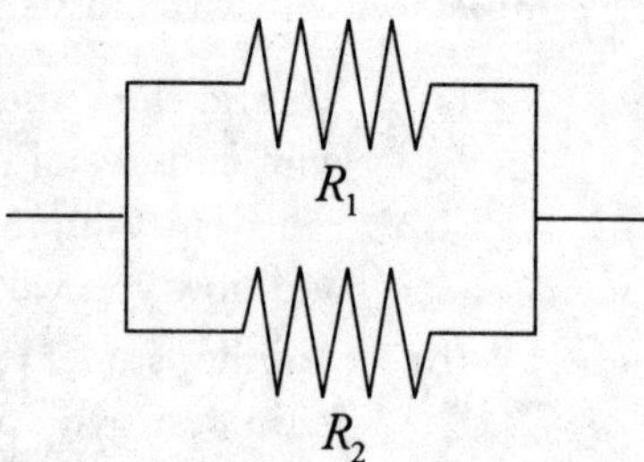

Resistors in Series

When two resistors called R_1 and R_2 are in series, they can be thought of as a single wire of a greater length. The "equivalent resistor" called R_S (where the subscript s means "in series") has a length of $L_{eq} = L_1 + L_2$. So the resistance of the equivalent resistor would be $R_S = \rho(L_1 + L_2)/A = \rho L_1/A + \rho L_2/A = R_1 + R_2$. This result holds for more than two resistors as well, so long as they are all arranged in series.

Equation Sheet

$$R_{eq,s} = \sum_i R_i$$

Resistors arranged in series result in an overall resistance that is greater than the resistance of any of the individual resistors in the arrangement.

One Over R_p
The equation for equivalent resistance in parallel has R_p in the denominator. Don't forget to take the reciprocal as the final step in calculating the equivalent resistance.

Resistors in Parallel

When the resistors are in parallel, they behave like a wire with a greater cross-sectional area. It would be equivalent to replace those two resistors with a single "equivalent resistor" called R_{eq} of area $A_p = A_1 + A_2$. Then the resistance of the equivalent resistor would be

$$R_p = \rho L/(A_1 + A_2) \rightarrow \frac{1}{R_p} = \frac{(A_1 + A_2)}{\rho L} = \frac{A_1}{\rho L} + \frac{A_2}{\rho L} = \frac{1}{R_1} + \frac{1}{R_2}$$

This result holds for more than two resistors as well, so long as they are all arranged in parallel.

$$\frac{1}{R_{\text{eq},p}} = \sum_i \frac{1}{R_i}$$

Equation Sheet

Resistors arranged in parallel result in an overall resistance that is less than the resistance of any of the resistors.

When dealing with arrangements of three or more resistors, it is possible for the arrangement to be a mixture of series and parallel.

(a) A pair of resistors R_A and R_B is in series. That combination is in parallel with a third resistor, R_C.

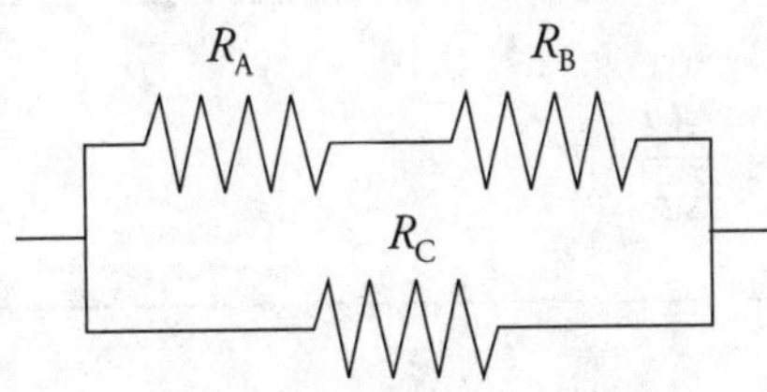

(b) A resistor R_D is in series with a pair of resistors R_E and R_F, which are in parallel.

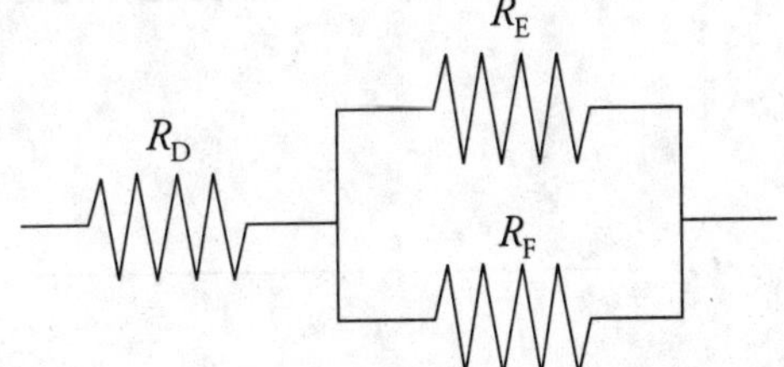

When resistors appear in a mixture of parallel and series, the equivalent resistance is calculated in multiple steps. First, the equivalent resistance of the part of the circuit that is purely series or parallel must be found. Then that resistance is used for further calculations. This process may need to be repeated several times in circuits with many resistors.

Example 3 Calculate the total equivalent resistance of the two arrangements in the figure above.

A Shortcut for Pairs of Parallel Resistors
The equivalent resistance of a pair of resistors in parallel is the product of the resistances over their sum. This shortcut does **not** work for combinations of more than two resistors.

Solution.

(a) The figure labeled (a) shows two resistors in series, R_A and R_B, which are in parallel with the third resistor R_C. The equivalent resistance of the arrangement is found by first combining the series resistors into $R_{\text{eq,AB}} = R_A + R_B$ and finally combining that with R_C as a parallel arrangement to get

$$\frac{1}{R_{\text{eq, total}}} = \frac{1}{R_A + R_B} + \frac{1}{R_C} \text{ so that } R_{\text{eq,total}} = \frac{R_C(R_A + R_B)}{R_A + R_B + R_C}$$

Wind as an Analogy to Current
Think of current as the wind. Blowing wind is comprised of many air molecules moving randomly and colliding with one another and with other things the air molecules encounter, but there is an overall flow of air in a particular direction. When talking about wind, the rate of motion of air molecules is considered. When talking about current, the rate of motion of electric charge is under consideration.

(b) The figure labeled (b) shows a resistor, R_D, in series with a pair of parallel resistors, R_E and R_F. The equivalent resistance of the arrangement is found by first combining the parallel resistors into $\frac{1}{R_{eq,EF}} = \frac{1}{R_E} + \frac{1}{R_F}$ so that $R_{eq,EF} = \frac{R_E R_F}{R_E + R_F}$. Then that is combined with R_D as a series arrangement to get $R_{eq,total} = R_D + \frac{R_E R_F}{R_E + R_F}$.

CURRENT

Establishing a potential difference across a wire creates an electric field in the wire, and as a result of that electric field, charges move. We can measure how much charge passes any particular point in the wire in a specified unit of time. When an amount of charge of magnitude ΔQ crosses an imaginary plane in a time interval Δt, then the **average current** is

Equation Sheet

$$I_{avg} = \frac{\Delta Q}{\Delta t}$$

Benjamin Franklin
Conventional direction of current is "wrong" because Benjamin Franklin did not have enough information to determine whether charge carriers were positive or negative charges (or both moving in opposite directions) when he was investigating electrostatics. He simply picked a direction based on the flow of positive charges.

Because current is charge per unit time, it's expressed in coulombs per second. One coulomb per second is an **ampere** (abbreviated **A**). Instead of calling the unit an ampere, it is often simply called an amp.

We know that the charge carriers that constitute the current within a metal are electrons. However, the direction of the current is taken to be the direction that positive charge carriers would move. This is referred to as *conventional current,* in contrast with *electron current,* which is the flow of electrons. On the AP Physics 2 Exam, assume all current flow is conventional current.

MEASURING CURRENT AND VOLTAGE IN A CIRCUIT

When a circuit is established, each electrical component in the circuit will have some electricity flowing through it and will have some energy being used up by it.

The current flowing past any point in a circuit is measured using an ammeter. An ammeter is a device with a very low resistance (essentially $R = 0\ \Omega$). To measure the current flowing through a resistor of resistance R, the ammeter is placed in series and the equivalent resistance of the resistor and ammeter is $R_S = R + 0 = R$. Because of the negligible resistance of the ammeter, the resistance of the pair is just R. The fact that this arrangement does not change the equivalent resistance means the ammeter reading is the same as the reading through the resistor when the ammeter is not present. Because current flows through the resistor, the reading will be the same no matter on which side of the resistor the meter is placed.

In a circuit diagram, an ammeter will be represented by

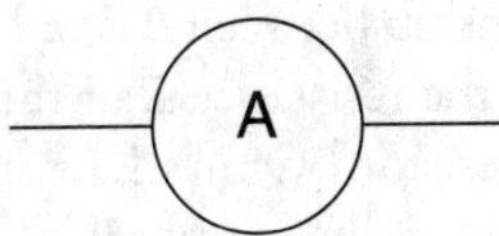

Current Is Not "Used Up"
Current flows through the devices in a circuit. The same amount of current flows into, and then back out of, each resistor.

The electrical energy is not measured directly, but instead the electric potential difference—commonly referred to as voltage or voltage drop—is measured using a voltmeter. A voltmeter has an extremely high resistance (essentially $R = \infty$). One lead is attached to each of the two points where the voltage difference is desired. To measure the voltage across a resistor of resistance R, the voltmeter must be in parallel with the resistor. Because they are in parallel, the equivalent resistance of the pair is found from $\frac{1}{R_P} = \frac{1}{R} + \frac{1}{\infty}$, so $R_P = R$. Because of the divergent resistance of the voltmeter, the resistance of the pair is just R. The fact that this arrangement does not change the equivalent resistance means the voltmeter reading is the same as the reading across the resistor when the meter is not present. In a circuit diagram, a voltmeter will be represented by

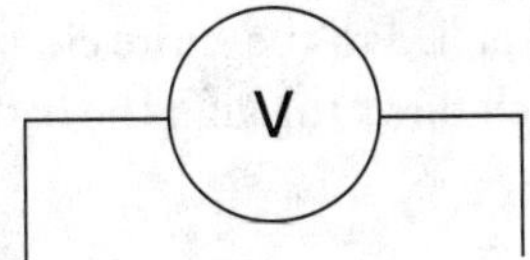

Example 4 A non-ideal ammeter with a resistance of 2 Ω used in series with a resistor with resistance of 25 Ω. By what percentage would adding the ammeter change the equivalent resistance of the circuit?

Solution. Since the two elements are in series, the equivalent resistance is found using

$$R_{eq} = R_1 + R_2 = 25\ \Omega + 2\ \Omega = 27\ \Omega$$

The percentage change is found by looking at the difference between the two things and dividing by the original.

$$\%_{change} = \frac{27\ \Omega - 25\ \Omega}{25\ \Omega} \times 100 = 8\%$$

So adding this non-ideal ammeter increases the equivalent resistance of the circuit by 8%.

OHM'S LAW

For many resistors, a plot of the current flowing through the resistor for various voltage drops results in a graph showing the relationship between ΔV and I is directly proportional. Moreover, the slope of such a graph is equal to the resistance R of the resistor. Whenever this is the case, the device is said to be an "Ohmic resistor." Ohm's Law states that the voltage drop across an Ohmic resistor is directly proportional to the current flowing through the resistor and that the proportionality constant is the resistance. Because the resistance of a component is determined by the geometric properties of its construction and the voltage supplied is determined by the power source, Ohm's Law tells the amount of current that will flow through the device and is written as

Equation Sheet

$$I = \frac{\Delta V}{R}$$

Not every device obeys Ohm's Law. However, unless a question explicitly states a device is non-Ohmic or asks to explain whether a device is Ohmic or not, every resistor you encounter in AP Physics 2 will be Ohmic. You can determine whether a resistive element in a circuit obeys Ohm's Law or not by making a plot of the voltage versus the current. A nonlinear plot indicates that the resistor is non-Ohmic, while a directly proportional graph will indicate that the resistor is Ohmic.

Ohm's Law can be applied to each individual resistor in a circuit as well as to the overall circuit. Applying Ohm's Law to an individual resistor yields the voltage drop across that resistor and the current flowing through the resistor. Applied to the entire circuit, Ohm's Law tells the voltage drop across the battery and the current which flows through the battery.

POWER DISSIPATION

The power dissipated by a circuit component is given by the product of the current through the component and the voltage drop across it.

Equation Sheet

$$P = I\Delta V$$

Heat and Power
Power in a circuit is very closely associated with heat given off. If we were to touch an incandescent light bulb, we would notice it becomes hot. The brighter the lightbulb is, the greater the heat given off.

The power dissipated by each individual component must sum to equal the power supplied by the battery (or other power source). For Ohmic resistors, the power equation may be combined with Ohm's Law to yield

$$P = I^2R \text{ and } P = V^2/R$$

KIRCHHOFF'S RULES

In addition to Ohm's Law, Kirchhoff's rules are used when analyzing circuits. These two laws are statements of conservation laws applied to circuit analysis. The **Loop Rule** states that the voltage drop across *any complete loop* in a circuit is 0 V. This statement follows from Conservation of Energy when applied to circuits. A corollary to the Loop Rule is that any pair of resistors in parallel will have identical voltage drops across them. The **Junction Rule** states that the sum of all current flowing into any junction is equal to the current flowing out of the junction. This statement follows from conservation of charge. A corollary of the Junction Rule is that for two resistors in series, the current must be the same through each because there is no junction.

A very simple circuit contains one battery with voltage V and one resistor of resistance R connected by wires.

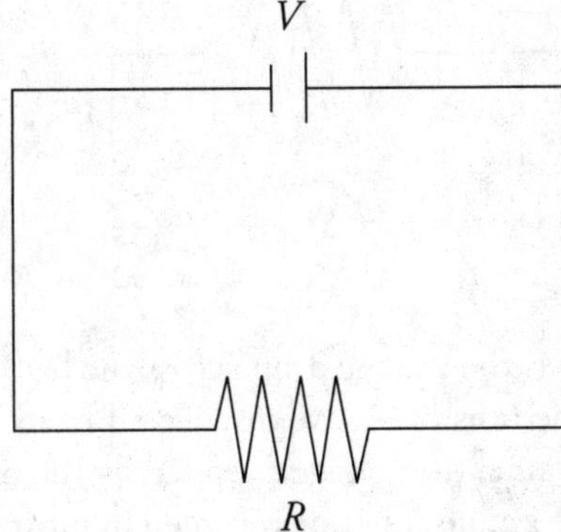

Let's consider how to apply Kirchhoff's Loop Rule to this circuit. First, assign the direction of current flow for all segments of the circuit. There are no junctions in this circuit, so one current flows throughout, and a current I flows from the positive terminal of the battery to the negative terminal, clockwise around the circuit. Then, choose a loop. In this simple circuit, there is only a single loop. Next, choose a starting point and a direction to traverse the loop. It is generally advisable to choose the direction of current, so start at point S and go clockwise.

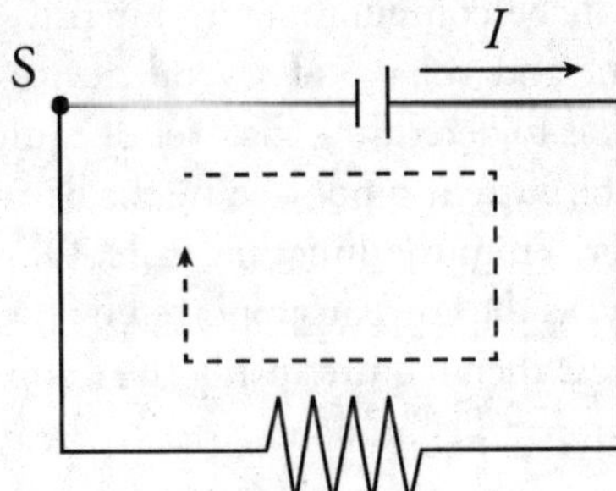

As we traverse the loop clockwise from point S, the first change in electric potential we experience is the increase in electric potential $+V$ from the battery. The next change in potential in the loop is a voltage drop across the resistor, $-IR$, after which we're back at starting point S. The loop must have an overall change in voltage of zero, so $V + (-IR) = 0$, and $V = IR$. (Note that if the loop went opposite the direction of current, the battery would correspond to a voltage drop and the resistor a voltage gain—this is why it's important to assign the direction of current first.) Add a pair of voltmeters and a pair of ammeters to the circuit in their proper positions to measure the current and voltage of the battery and the resistor.

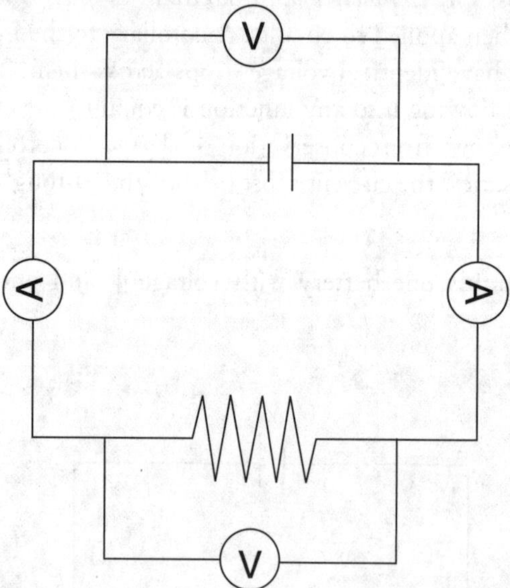

The reading in the top voltmeter is the voltage drop across the battery. The reading in the lower voltmeter is the voltage across the resistor. As we discovered from the Loop Rule, the readings in these voltmeters must be the same; the voltage supplied by the battery is equal to the voltage drop across the resistor. As there are no junctions in the circuit (no current flows through the voltmeters due to their nearly infinite resistance), each ammeter will also read an equal value for the current. In this simple circuit, all the current that flows out of the battery flows into the resistor, and all the current that flows out of the resistor flows back into the battery.

ANALYSIS OF CIRCUITS WITH RESISTORS

Circuits with several resistors can be computationally intensive. Problems frequently involve iteratively solving for the current and voltage of several devices in several steps. The typical approach to quantitative analysis is to determine the overall equivalent resistance of the circuit. Then given the battery voltage, the current supported by the battery can be calculated as well as the power supplied by the battery. From the Junction Rule, Ohm's Law is then applied to any resistors in series with the battery, as the current supported by the battery flows through each of those resistors. The iterative process then requires using the Loop Rule to determine the voltage drop across the paths that contain parallel branches.

Example 5 The circuit below shows a 12 V battery connected to three resistors. The resistances include $R_1 = 4\ \Omega$, $R_2 = 3\ \Omega$, and $R_3 = 6\ \Omega$. Determine the current supported by the battery and the voltage drop across each resistor.

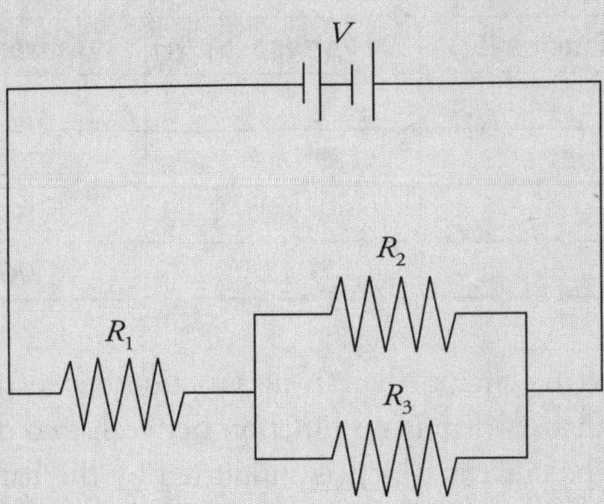

Solution. To begin, it is helpful to make a chart of the resistance, voltage, current, and power for each device. We need a row for each resistor and an additional row for the entire circuit. Any time that any two of the columns are known in a row, Ohm's Law and the power equation can be used to solve for the other two columns. We can immediately fill out any known resistance, current, voltage, or power values. Here, we have three resistances and one voltage as given.

	Resistance (Ω)	Voltage (V)	Current (A)	Power (W)
R_1	4			
R_2	3			
R_3	6			
Entire Circuit		12		

Next, we can use the concept of equivalent resistance to get resistance of the entire circuit, which will give us two values in the bottom row.

To begin, we combine R_2 and R_3 to get an equivalent resistor $R_{2,3}$. Since R_2 and R_3 are in parallel,

$$\frac{1}{R_{2,3}} = \frac{1}{R_2} + \frac{1}{R_3} = \frac{1}{3} + \frac{1}{6} = \frac{2}{6} + \frac{1}{6} = \frac{3}{6} = \frac{1}{2}$$

$$R_{2,3} = 2\ \Omega$$

We now have R_1 in series with $R_{2,3}$. The equivalent resistance $R_{1,2,3} = R_1 + R_{2,3} = 4 + 2 = 6\ \Omega$. The total resistance of the circuit is $R_{total} = 6\ \Omega$. We add that to the bottom row of the table. Then, using Ohm's Law and the power equation, we can complete the bottom row of the table.

	Resistance (Ω)	Voltage (V)	Current (A)	Power (W)
R_1	4			
R_2	3			
R_3	6			
Entire Circuit	**6**	12	12/6 = 2	$12^2/6 = 24$

According to the image of the circuit, there is no junction between V and R_1. The corollary to the Junction Rule states that if there is no junction between two devices, the current through each must be equal. All the current which is supported by the battery must flow through R_1. This allows us to fill in the current through R_1. Then, using Ohm's Law and the power equation, we can complete the R_1 row of the table.

	Resistance (Ω)	Voltage (V)	Current (A)	Power (W)
R_1	4	2 * 4 = 8	**2**	$2^2 * 4 = 16$
R_2	3			
R_3	6			
Entire Circuit	6	12	2	24

After the Junction Rule is used to complete a row, the Loop Rule is typically used to complete the next step. We know that the total voltage drop through any complete loop must be equal to the voltage supplied by the battery. This circuit has two completed loops: $V = V_1 + V_2$ and $V = V_1 + V_3$. We know V is 12 V and we know the voltage drop across R_1 is $V_1 = 8$ V, so the voltage drop across either R_2 or R_3 must be 4 V. This conclusion is consistent with the corollary to the Loop Rule that the voltage drops across resistors which are in parallel are equal. We can now complete the rest of the table.

	Resistance (Ω)	Voltage (V)	Current (A)	Power (W)
R_1	4	8	2	16
R_2	3	**4**	4/3 = 1.33	$4^2/3 = 5.33$
R_3	6	**4**	4/6 = 0.67	$4^2/6 = 2.67$
Entire Circuit	6	12	2	24

This example did not ask for the power dissipated by the circuit or by any component, but it is useful to have that column. As you will note, the resistances do not sum to the entire resistance. The voltages do not sum to the entire voltage. The currents do not sum to the entire current. However, the power will always sum to the total power. Here, 24 = 16 + 5.33 + 2.67 is true. This simple check allows you to verify the likelihood that your answer is correct and that you didn't make any algebra mistakes in calculating the voltages or currents.

Example 6 In this circuit, the voltage of the battery and the resistance of three of the resistors is known. An ammeter placed at P_2 reads 6 A.

(a) What is the value of the unknown resistor?

(b) What will a voltmeter read that is placed between P_1 and P_2?

(c) What percentage of the power dissipated by the circuit is dissipated by the 150 Ω resistor?

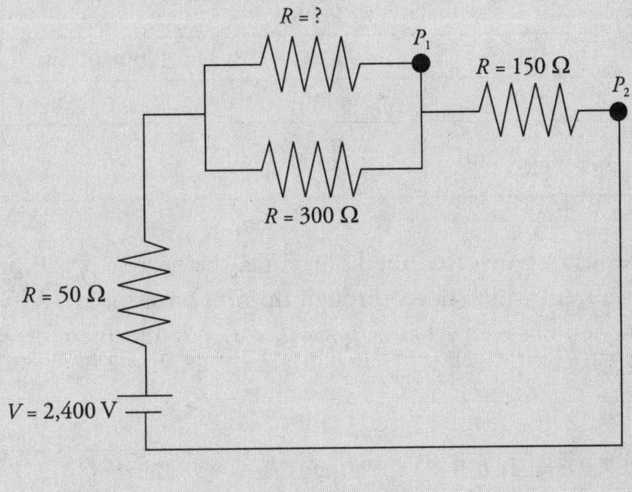

Solution. For the chart, we will need to label each resistor. Going in the direction of current flow from the battery, the 50 Ω resistor is R_1, the unknown resistor is R_2, the 300 Ω resistor is R_3, and the 150 Ω resistor is R_4. Fill out the given three resistances and one voltage.

	Resistance (Ω)	Voltage (V)	Current (A)	Power (W)
R_1	50			
R_2				
R_3	300			
R_4	150			
Entire Circuit		2400		

Additionally, we are given the current at point P_2 is 6 A. Whenever there is no junction between two devices, the current through each must be equal, so any current through an ammeter at P_2 must flow through the battery, R_1, and R_4.

	Resistance (Ω)	Voltage (V)	Current (A)	Power (W)
R_1	50	$50 * 6 = 300$	6	$6^2 * 50 = 1800$
R_2				
R_3	300			
R_4	150	$150 * 6 = 900$	6	$6^2 * 150 = 5400$
Entire Circuit	$2400/6 = 400$	2400	6	$2400 * 6 = 14,400$

Since the last step involved the Junction Rule, the next step will involve the Loop Rule. The voltage across R_2 (which we know is equal to the voltage across R_3) is $V - V_1 - V_4 = 2{,}400 - 900 - 300 = 1{,}200$ V.

	Resistance (Ω)	Voltage (V)	Current (A)	Power (W)
R_1	50	300	6	1800
R_2		**1200**		
R_3	300	**1200**	1200/300 = 4	$1200^2/300 = 4800$
R_4	150	900	6	5400
Entire Circuit	2400/6 = 400	2400	6	14,400

Now, since the previous step involved the Loop Rule, the next step will involve the Junction Rule. We know the current which flows through the 50 Ω resistor is 6 A. We know this must all flow into the junction between R_2 and R_3. So $I_1 = I_2 + I_3$ and therefore $6 = I_2 + 4$. We can conclude that I_2 is 2 A.

	Resistance (Ω)	Voltage (V)	Current (A)	Power (W)
R_1	50	300	6	1800
R_2	1200/2 = 600	1200	**2**	2 * 1200 = 2400
R_3	300	1200	1200/300 = 4	$1200^2/300 = 4800$
R_4	150	900	6	5400
Entire Circuit	2400/6 = 400	2400	6	14,400

Finally, we can check to see if the sum of the power of each resistor correctly adds to the entire power: 1800 + 2400 + 4800 + 5400 = 14,400 is correct.

With the analysis complete, we can address the questions that were asked in the problem.

(a) The unknown resistance (R_2) is 600 Ω.

(b) A voltmeter placed between P_1 and P_2 will be in parallel with R_3 and will therefore have a reading of 1200 V.

(c) The 150 Ω resistor dissipates 5400 W of the entire 14,400 W, which is 37.5%.

EMF and Internal Resistance

The voltage supplied by an ideal battery is often, for historical reasons, referred to as EMF (electromotive force, ε). This is simply another name for voltage and electric potential difference, but it is almost always encountered in the context of "voltage supplied by the battery" (or in the context of Faraday's Law in the magnetism chapter).

A battery in a circuit can be modeled as an ideal EMF in series with a resistor, *r*. In this model, *r* is called the "internal resistance" of the battery and the voltage measured across the EMF source, and *r* is the "terminal voltage." This would be the measurement reading from a voltmeter placed across the terminals of the battery.

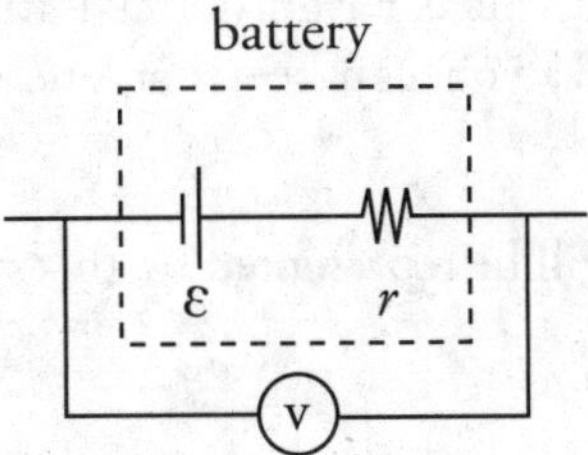

> **Example 7** If the 50 Ω resistor in the previous example were an internal resistor, what would the terminal voltage of the battery in that example equal?

Solution. The terminal voltage would be the voltage measured from one side of the battery across the internal resistor. Here, that would just be $V_{\text{terminal}} = V - V_1 = 2400 - 300 = 2100$ V.

CAPACITORS IN CIRCUITS

In the previous chapter, we looked at the electric field generated by parallel-plate capacitors. The electric field between the plates of a capacitor allows capacitors to be used in circuits as devices to store energy. They also have interesting properties related to the amount of time they take to charge, but those non-steady-state conditions are beyond the scope of the AP Physics 2 Exam, as their analysis requires calculus.

A capacitor stores energy in an electric field. A parallel-plate capacitor is comprised of two plates of area *A* and separation *d*. An insulator (called a dielectric in this context) may fill the area between the plates, or the space between them may be in vacuum.

The capacitance, which determines the amount of charge that can be stored on the plates for a given voltage, depends on the plate area, separation, and the dielectric constant, κ, of whatever material is between the plates (for vacuum, κ = 1).

$$C = \kappa\varepsilon_0 \frac{A}{d}$$

Equation Sheet

As current flows in a circuit with a capacitor, charges move onto one plate and off of the other plate. As a charge excess builds up on one plate, the electric field between the plates builds. The ratio of the charge on either plate to the voltage across the plates is the capacitance.

Equation Sheet

$$\Delta V = Q/C$$

Capacitance is measured in a unit called **Farad** (named after Michael Faraday). Rearranging the equation above to $C = Q/\Delta V$, you can see that one Farad is one Coulomb per Volt, 1F = 1C/V.

In a circuit diagram, a capacitor will be represented by this symbol:

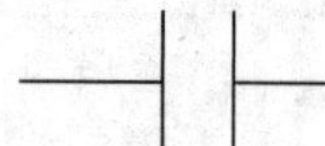

Combinations of Capacitors and Equivalent Capacitance

Multiple capacitors may be in a circuit in which the equivalent capacitance may be calculated. Two capacitors which are arranged in parallel are equivalent to a single capacitor of area $A_p = A_1 + A_2$. Following a similar analysis to a pair of resistors, we see that

$$C_{\text{eq},p} = \kappa\varepsilon_0 \frac{A_1 + A_2}{d} = \kappa\varepsilon_0 \frac{A_1}{d} + \kappa\varepsilon_0 \frac{A_2}{d} = C_1 + C_2$$

This result holds for more than two capacitors as well, so long as they are all arranged in parallel.

Equation Sheet

$$C_{\text{eq},p} = \sum_i C_i$$

Similarly, the capacitors may be arranged in series. That arrangement is equivalent to a single capacitor of separation: $d_s = d_1 + d_2$. Following a similar analysis to a pair of resistors in parallel, we see that for capacitors in series

$$C_{\text{eq},s} = \kappa\varepsilon_0 \frac{A}{d_1 + d_2} \rightarrow \frac{1}{C_{\text{eq},s}} = \frac{(d_1 + d_2)}{\kappa\varepsilon_0 A} = \frac{d_1}{\kappa\varepsilon_0 A} + \frac{d_2}{\kappa\varepsilon_0 A} = \frac{1}{C_1} + \frac{1}{C_2}$$

This result holds for more than two capacitors as well, so long as they are all arranged in series.

Equation Sheet

$$\frac{1}{C_{\text{eq},s}} = \sum_i \frac{1}{C_i}$$

Example 8 Calculate the equivalent capacitance of a pair of capacitors that are in series when that combination is in parallel with a third capacitor. All the capacitors have capacitance C.

Solution. In the same way we combined part of the resistor circuits, we first combine only the portion of this mixed series and parallel capacitor arrangement in which the capacitors are in series: $\frac{1}{C_{1,2}} = \frac{1}{C_1} + \frac{1}{C_2}$ so $C_{1,2} = \frac{C_1 C_2}{C_1 + C_2}$. Because all the capacitors have capacitance C, $C_{1,2} = \frac{C^2}{2C} = \frac{C}{2}$. Then combining that equivalent capacitor with the third capacitor in parallel, $C_{1,2,3} = C_{1,2} + C_3 = \frac{C}{2} + C = \frac{3C}{2}$.

Charging and Discharging Capacitors

An uncharged capacitor may become charged by connecting it to a battery. The instant that connection is made, the uncharged capacitor has a potential difference of 0 V across it and current flows into the capacitor just like current would flow through a wire. For some amount of time, the capacitor is partially charged and the potential difference across the plates grows as more and more charge builds up on the plate. Eventually, the capacitor reaches a state at which the potential difference across the plates is equal to the voltage of the battery and the current flow onto the capacitor plates ceases. In this state, the capacitor acts like an open switch and no current flows onto or off of the plates.

A charged capacitor may be discharged by connecting the two plates to a resistor. The voltage between the two plates causes current to flow across the resistor. As the current flows, the amount of charge on the plates, Q, decreases and the voltage ΔV also decreases. Eventually, when the capacitor is fully discharged so that $Q = 0$ C, the voltage across the resistor is $\Delta V = 0$ V, from $\Delta V = Q/C$. Ohm's Law, $V = IR$, indicates that no more charge flows across the resistor.

Calculating the amount of time required for a capacitor to charge and discharge requires calculus. It is not part of the AP Physics 2 Exam. Instead, the exam focuses on capacitors in "steady state," in which enough time has passed for the capacitor to begin each problem in a state which is either completely discharged or completely charged.

Altering the Capacitance of a Capacitor

The capacitance is given by

Equation Sheet

$$C = \kappa\varepsilon_0 \frac{A}{d}$$

Three things may be done to alter the capacitance of a capacitor: a dielectric may be inserted (altering κ), the area of the plates may change (changing A), or the spacing between the plates may change (which changes d).

There are two scenarios for making these changes to a charged capacitor in a circuit:

1. The battery may remain connected, causing the potential difference to remain constant.
2. The battery may be disconnected, causing the charge on each plate to remain constant.

Which of these scenarios unfolds will influence an analysis of the capacitor in terms of its charge on each plate, potential difference, stored electric field, and stored energy.

Let's look at what happens when the capacitance increases as a result of adding a dielectric (κ increases).

Capacitance is increased with a battery connected. *V* is constant and *Q* changes.	
V	Constant
$Q = CV$	Increases
$E = V/d$	Constant
$U_C = \frac{1}{2}QV$	Increases

Capacitance is increased with a battery disconnected. *Q* is constant and *V* changes.	
$V = Q/C$	Decreases
Q	Constant
$E = V/d$	Decreases
$U_C = \frac{1}{2}QV$	Decreases

If the capacitance were decreased instead of increased, say by removing a dielectric from a charged capacitor (κ decreases), the same analysis would apply. The results would be as follows.

Capacitance is decreased with a battery connected. *V* is constant and *Q* changes.	
V	Constant
$Q = CV$	Decreases
$E = V/d$	Constant
$U_C = \frac{1}{2}QV$	Decreases

Capacitance is decreased with a battery disconnected. *Q* is constant and *V* changes.	
$V = Q/C$	Increases
Q	Constant
$E = V/d$	Increases
$U_C = \frac{1}{2}QV$	Increases

RC CIRCUITS WITH CAPACITORS IN STEADY STATE

The two possible steady state conditions for a capacitor in a circuit occur when the capacitor is discharged or when it is fully charged.

When a capacitor is discharged, the voltage between the plates is 0 V. As a result, the discharged capacitor *acts like a wire*. This causes all of the current from any junction where one parallel path has the discharged capacitor and the other has a resistor to flow into the capacitor and none to flow into the resistor. Essentially, the discharged capacitor short circuits any resistors arranged in parallel with the capacitor.

The situation is reversed when a capacitor is fully charged. The voltage between the plates is the maximum amount that can be supplied by the battery. As a result, the fully charged capacitor *acts like a broken wire*. This causes all of the current from any junction where one parallel path has the fully charged capacitor and the other has a resistor to flow into the resistor and none to flow into the capacitor. Essentially, the fully charged capacitor short circuits any resistors arranged in series with the capacitor.

When we analyze circuits with capacitors, we still use a chart of the resistance, voltage, current, and power. However, because our analysis is limited to one of two particular instants in time (the discharged capacitor or the fully charged capacitor), both the "current" and the "power" columns are not applicable to the capacitor, as both of those quantities explicitly depend on Δt.

Example 9 The image shows an RC circuit with an uncharged capacitor. Determine the following:

(a) the current through the capacitor when the switch has just been closed

(b) the voltage across the capacitor after the switch has been closed for a long time

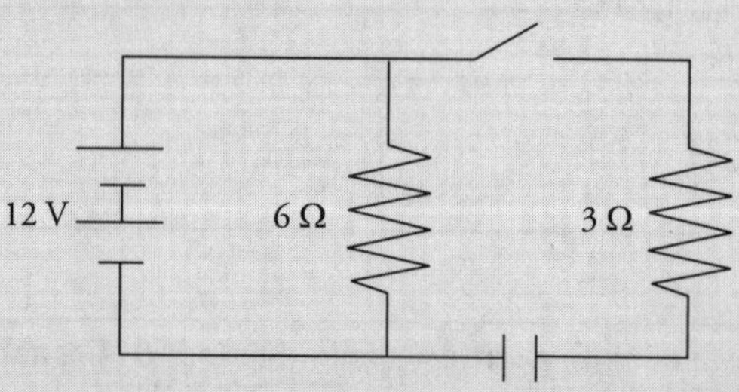

Solution.

(a) Immediately after the switch is closed, the uncharged capacitor acts like a wire.

	Resistance (Ω)	Voltage (V)	Current (A)	Power (W)
R_1	6			
R_2	3			
C	**0**	**0**	N/A	N/A
Entire Circuit		12		

We can find the resistance of the entire circuit by noting the 6 Ω and 3 Ω resistors are in parallel, so the equivalent resistance is $\frac{1}{R} = \frac{1}{3} + \frac{1}{6}$ so that $R = 2\ \Omega$.

	Resistance (Ω)	Voltage (V)	Current (A)	Power (W)
R_1	6			
R_2	3			
C	0	0	N/A	N/A
Entire Circuit	**2**	12	12/2 = 6	$12^2/2 = 72$

Now that we have used the current, we look to the Loop Rule. Since the 6 Ω and 3 Ω resistors are both in parallel with the battery, they all have equal voltages.

	Resistance (Ω)	Voltage (V)	Current (A)	Power (W)
R_1	6	**12**	12/6 = 2	$12^2/6 = 24$
R_2	3	**12**	12/3 = 4	$12^2/3 = 48$
C	0	0	N/A	N/A
Entire Circuit	2	12	6	72

Finally, we return to the Junction Rule and see that the capacitor is in series with the 3 Ω resistor, so the current flowing to the capacitor must be 4 A.

Note that as with wires, we cannot apply Ohm's Law to the capacitor because we would be dividing by 0 when calculating the current.

(b) After the switch is closed for a long time, the capacitor has filled. The capacitor acts like a broken wire. Now, R_2 is in series with the fully charged capacitor, so it has been shorted out of the circuit. Therefore, when we complete the table, we place a value of 0 Ω in for R_2 instead of 3 Ω. The resistance of the resistor is still 3 Ω, but because it is shorted out of the circuit, its effect is 0 Ω.

	Resistance (Ω)	**Voltage (V)**	**Current (A)**	**Power (W)**
R_1	6			
R_2	**0**			
C	∞		N/A	N/A
Entire Circuit		12		

The equivalent resistance of the circuit is just 6 Ω.

	Resistance (Ω)	**Voltage (V)**	**Current (A)**	**Power (W)**
R_1	6			
R_2	0			
C	∞		N/A	N/A
Entire Circuit	**6**	12	12/6 = 2	$12^2/6 = 24$

All of the current supported by the battery flows through R_1 and none through R_2. Because no current flows across R_2, there must be 0 V voltage between the two ends of the resistor and it must be dissipating 0 W of power.

	Resistance (Ω)	**Voltage (V)**	**Current (A)**	**Power (W)**
R_1	6	6 * 2 = 12	2	$12^2/6 = 24$
R_2	0	**0**	**0**	**0**
C	∞		N/A	N/A
Entire Circuit	**6**	12	2	24

We know that each parallel branch has to have the same voltage. Since the current, voltage, and resistance of R_2 are all 0, then the voltage across the filled capacitor must be 12 V.

Example 10 Determine the current through and the voltage across each resistor and the battery in the following circuit when

(a) the switch has just been closed

(b) the switch has been closed for a long period of time

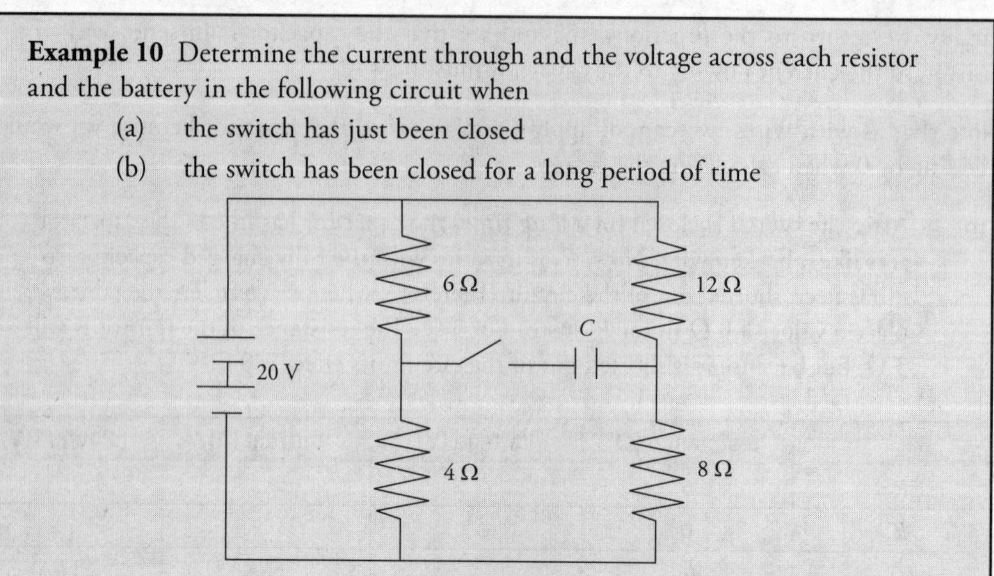

Solution.

(a) If we imagine the capacitor as a wire, the 6 Ω and 12 Ω resistors are in parallel, with an equivalent resistance of 4 Ω. The 4 Ω and 8 Ω resistors are in parallel, with an equivalent resistance of $\frac{8}{3}$ Ω. The equivalent 4 Ω resistor and $\frac{8}{3}$ Ω resistor are in series. The total resistance of the circuit is $\frac{20}{3}$ Ω.

	Resistance (Ω)	Voltage (V)	Current (A)	Power (W)
R_1	6			
R_2	12			
R_3	4			
R_4	8			
C	**0**	**0**	N/A	N/A
Entire Circuit	**20/3**	20	20/6.67 = 3	20^2/6.67 = 60

Now, we again have a mixture of parallel and series. We could continue our methodical approach and find the equivalent resistance combination of R_1 and R_2 since we know the current through that combination is 3 A. That would allow us to calculate the voltage across each of those resistors. However, since we know the 3 A of total current divides between the 6 Ω and 12 Ω resistors, and since those are in parallel, they must have the same voltage, so $I_1 * 6 = I_2 * 12$, and we can conclude that the current in R_1 is 2 A while the current in R_2 is 1 A. Similarly, the 3 A of total current divides between the 4 Ω and 8 Ω resistors. Since those are in parallel, they must have the same voltage, so $I_3 * 4 = I_4 * 8$, and we can conclude that the current in R_3 is 2 A while the current in R_4 is 1 A.

	Resistance (Ω)	Voltage (V)	Current (A)	Power (W)
R_1	6	6 * 2 = 12	**2**	$2^2 * 6 = 24$
R_2	12	12 * 1 = 12	**1**	$1^2 * 12 = 12$
R_3	4	4 * 2 = 8	**2**	$2^2 * 4 = 16$
R_4	8	8 * 1 = 8	**1**	$1^2 * 8 = 8$
C	**0**	**0**	N/A	N/A
Entire Circuit	20/3	20	3	60

(b) When the switch has been closed for a long time, it acts like a broken wire. This makes the 6 Ω and 4 Ω resistors in series, acting as a single 10 Ω resistor, and that combination is in parallel with the combination of the 12 Ω and 8 Ω resistors in series, acting as a single 20 Ω resistor. The equivalent resistance of the entire circuit is $\frac{1}{R} = \frac{1}{10} + \frac{1}{20}$, so $R = 6.67$.

	Resistance (Ω)	Voltage (V)	Current (A)	Power (W)
R_1	6			
R_2	12			
R_3	4			
R_4	8			
C	∞		N/A	N/A
Entire Circuit	**6.67**	20	20/6.67 = 3	$20^2/6.67 = 60$

Now, we again have a mixture of parallel and series. We could continue our methodical approach and find the equivalent resistance combination of R_1 and R_3 since we know the voltage across that combination is 20 V. That would allow us to calculate the current through each of those resistors. However, since we know the 6 A of total current divides between the 10 Ω and 20 Ω paths, we know that 2 A goes through R_1 and R_3 while 1 A goes through R_2 and R_4.

	Resistance (Ω)	Voltage (V)	Current (A)	Power (W)
R_1	6	6 * 2 = 12	**2**	$12^2/6 = 24$
R_2	12	12 * 1 = 12	**1**	$12^2/2 = 12$
R_3	4	4 * 2 = 8	**2**	$8^2/4 = 16$
R_4	8	8 * 1 = 8	**1**	$8^2/8 = 8$
C	∞		N/A	N/A
Entire Circuit	6.67	20	20/6.67 = 3	$20^2/6.67 = 60$

Chapter 6 Review Questions

Answers and explanations can be found in Chapter 11.

Section I: Multiple Choice

1 Mark for Review

A wire made of brass and a wire made of silver have the same length, but the diameter of the brass wire is 4 times the diameter of the silver wire. The resistivity of brass is 5 times greater than the resistivity of silver. If R_B denotes the resistance of the brass wire and R_S denotes the resistance of the silver wire, which of the following is true?

(A) $R_B = \frac{5}{16} R_S$

(B) $R_B = \frac{4}{5} R_S$

(C) $R_B = \frac{5}{4} R_S$

(D) $R_B = \frac{5}{2} R_S$

2 Mark for Review

For an Ohmic conductor, doubling the voltage without changing the resistance will cause the current to

(A) decrease by a factor of 4

(B) decrease by a factor of 2

(C) increase by a factor of 2

(D) increase by a factor of 4

3 Mark for Review

A battery, an uncharged capacitor, and resistor are all placed in parallel. Which answer best describes the current flowing in this circuit?

(A) The capacitor and resistor both experience the same current.

(B) Initially, all the current flows through the path with the resistor, then, after a long time, all the current flows through the path with the capacitor.

(C) Initially, all the current flows through the path with the capacitor, then, after a long time, all the current flows through the path with the resistor.

(D) No current will flow in this circuit.

4 Mark for Review

A student wants to determine the resistivity of copper. She has a voltmeter reading for a copper wire of known length. What other information will she need?

(A) An ammeter reading

(B) The diameter of the wire

(C) Both an ammeter reading and the diameter of the wire

(D) Neither an ammeter reading nor the diameter of the wire

5 Mark for Review

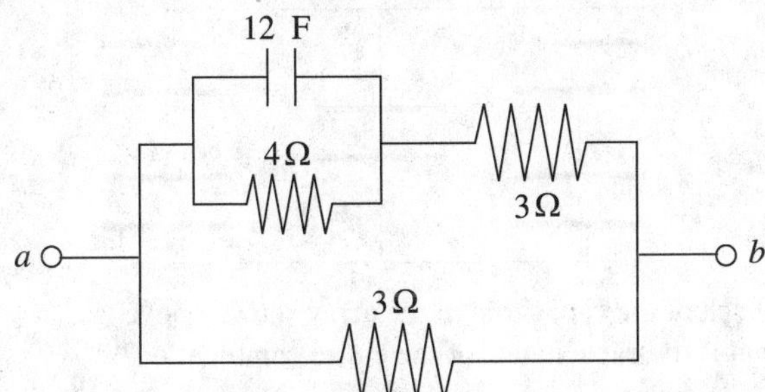

The circuit shown above has a constant voltage between points *a* and *b*. The voltage has been applied for a long time. What is the resistance of the circuit?

(A) 0.47 Ω

(B) 1.5 Ω

(C) 2.1 Ω

(D) 10 Ω

6 Mark for Review

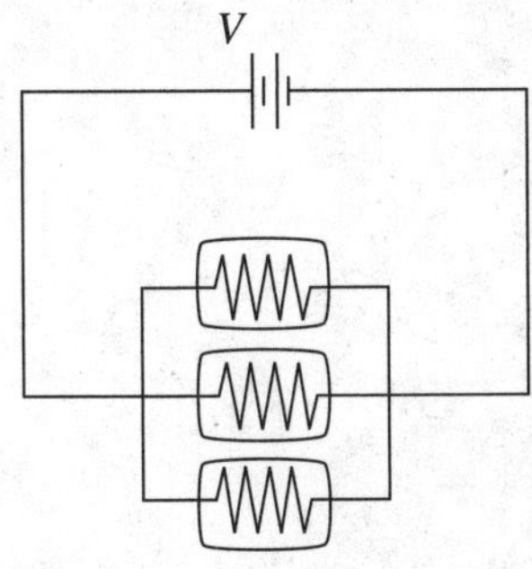

Three identical light bulbs are connected to a source of emf, as shown in the diagram above. What will happen if the middle bulb burns out?

(A) The light intensity of the other two bulbs will decrease (but they won't go out).

(B) The light intensity of the other two bulbs will increase.

(C) The light intensity of the other two bulbs will remain the same.

(D) More current will be supported by the source of emf.

7 Mark for Review

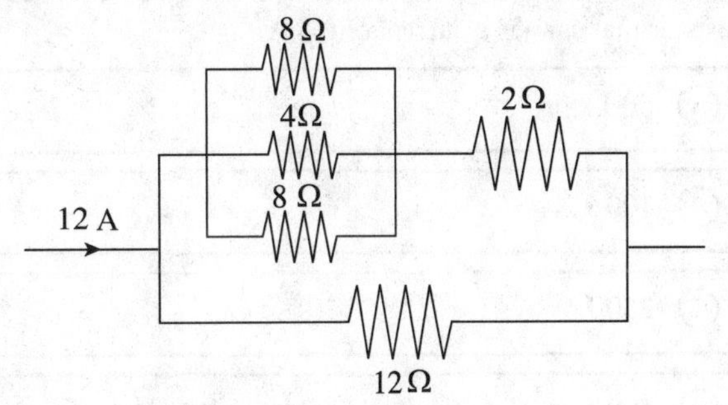

What is the voltage drop across the 12 Ω resistor in the portion of the circuit shown above?

(A) 24 V

(B) 36 V

(C) 48 V

(D) 72 V

8 Mark for Review

A simple DC circuit with a single ohmic resistor is constructed. A graph plots the voltage drop across the resistor versus the current flowing through the resistor. For an ideal battery, the graph shows a directly proportional relationship. If the battery has some internal resistance, how, if at all, would such a graph change?

(A) The graph would continue to show a directly proportional relationship.

(B) The graph would be linear, but would have a negative *y*-intercept.

(C) The graph would be linear, but would have a positive *y*-intercept.

(D) The graph would be nonlinear.

9 Mark for Review

How much energy is dissipated as heat in 20 s by a 100 Ω resistor that carries a current of 0.5 A?

(A) 50 J

(B) 100 J

(C) 250 J

(D) 500 J

10 Mark for Review

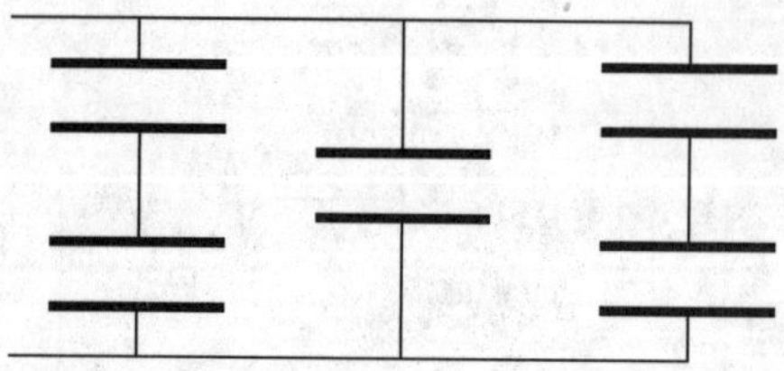

If each of the capacitors in the array shown above is C, what is the capacitance of the entire combination?

(A) $C/2$

(B) $2C/3$

(C) $5C/6$

(D) $2C$

Section II: Free Response

1 Mark for Review

Consider the following circuit in Figure 1:

Figure 1

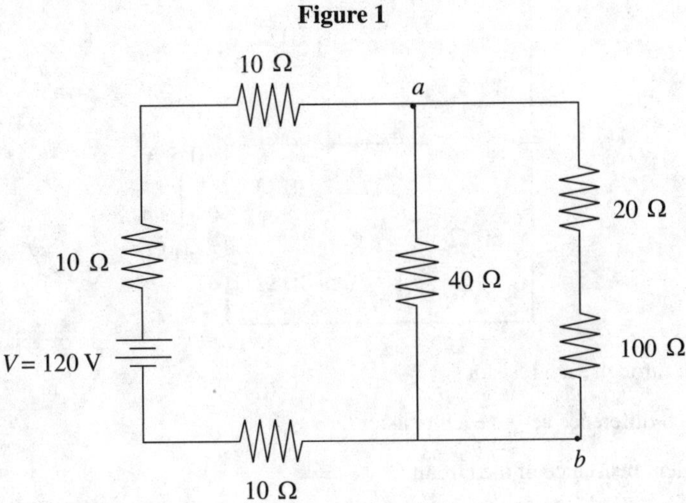

A. At what rate does the battery deliver energy to the circuit?

B. Find the current through the 40 Ω resistor.

C.

i. Determine the potential difference between points *a* and *b*.

ii. At which of these two points is the potential higher?

D. Given that the 100 Ω resistor is a solid cylinder that's 4 cm long, composed of a material whose resistivity is 0.45 Ω • m, determine the radius of its cross-section.

E. Which resistor or resistors, if any, could be replaced with a capacitor so that when the capacitor is fully charged, no current flow is supported by the battery? Explain your answer.

2 Mark for Review

Consider the following circuit in Figure 1:

Figure 1

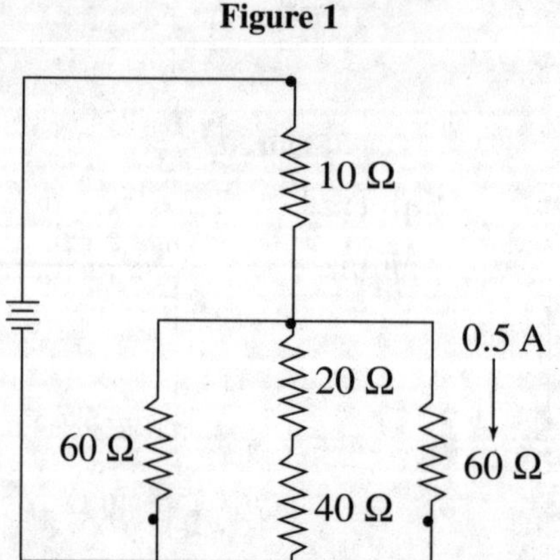

A. What is the current through each resistor?

B. What is the potential difference across each resistor?

C. What is the equivalent resistance of the circuit?

D. The leftmost 60 Ω resistor is replaced with a capacitor.

   i. When the capacitor is uncharged and just beginning to charge, will the current in the rightmost 60 Ω resistor be greater than, equal to, or less than the 0.5 A it was before the capacitor was inserted into the circuit?

   ii. When the capacitor is fully charged, will the current in the rightmost 60 Ω resistor be greater than, equal to, or less than the 0.5 A it was before the capacitor was inserted into the circuit?

3 Mark for Review

Figure 1

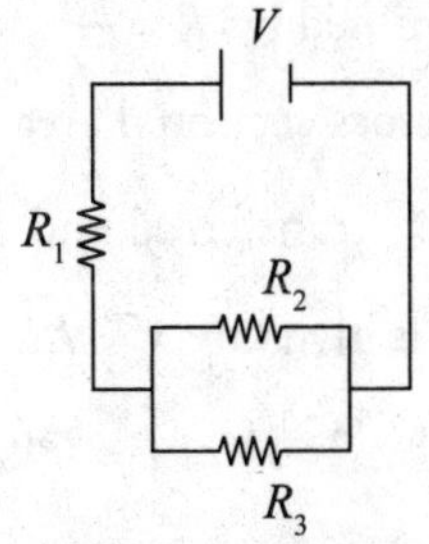

The circuit in Figure 1 contains a battery of voltage V and three resistors with resistances R_1, R_2, and R_3, respectively. As part of an experiment, a student has been given two measuring devices: a voltmeter and an ammeter. The first can be used to measure the changes in voltage of a circuit. The second can be used to measure the current flowing through a particular segment of wire. For answering the questions below, a voltmeter and ammeter look like —(V)— and —(A)—, respectively, when drawn in a circuit diagram.

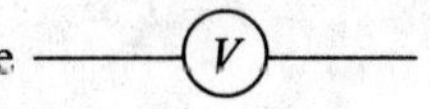

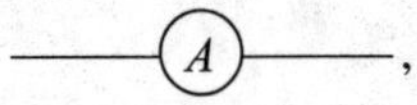

A. In terms of the known variables, what is the voltage drop across the first resistor?

B. Draw a diagram showing how you would integrate the voltmeter to measure the voltage drop across resistor labeled R_1. Explain the reasoning behind your decision.

C. Draw a diagram showing how you would integrate the ammeter to measure the current passing through the resistor labeled R_1. Explain the reasoning behind your decision.

D. What would be the ideal resistances for each device to have? Explain why each would be ideal for that device.

Chapter 6 Summary

- The resistance of an object can be determined by $R = \frac{\rho \ell}{A}$, where ρ is the resistivity (a property of the material), ℓ is the length, and A is the cross-sectional area.
- The current is the rate at which charge is transferred and given by $I_{avg} = \frac{\Delta Q}{\Delta t}$.
- Many objects obey Ohm's Law, which is given by $V = IR$.
- The electrical power in a circuit is given by $P = IV$ or $P = I^2R$ or $P = \frac{V^2}{R}$.
  This is the same power we've encountered in our discussion of energy $P = \frac{W}{t}$.
- In a series circuit / In a parallel circuit

In a series circuit	In a parallel circuit
$V_{Bat} = V_1 + V_2 + \ldots$	$V_{Bat} = V_1 = V_2 = \ldots$
$I_{Bat} = I_1 = I_2 = \ldots$	$I_{Bat} = I_1 + I_2 + \ldots$
$R_{eq,s} = R_1 + R_2 + \ldots$	$\frac{1}{R_{eq,p}} = \frac{1}{R_1} + \frac{1}{R_2} + \ldots$
$R_{eq,s} = \sum_i R_i$	$\frac{1}{R_{eq,p}} = \sum_i \frac{1}{R_i}$

- Kirchhoff's Loop Rule tells us that the sum of the potential differences in any closed loop in a circuit must be zero.
- Kirchhoff's Junction Rule (Node Rule) tells us the total current that enters a junction must equal the total current that leaves the junction.

- To find the combined capacitance of capacitors in parallel (side by side), simply add their capacitances:

  $$C_{\text{eq,p}} = C_1 + C_2 + \ldots \text{ or } C_{\text{eq},p} = \sum_i C_i$$

- To find the capacitance of capacitors in series (one after another), add their inverses:

  $$\frac{1}{C_{\text{eq,s}}} = \frac{1}{C_1} + \frac{1}{C_2} + \ldots \text{ or } \frac{1}{C_{\text{eq},s}} = \sum_i \frac{1}{C_i}$$

- When a capacitor is in a circuit and is completely discharged, it will act like a closed switch (equivalent to a resistor with $R = 0\ \Omega$). When the capacitor is fully charged, it acts like an open switch (equivalent to an infinite resistor).

Chapter 7
Magnetism and Electromagnetism

INTRODUCTION

In a previous chapter, we learned that electric charges are the sources of electric fields and that other charges experience an electric force in those fields. The charges generating the field were assumed to be at rest, because if they weren't, then another force field would have been generated in addition to the electric field. Electric charges *that move* are the sources of **magnetic fields**, and other charges that move can experience a magnetic force in these fields.

MAGNETIC FIELDS

Similar to our discussion about electric fields, the space surrounding a magnet is permeated by a magnetic field. The direction of the magnetic field is defined as pointing out of the north end of a magnet and into the south end of a magnet, as illustrated below.

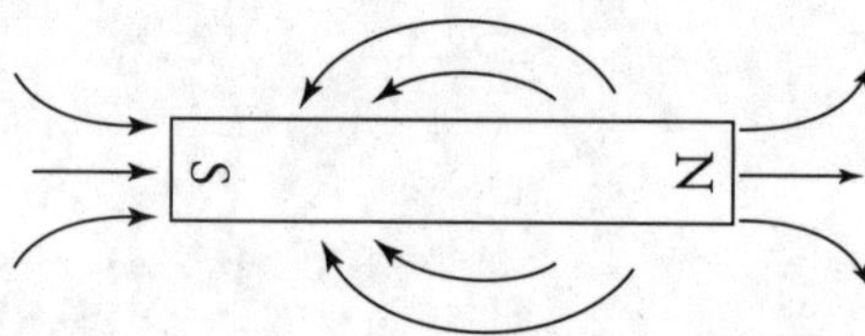

Magnetic fields have always been found to have both a north and a south pole, defined so that the field comes out of the north pole and enters back into the magnet at the south pole. If you look back at Chapter 5, electric fields could be created from a single positive or single negative charge. You will also see that a pair of charges, one positive and one negative, generated an electric dipole. Magnetic fields always have both a north and a south pole.

When two magnets get near each other, the magnetic fields interfere with each other and can be drawn as follows. Note that, for simplicity's sake, only the field lines closest to the poles are shown.

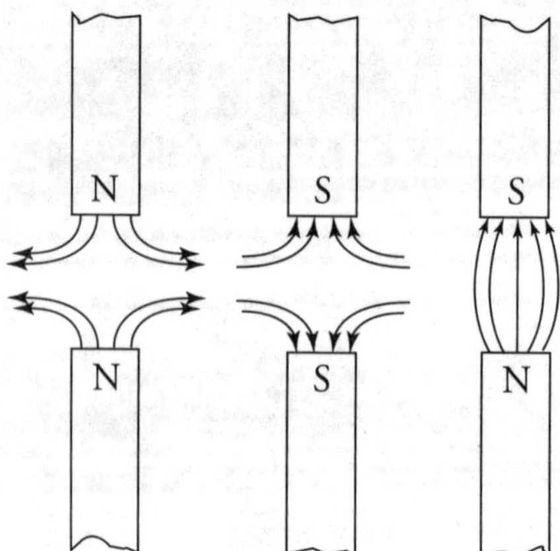

Notice there is a curve to the previous fields. We call the field uniform if the field lines are parallel and of equal strength. It is easy to recognize a uniform magnetic field to the right, left, top of the page, or bottom of the page. But you will also see a field going into or out of the page. A field into the page looks as if there were a north pole of a magnet above the page pointing down at the south pole of a magnet that is below the page. It is represented by an area with X's going into the page. A field coming out of the page looks as if there were a north pole of a magnet below the page pointing up at the south pole of a magnet that is above the page. It is represented by an area with dots (•) coming out of the page.

INTO	OUT OF
× × × ×	• • • •
× × × ×	• • • •
× × × ×	• • • •
× × × ×	• • • •

BAR MAGNETS

A permanent bar magnet creates a magnetic field that closely resembles the magnetic field produced by a circular loop of current-carrying wire:

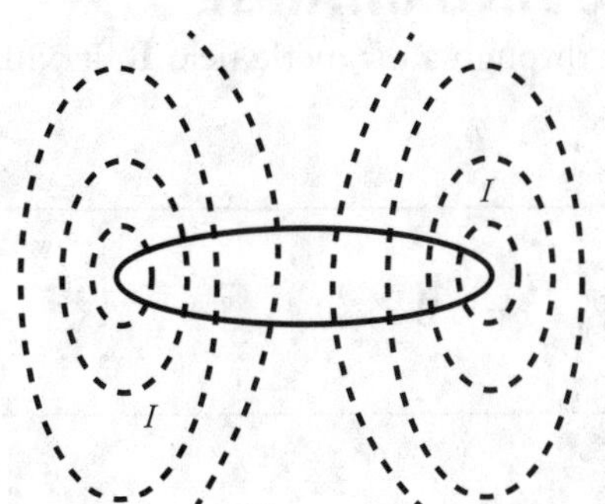

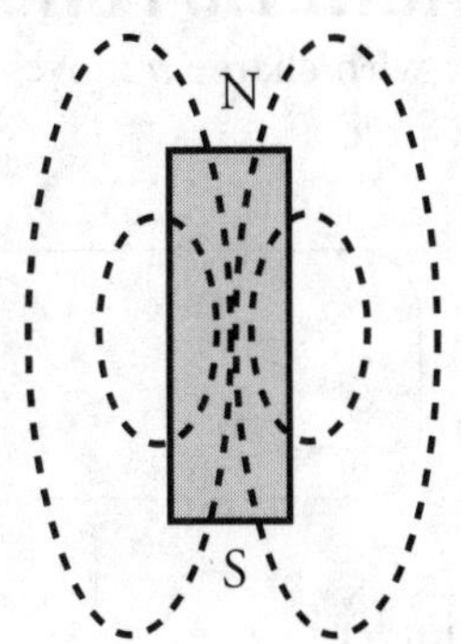

By convention, the magnetic field lines emanate from the end of the magnet designated the north pole (N) and then curl around and reenter the magnet at the end designated the south pole (S). The magnetic field decreases in strength as distance from the magnet increases. The magnetic field created by a permanent bar magnet is due to the electrons, which have an intrinsic spin as they orbit the nuclei. They are literally charges in motion, which is the ultimate source of any and all magnetic fields. If a piece of iron is placed in an external magnetic field (for example, one created by a current-carrying solenoid), the individual magnetic dipole moments of the electrons will be forced to more or less line up. Because iron is ferromagnetic, these now-aligned magnetic dipole moments tend to retain this configuration, thus permanently magnetizing the bar and causing it to produce its own magnetic field. All materials have some **magnetic permeability**, μ, which determines how great a magnetic field an object will develop when placed in an external field.

As with electric charges, like magnetic poles repel each other, while opposite magnetic poles attract each other.

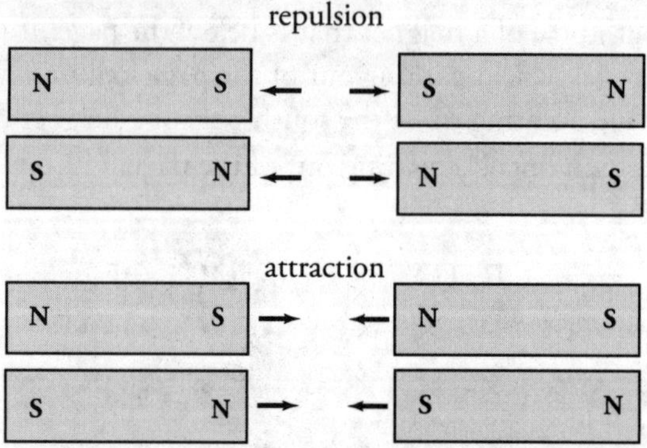

However, while you can have a positive electric charge all by itself, you can't have a single magnetic pole all by itself: remember, the existence of a lone magnetic pole has never been confirmed. That is, there are no magnetic monopoles; magnetic poles always exist in pairs. If you break a bar magnet into two pieces, it does not produce one piece with just an N and another with just an S; it produces two separate, complete magnets, each with an N-S pair.

THE MAGNETIC FORCE ON A MOVING CHARGE

If a particle with charge q moves with velocity $\mathbf{v}$ through a magnetic field $\mathbf{B}$, it will experience a magnetic force, $\mathbf{F}_B$:

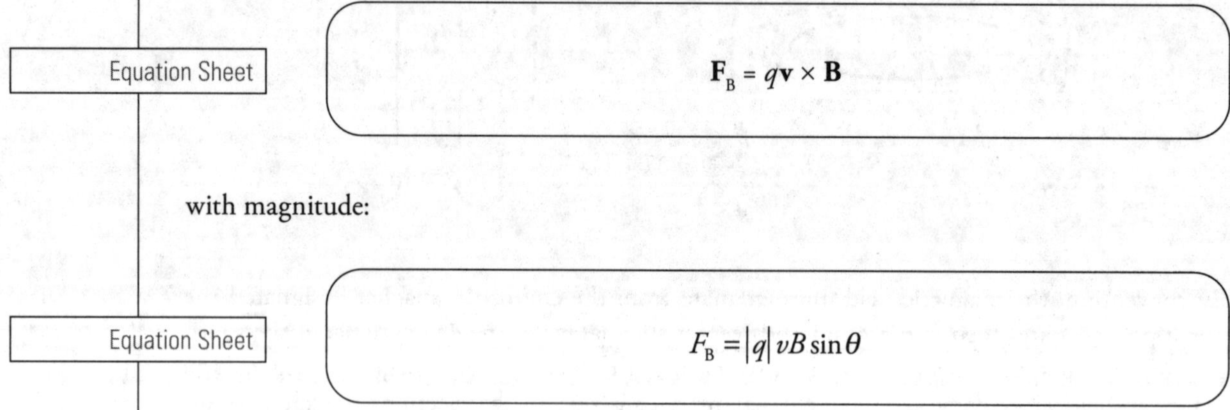

$$\mathbf{F}_B = q\mathbf{v} \times \mathbf{B}$$

with magnitude:

$$F_B = |q| vB \sin\theta$$

where θ is the angle between $\mathbf{v}$ and $\mathbf{B}$. From this equation, we can see that if the charge is at rest, then $v = 0$ immediately gives us $F_B = 0$. This tells us that magnetic forces act only on moving charges. Also, if $\mathbf{v}$ is parallel (or antiparallel) to $\mathbf{B}$, then $F_B = 0$ since, in either of these cases, $\sin\theta = 0$. So, only charges that cut across the magnetic field lines will experience a magnetic force. Furthermore, the magnetic force is maximized when $\mathbf{v}$ is perpendicular to $\mathbf{B}$, since if $\theta = 90°$, then $\sin\theta$ is equal to 1, its maximum value.

The direction of $\mathbf{F}_B$ is always perpendicular to both **v** and **B** and depends on the sign of the charge q and the direction of $\mathbf{v} \times \mathbf{B}$ (which can be found by using the Right-Hand Rule).

> **0 or 180**
> If the velocity **v** and the magnetic field **B** are parallel or antiparallel, which means oriented so the angle between the vectors is 180 degrees, the magnetic force $\mathbf{F}_B = 0$.

If q is *positive*, use your *right* hand and the *Right*-Hand Rule.

If q is *negative*, you need your thumb to point in the direction of the product of q and v (which is the direction opposite of v), not just the direction of the velocity when you apply the Right-Hand Rule.

Whenever you use the Right-Hand Rule, keep your hand flat, and follow these steps:

1. Orient your hand so that your thumb points in the direction of the velocity **v**. If the charge is negative, turn your thumb by 180 degrees.
2. Without changing the direction of your thumb, rotate your hand to point your fingers in the direction of **B**.
3. The direction of $\mathbf{F}_B$ will then be perpendicular to your palm.

Think of your palm pushing with the force $\mathbf{F}_B$; the direction it pushes is the direction of $\mathbf{F}_B$.

Right-Hand Rule:

For determining the direction of the magnetic force, $\mathbf{F}_B$, on a *positive* charge

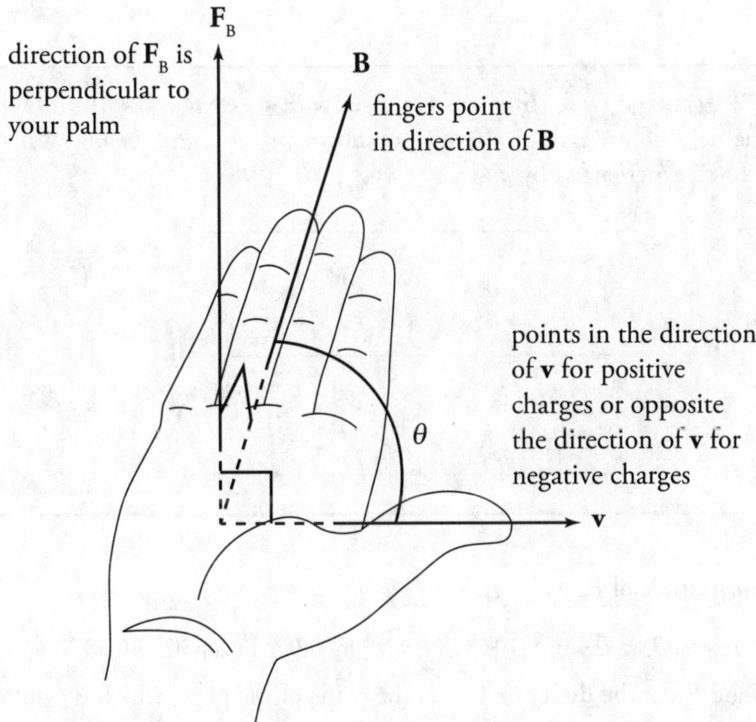

Note that there are fundamental differences between the electric force and magnetic force on a charge. First, a magnetic force acts on a charge only if the charge is moving; the electric force acts on a charge whether it moves or not. Second, the direction of the magnetic force is always perpendicular to the magnetic field, while the electric force is always parallel (or antiparallel) to the electric field.

Example 1 For each of the following charged particles moving through a magnetic field, determine the direction of the force acting on the charge.

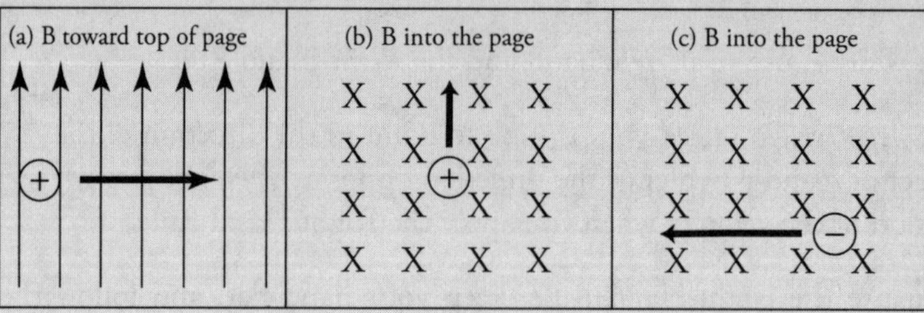

Solution.

(a) If you point your fingers to the top of the page and thumb to the right of the page, your palm should point out of the page. The force is out of the page.

(b) If you point your fingers into the page and thumb to the top of the page, your palm should point to the left of the page. The force is to the left of the page.

(c) If you point your fingers into the page and thumb to the right of the page (remember, when the charge is negative, your thumb points in the opposite direction from the velocity), your palm should point to the top of the page. The force points up toward the top of the page.

The SI unit for the magnetic field is the **tesla** (abbreviated **T**), which is one newton per amperemeter.

Example 2 A charge $+q = +6 \times 10^{-6}$ C moves with speed $v = 4 \times 10^5$ m/s through a magnetic field of strength $B = 0.4$ T, as shown in the figure below. What is the magnetic force experienced by q?

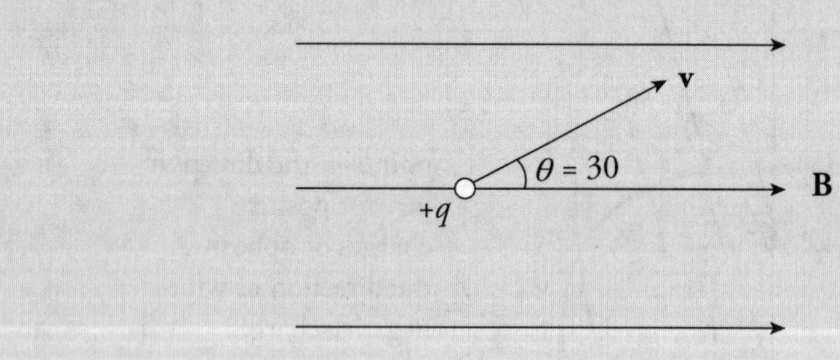

Solution. The magnitude of F_B is

$$F_B = qvB \sin \theta = (6 \times 10^{-6} \text{ C})(4 \times 10^5 \text{ m/s})(0.4 \text{ T}) \sin 30° = 0.48 \text{ N}$$

By the Right-Hand Rule, the direction is into the plane of the page, which is symbolized by ×.

Example 3 A particle of mass m and charge $+q$ is projected with velocity **v** (in the plane of the page) into a uniform magnetic field **B** that points into the page. How will the particle move?

Solution. Since **v** is perpendicular to **B**, the particle will feel a magnetic force of strength qvB, which will be directed perpendicular to **v** (and to **B**) as shown:

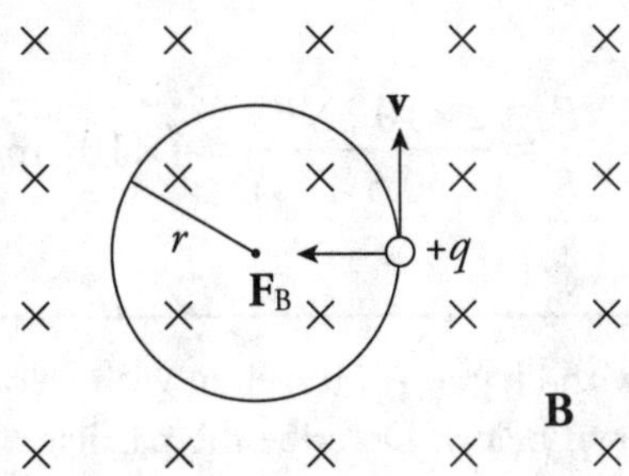

Since $\mathbf{F}_B$ is always perpendicular to **v**, the particle will undergo uniform circular motion; $\mathbf{F}_B$ will provide the centripetal force. Notice that, because $\mathbf{F}_B$ is always perpendicular to **v**, the magnitude of **v** will not change, just its direction. *Magnetic forces alone cannot change the speed of a charged particle; they can change only its direction of motion.* The radius of the particle's circular path is found from the equation $F_B = F_C$:

$$qvB = \frac{mv^2}{r} \Rightarrow r = \frac{mv}{qB}$$

Magnetic Fields

$\mathbf{F}_B$ is always perpendicular to both **v** and **B**. Magnetic forces cannot change the speed of an object, only its direction. The magnetic field does no work on any charges.

Example 4 A particle of charge $-q$ is shot into a region that contains an electric field, **E**, crossed with a perpendicular magnetic field, **B**. If $E = 2 \times 10^4$ N/C and $B = 0.5$ T, what must be the speed of the particle if it is to cross this region without being deflected?

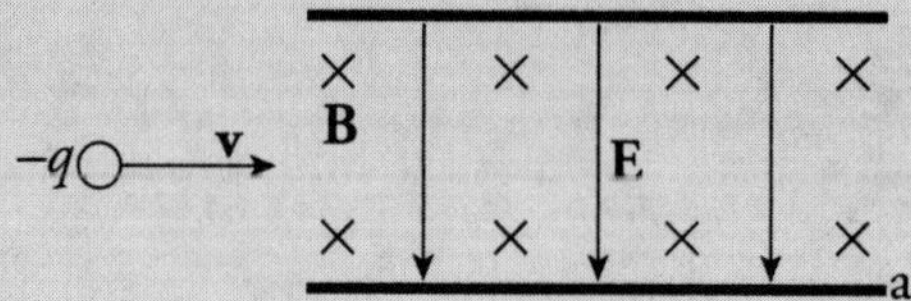

Solution. If the particle is to pass through undeflected, the electric force it feels has to be canceled by the magnetic force. In the diagram above, the electric force on the particle is directed upward (since the charge is negative and E is downward), and the magnetic force is directed downward by the Right-Hand Rule. So F_E and F_B point in opposite directions, and in order for their magnitudes to balance, qE must equal qvB, so v must equal E/B, which in this case gives

$$v = \frac{E}{B} = \frac{2 \times 10^4 \text{ N/C}}{0.5 \text{ T}} = 4 \times 10^4 \text{ m/s}$$

Example 5 A particle with charge $+q$, traveling with velocity **v**, enters a uniform magnetic field **B**, as shown below. Describe the particle's subsequent motion.

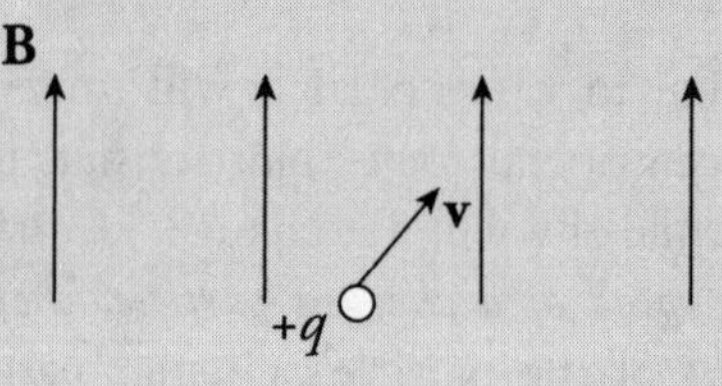

Solution. If the particle's velocity were parallel to **B**, then it would be unaffected by **B**. If **v** were perpendicular to **B**, then it would undergo uniform circular motion (as we saw in Example 2). In this case, **v** is neither purely parallel nor perpendicular to **B**. It has a component ($\mathbf{v}_1$) that's parallel to **B** and a component ($\mathbf{v}_2$) that's perpendicular to **B**.

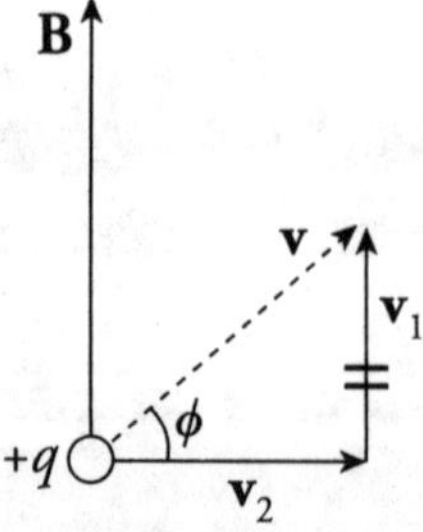

Component $\mathbf{v}_1$ will not be changed by $\mathbf{B}$, so the particle will continue upward in the direction of $\mathbf{B}$. However, the presence of $\mathbf{v}_2$ will create circular motion. The superposition of these two types of motion will cause the particle's trajectory to be a *helix*; it will spin in circular motion while traveling upward with the speed $v_1 = v \sin \phi$:

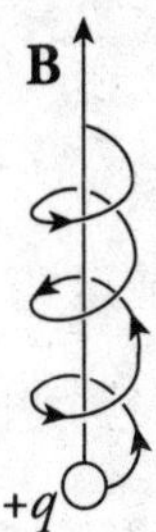

THE MAGNETIC FORCE ON A CURRENT-CARRYING WIRE

Since magnetic fields affect moving charges, they should also affect current-carrying wires. After all, a wire that contains a current contains charges that move. Remember, a current is the flow of positive charges.

Let a wire of length ℓ be immersed in magnetic field $\mathbf{B}$. If the wire carries a current I, then the force on the wire is

$$\mathbf{F}_B = I\boldsymbol{\ell} \times \mathbf{B}$$

Equation Sheet

with magnitude

$$F_B = BI\ell \sin \theta$$

Equation Sheet

where θ is the angle between ℓ and $\mathbf{B}$. Here, the direction of ℓ is the direction of the current, I. The direction of $\mathbf{F}_B$ can be found using the Right-Hand Rule and by letting your thumb point in the direction in which the current flows. Remember, conventional current is the flow of positive charges.

Another Way of Seeing It

Current I is charge over time (q/t) and the length of a wire I is a distance d. Hence, $I\ell = q(d/t)$. This gives us the same formula as before: $F_B = qvB\sin\theta = I\ell B\sin\theta$.

Example 6 A U-shaped wire of mass m is placed in a magnetic field **B** that points out of the plane of the page. How much current I must pass through the wire in order to cause the net force on the wire to be zero?

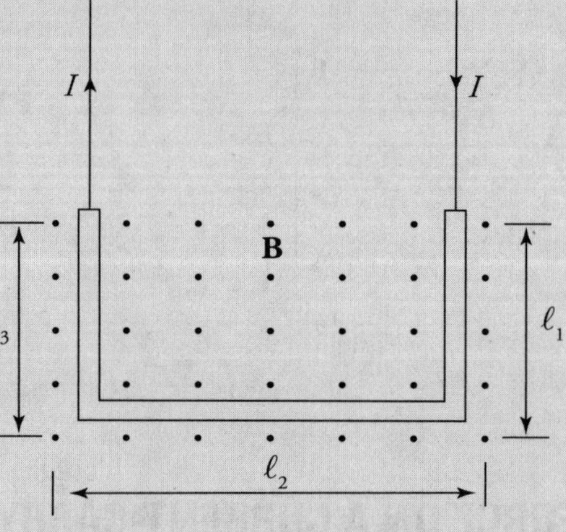

Solution. The total magnetic force on the wire is equal to the sum of the magnetic forces on each of the three sections of wire. The force on the first section (the right, vertical one), $\mathbf{F}_{B1}$, is directed to the left (applying the Right-Hand Rule), and the force on the third piece (the left, vertical one), $\mathbf{F}_{B3}$, is directed to the right. Since these pieces are the same length, these two oppositely directed forces have the same magnitude, $I\ell_1\text{B} = I\ell_3\text{B}$, and they cancel. So, the net magnetic force on the wire is the magnetic force on the middle piece. Since I points to the left and **B** is out of the page, the Right-Hand Rule tells us the force is upward.

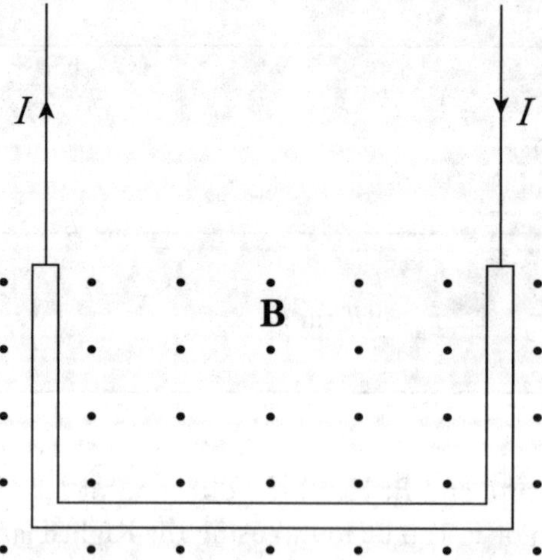

Since the magnetic force on the wire is $I\ell_2B$, directed upward, the amount of current must create an upward magnetic force that exactly balances the downward gravitational force on the wire. Because the total mass of the wire is m, the resultant force (magnetic + gravitational) will be zero if

$$I\ell_2\text{B} = mg \quad \Rightarrow \quad I = \frac{mg}{\ell_2\text{B}}$$

Example 7 A rectangular loop of wire that carries a current I is placed in a uniform magnetic field, **B**, as shown in the diagram below and is free to rotate. What torque does it experience?

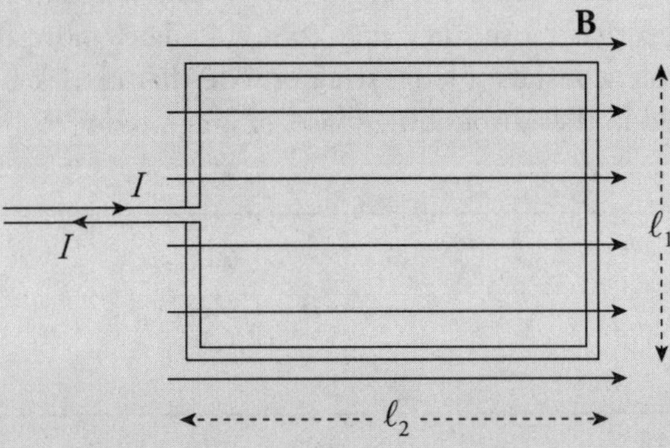

Solution. Ignoring the tiny gap in the vertical left-hand wire, we have two wires of length ℓ_1 and two of length ℓ_2. There is no magnetic force on either of the sides of the loop of length ℓ_2 because the current in the top side is parallel to **B** and the current in the bottom side is antiparallel to **B**. The magnetic force on the right-hand side points out of the plane of the page, while the magnetic force on the left-hand side points into the plane of the page.

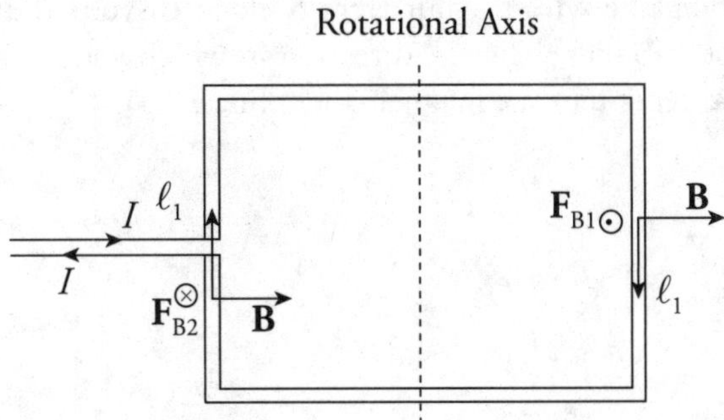

Each of these two forces exerts a torque that tends to turn the loop in such a way that the right-hand side rises out of the plane of the page while the left-hand side rotates into the page. Relative to the axis shown above (which cuts the loop in half), the torque of $\mathbf{F}_{B1}$ is

$$\tau_1 = rF_{B1} \sin\theta = (\tfrac{1}{2}\ell_2)(I\ell_1 B)\sin 90° = \tfrac{1}{2}I\ell_1\ell_2 B$$

and the torque of $\mathbf{F}_{B2}$ is

$$\tau_2 = rF_{B2} \sin\theta = (\tfrac{1}{2}\ell_2)(I\ell_1 B)\sin 90° = \tfrac{1}{2}I\ell_1\ell_2 B$$

Since both these torques rotate the loop in the same direction, the net torque on the loop is

$$\tau_1 + \tau_2 = I\ell_1\ell_2 B$$

MAGNETIC FIELDS CREATED BY CURRENT-CARRYING WIRES

In the previous section, we examined the force that a current-carrying wire experiences when it is subjected to an external magnetic field. The source of the magnetic field in the previous section was unspecified. As we said at the beginning of this chapter, the sources of magnetic fields are electric charges that move; they may spin, circulate, move through space, or flow down a wire. For example, consider a long, straight wire that carries a current *I*. The current generates a magnetic field in the surrounding space, of magnitude

Equation Sheet

$$B = \frac{\mu_0}{2\pi}\frac{I}{r}$$

Magnetic Fields vs. Electric Fields
Unlike electric fields that start at a point charge and end at another point charge, magnetic fields are unending loops. Electric charges can be positive or negative and exist by themselves. There is no such thing as a monopole for magnets.

where r is the distance from the wire. The symbol μ_0 denotes a fundamental constant called the permeability of free space. Its value is

$$\mu_0 = 4\pi \times 10^{-7}\ \text{N/A}^2 = 4\pi \times 10^{-7}\ \text{T}\cdot\text{m/A}$$

The magnetic field lines are actually circles whose centers are on the wire. The direction of these circles is determined by a variation of the Right-Hand Rule. Imagine grabbing the wire in your right hand with your thumb pointing in the direction of the current. Then, the direction in which your fingers curl around the wire gives the direction of the magnetic field lines.

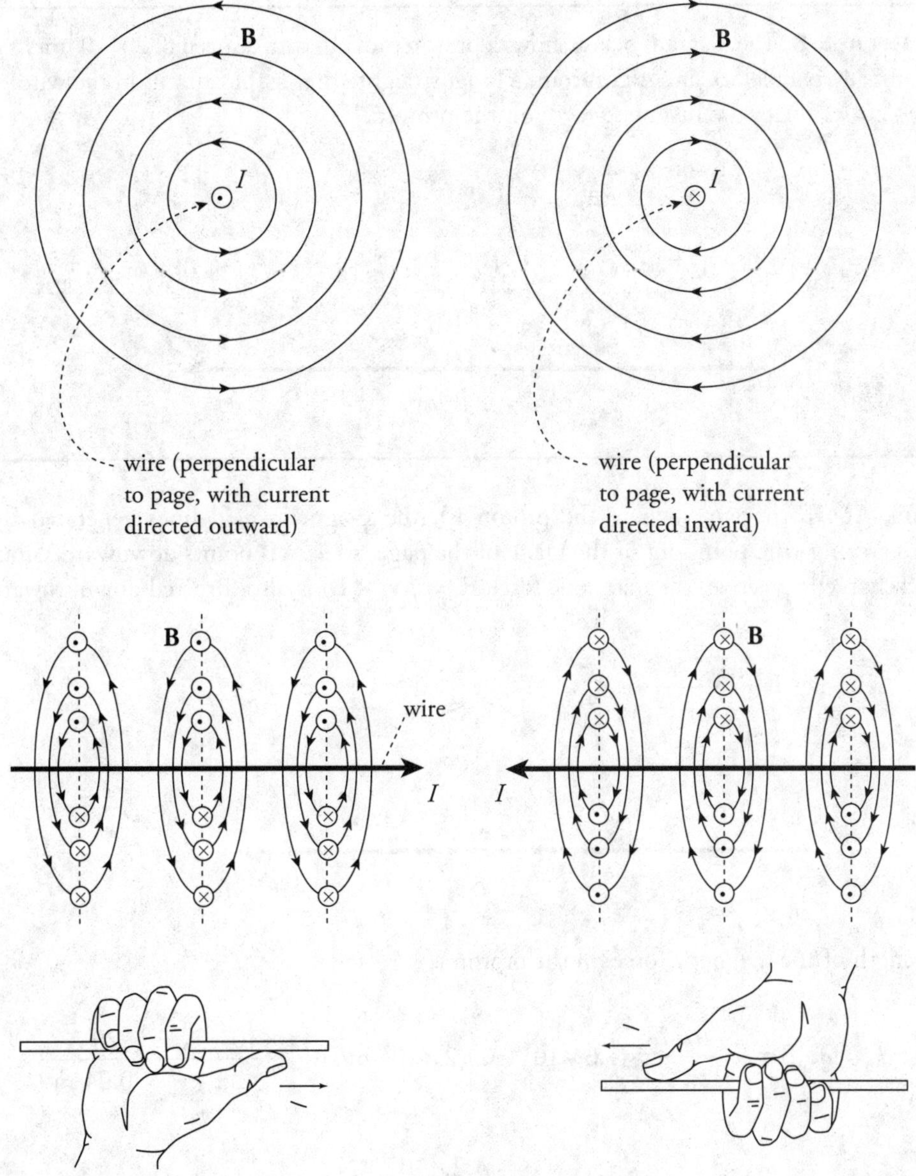

Right-Hand Rule for the Magnetic Field Created by a Current-Carrying Wire:

1. Put your thumb in the direction of the current or in the direction of a positive traveling charge.
2. Grab the wire/path.
3. As the fingers curl around your thumb, it represents the magnetic field going around the wire/path.

Example 8 The diagram below shows a proton moving with a speed of 2×10^5 m/s, initially parallel to, and 4 cm from, a long, straight wire. If the current in the wire is 20 A, what's the magnetic force on the proton?

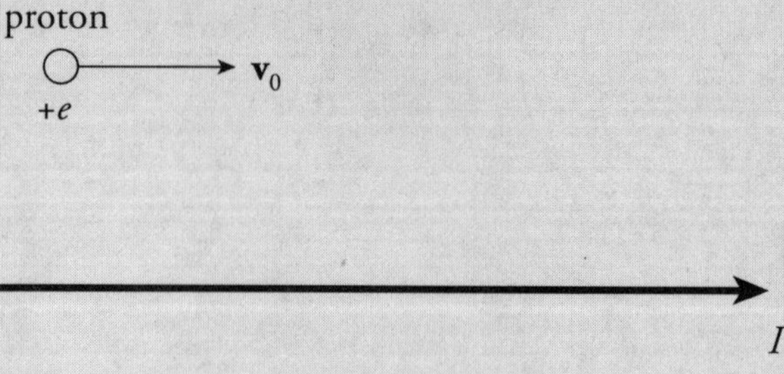

Solution. Above the wire (where the proton is), the magnetic field lines generated by the current-carrying wire point out of the plane of the page, so $\mathbf{v}_0 \times \mathbf{B}$ points downward. Since the proton's charge is positive, the magnetic force $\mathbf{F}_B = q(\mathbf{v}_0 \times \mathbf{B})$ is also directed down, toward the wire.

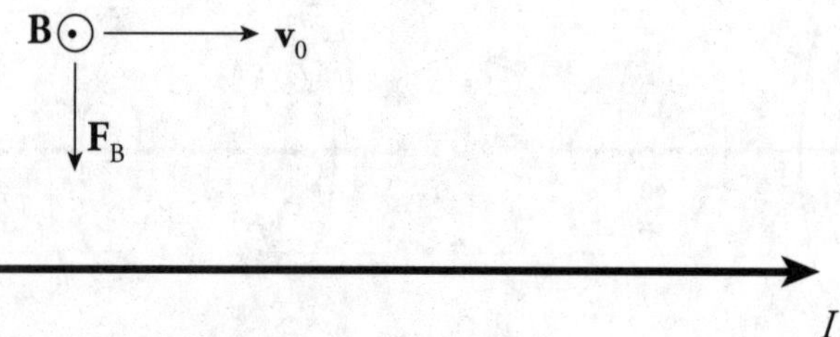

The strength of the magnetic force on the proton is

$$F_B = qv_0 B = ev_0 \frac{\mu_0}{2\pi}\frac{I}{r} = (1.6\times10^{-19}\text{ C})(2\times10^5\text{ m/s})\frac{4\pi\times10^{-7}\text{ N/A}^2}{2\pi}\frac{20\text{ A}}{0.04\text{ m}}$$

$$= 3.2\times10^{-18}\text{ N}$$

Example 9 The diagram below shows a pair of long, straight, parallel wires, separated by a small distance, r. If currents I_1 and I_2 are established in the wires, what is the magnetic force per unit length they exert on each other?

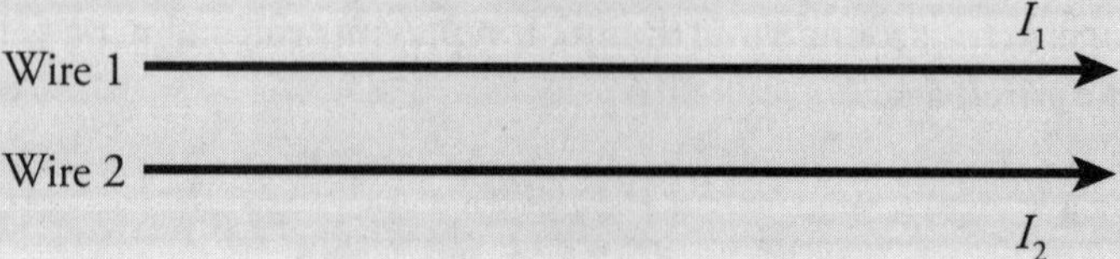

Solution. To find the force on Wire 2, consider the current in Wire 1 as the source of the magnetic field. Below Wire 1, the magnetic field lines generated by Wire 1 point into the plane of the page. Therefore, the force on Wire 2, as given by the equation $\mathbf{F}_{B2} = I_2(\ell_2 \times \mathbf{B}_1)$, points upward.

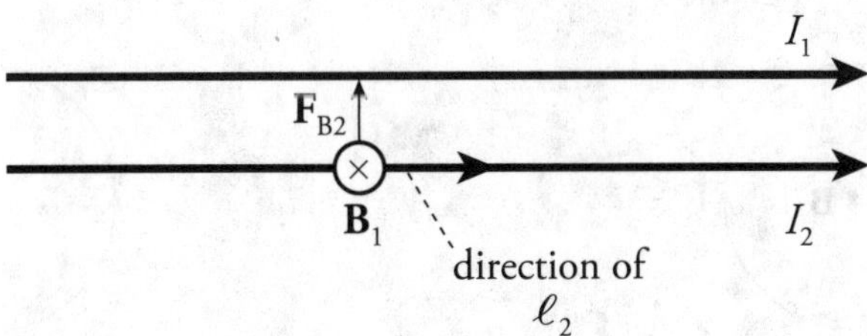

The magnitude of the magnetic force per unit length felt by Wire 2, due to the magnetic field generated by Wire 1, is found this way:

$$F_{B2} = I_2\ell_2 B_1 = I_2\ell_2 \frac{\mu_0}{2\pi}\frac{I_1}{r} \quad \Rightarrow \quad \frac{F_{B2}}{\ell_2} = \frac{\mu_0}{2\pi}\frac{I_1 I_2}{r}$$

By Newton's Third Law, this is the same force that Wire 1 feels due to the magnetic field generated by Wire 2. The force is attractive because the currents point in the same direction; if one of the currents were reversed, then the force between the wires would be repulsive.

SOLENOIDS CREATE UNIFORM FIELDS

The parallel-plate capacitor that was discussed in the chapter on electric fields appears frequently on the AP Physics 2 Exam because it generates a constant electric field in the region between the plates. Knowing that a current-carrying wire creates a magnetic field, if you look back at the section on bar magnets, you see that the magnetic field produced by a bar magnet closely resembles that produced by a single closed loop of current-carrying wire.

A device called a solenoid is constructed by a series of coaxial loops of wires through which a continuous current flows. The magnetic fields from each of the individual loops of wire add together and within the region inside the solenoid, the magnetic field becomes uniform. If the solenoid is very long in comparison to its diameter and the coils are tightly packed together, the field inside a solenoid created in a laboratory is very nearly uniform in both direction and strength.

Unlike a bar magnet, a solenoid is an electromagnet. When there is a current flowing through the wires of the solenoid, the solenoid creates a magnetic field. When there is no current flowing through the wire, there is not a magnetic field. The bar magnet, on the other hand, always produces a magnetic field.

MAGNETIC FIELDS CREATED BY CURRENT LOOPS

When the magnetic field is created by a current that is flowing through a loop of wire or a solenoid, the above Right-Hand Rule still applies. However, there's another variation that can be used:

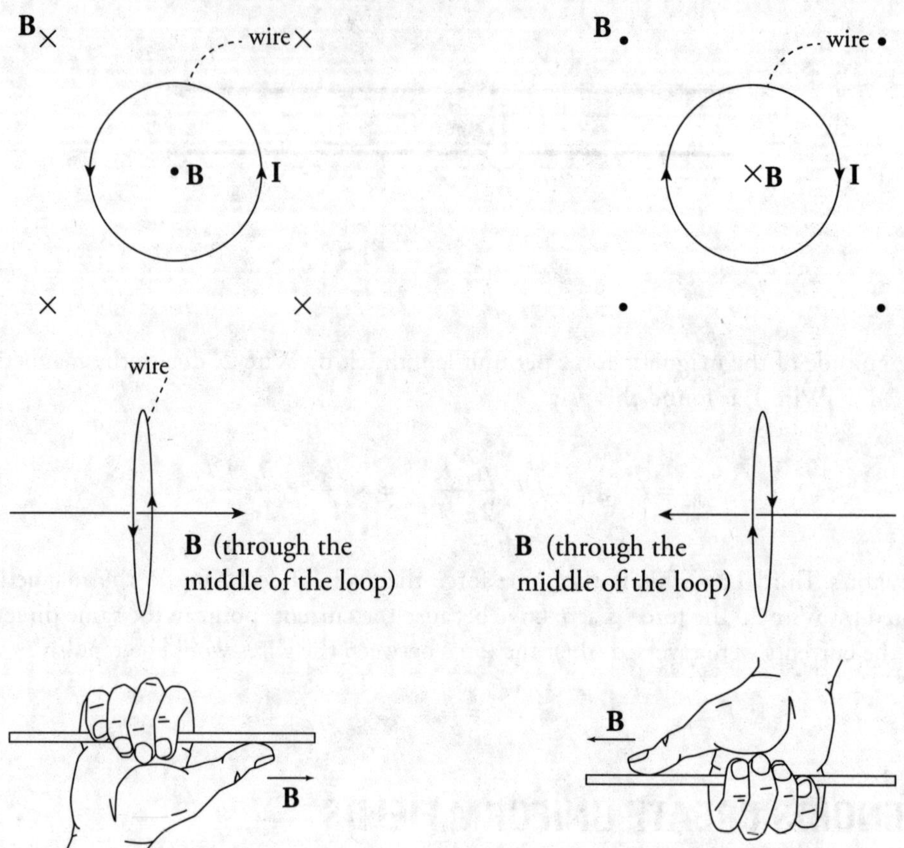

Right-Hand Rule for the Magnetic Field Created by a Current Loop or Solenoid:

1. Curl your fingers in the direction of the current loop.
2. Make a thumbs up.
3. Your thumb is pointing in the direction of the magnetic field in the center of the loop/solenoid. Outside the loop, the magnetic field goes in the opposite direction.

You'll notice that it is very similar to the other Right-Hand Rule, except now your fingers are the current and your thumb is the magnetic field.

MOTIONAL EMF

The simple act of moving a conducting rod in the presence of an external magnetic field *creates an electric field* within the rod. The figure below shows a conducting wire of length ℓ moving with constant velocity v in the plane of the page through a uniform magnetic field **B** that's perpendicular to the page. The presence of a moving conductor (recall there are many electrons present in the wire) inside a magnetic field results in the creation of an electric field that points in the same direction as the magnetic force on the electrons. (We know that the magnetic force is not directly responsible for the movement of the electrons in the rod because the magnetic force is *always* perpendicular to the direction of the motion of charges and therefore does no work. The magnetic field cannot be the source of the energy that causes the charges to move.) We can find the direction of the magnetic field using the Right-Hand Rule the instant the charges begin to move. The charges are negative and moving to the right, so $q\mathbf{v}$ points to the left (thumb of your right hand). **B** points into the page (fingers of your right hand). Your palm shows the direction of the magnetic force is downward. Therefore, the motion of the rod through the external magnetic field generates an electric field within the rod that causes electrons to move toward the bottom of the rod. This *induced electric field* points downward.

Vice-Versa
Earlier, we learned that moving charges generate magnetic fields. If we think about this backward, magnetic fields then can generate moving charges or current.

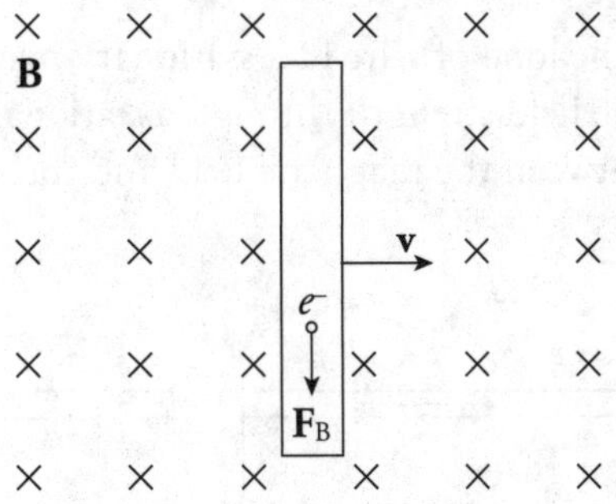

As long as the rod continues to move at velocity **v**, the electric field will be maintained within the rod. Electric fields are generated by charge separations, and therefore there will be a surfeit of negative charges on the bottom of the rod and a deficit of electrons at the top of the rod.

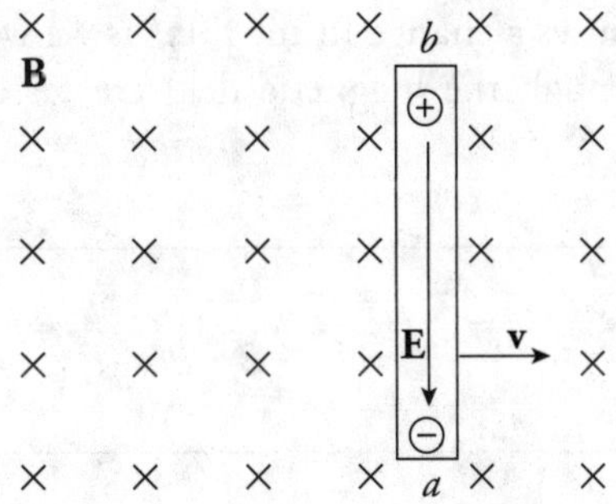

Right-Hand Rules
You will have to use a combination of your Right-Hand Rules in order to find the direction of current. Just pay attention to orientation.

A charge q in the wire feels two forces: an electric force, $\mathbf{F}_E = q\mathbf{E}$, and a magnetic force, $F_B = qvB\sin\theta = qvB$, because $\theta = 90°$.

If q is negative, $\mathbf{F}_E$ is upward and $\mathbf{F}_B$ is downward; if q is positive, $\mathbf{F}_E$ is downward and $\mathbf{F}_B$ is upward. So, in both cases, the forces act in opposite directions. Once the magnitude of $\mathbf{F}_E$ equals the magnitude of $\mathbf{F}_B$, the charges in the wire are in electromagnetic equilibrium. This occurs when $qE = qvB$; that is, when $E = vB$.

The presence of the electric field requires a potential difference between the ends of the rod. Since negative charge accumulates at the lower end (which we'll call point *a*) and positive charge accumulates at the upper end (point *b*), point *b* is at a higher electric potential. The potential difference V_{ba} is equal to $E\ell$ and, since $E = vB$, the potential difference can be written as $vB\ell$.

Now, imagine that the rod is sliding along a pair of conducting rails connected at the left by a stationary bar. The sliding rod now completes a rectangular circuit, and the potential difference V_{ba} causes current to flow.

Induced Current

An induced current can be created in three different ways:

1. changing the area of the loop of wire in a stationary magnetic field
2. changing the magnetic field strength through a stationary circuit
3. changing the angle between the magnetic field and the wire loop

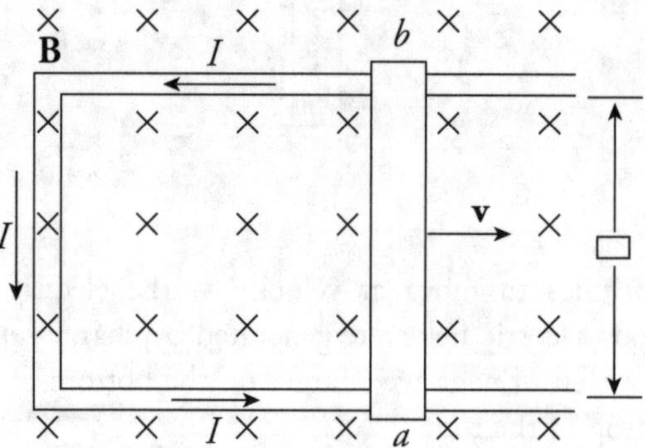

Each of these three changes causes a change in the flux, as we will see in the next section. The motion of the sliding rod through the magnetic field creates an electromotive force, called **motional emf**:

Equation Sheet

$$\mathcal{E} = B\ell v$$

The existence of a current in the sliding rod causes the magnetic field to exert a force on it. Using the formula $F_B = BI\ell$, the fact that ℓ points upward (in the direction of the current) and $\mathbf{B}$ is into the page tells us that the direction of $\mathbf{F}_B$ on the rod is to the left. An external agent must provide this same amount of force to the right to maintain the rod's constant velocity and keep the current flowing. The power that the external agent must supply is $P = Fv = I\ell Bv$, and the electrical power delivered to the circuit is $P = IV_{ba} = I\mathcal{E} = IvB\ell$. Notice that these two expressions are identical. The energy provided by the external agent is transformed first into electrical energy and then thermal energy as the conductors making up the circuit dissipate heat.

This relationship between current in a coil of wire and magnetic fields sets the basis for Faraday's Law.

FARADAY'S LAW OF ELECTROMAGNETIC INDUCTION

Electromotive force can be created by the motion of a conducting wire through a magnetic field, but there is another way to generate an emf from a magnetic field.

Faraday discovered that a current is induced when the magnetic flux passing through the coil or loop of wire changes. Magnetic flux helps us define what that amount of field passing into the loop means.

Imagine holding a loop of wire in front of a fan as shown:

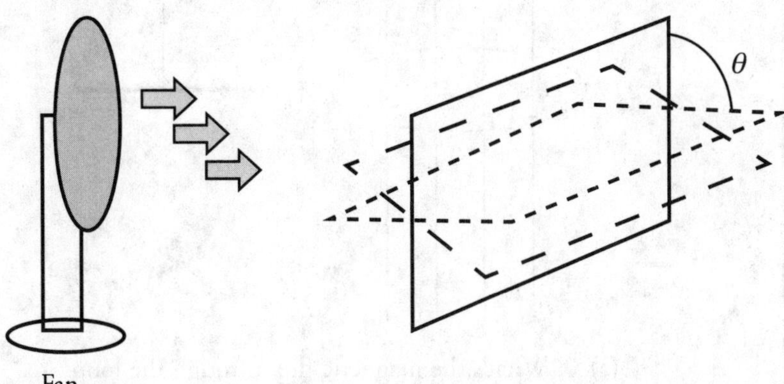

The amount of air that flows through the loop depends on the area of the loop as well as its orientation (tilt angle θ).

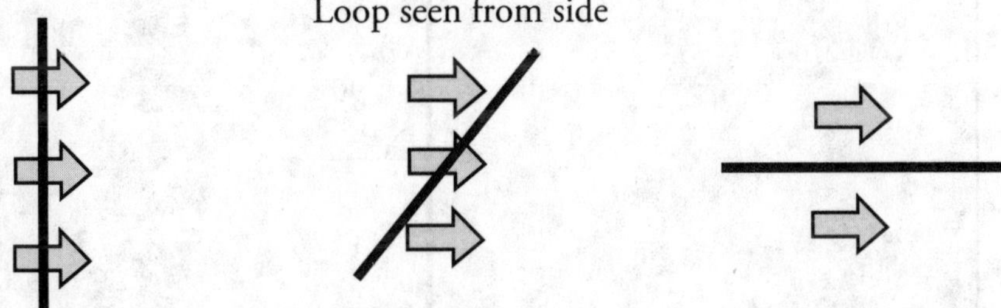

The most effective airflow occurs when the loop is completely perpendicular, as in the situation to the left. The least effective occurs when the airflow and loop are in the situation to the right.

We can apply this idea to a magnetic field passing through a loop. The magnetic flux, Φ_B, through an area A is equal to the product of **A** and the magnetic field parallel to the area vector. The area vector points normal to the surface.

Equation Sheet

Faraday's Law

Course Description version:	Equation Sheet version:
$\Phi_B = BA\cos\theta$	$\Phi_B = \vec{B}\bullet\vec{A} = \lvert\vec{B}\rvert\cos\theta\lvert\vec{A}\rvert$

Magnetic flux measures the density of magnetic field lines that cross through an area. (Note that the direction of A is taken to be perpendicular to the plane of the loop.)

Magnetic Flux Units
The SI unit for magnetic flux is called the weber (Wb), which is equivalent to one Tesla meter-squared (T·m²).

Example 10 The figure below shows two views of a circular loop of radius 3 cm placed within a uniform magnetic field, **B** (magnitude 0.2 T).

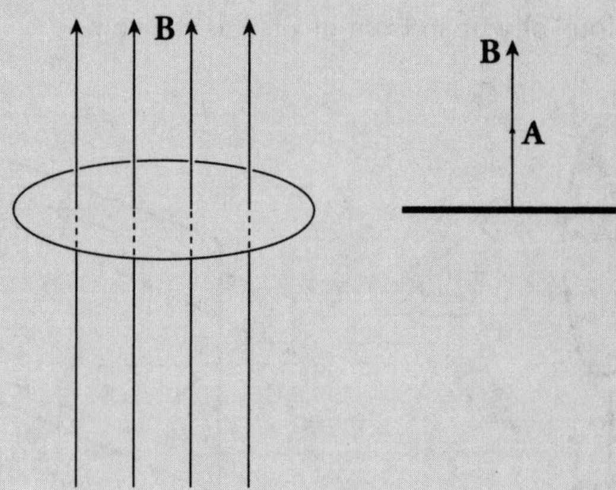

(a) What's the magnetic flux through the loop?
(b) What would be the magnetic flux through the loop if the loop were rotated 45°?
(c) What would be the magnetic flux through the loop if the loop were rotated 90°?

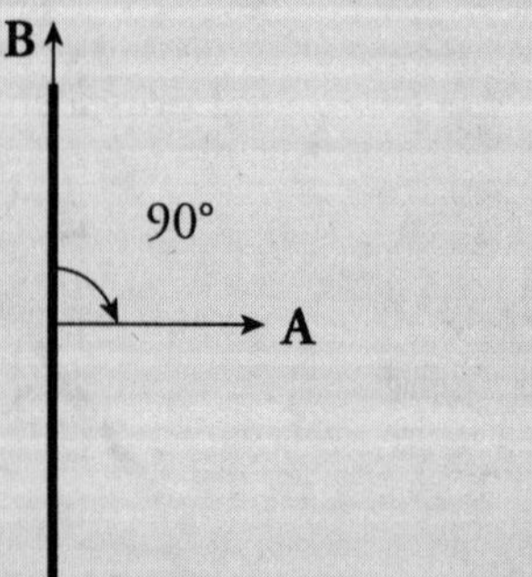

Solution.

(a) Since **B** is parallel to **A**, the magnetic flux is equal to BA:

$$\Phi_B = BA = B \cdot \pi r^2 = (0.2 \text{ T}) \cdot \pi(0.03 \text{ m})^2 = 5.7 \times 10^{-4} \text{ T·m}^2$$

The SI unit for magnetic flux, the tesla meter², is called a **weber** (abbreviated **Wb**). So $\Phi_B = 5.7 \times 10^{-4}$ Wb.

(b) Since the angle between **B** and **A** is 45°, the magnetic flux through the loop is

$$\Phi_B = BA \cos 45° = B \cdot \pi r^2 \cos 45° = (0.2 \text{ T}) \cdot \pi(0.03 \text{ m})^2 \cos 45° = 4.0 \times 10^{-4} \text{ Wb}$$

(c) If the angle between **B** and **A** is 90°, the magnetic flux through the loop is zero, since cos 90° = 0.

The concept of magnetic flux is crucial because changes in magnetic flux induce emf. According to **Faraday's Law of Electromagnetic Induction**, the magnitude of the emf induced in a circuit is equal to the rate of change of the magnetic flux through the circuit. This can be written mathematically in the form

$$\left|\mathcal{E}_{\text{avg}}\right| = \left|\frac{\Delta\Phi_B}{\Delta t}\right|$$

This induced emf can produce a current, which will then create its own magnetic field. The direction of the induced current is determined by the polarity of the induced emf and is given by **Lenz's Law**: the induced current will always flow in the direction that opposes the change in magnetic flux that produced it. If this were not so, then the magnetic flux created by the induced current would magnify the change that produced it, and energy would not be conserved. Lenz's Law can be included mathematically with Faraday's Law by the introduction of a minus sign; this leads to a single equation that expresses both results:

$$|\varepsilon| = \left|\frac{\Delta\Phi_B}{\Delta t}\right|$$

Equation Sheet

Example 11 The circular loop of Example 10 rotates at a constant angular speed through 45° in 0.5 s.

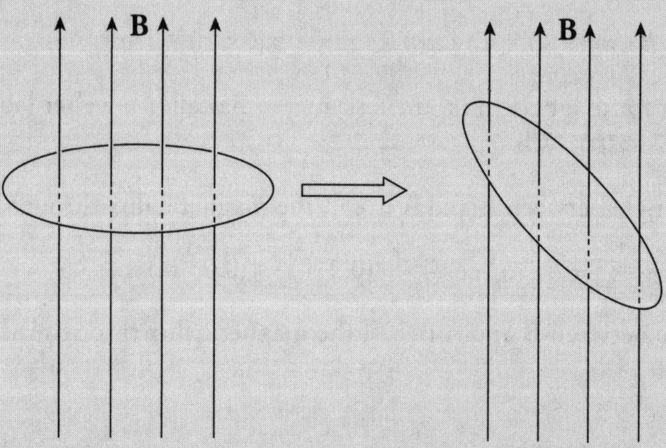

(a) What's the induced emf in the loop?

(b) In which direction will current be induced to flow?

Solution.

(a) As we found in Example 10, the magnetic flux through the loop changes when the loop rotates. Using the values we determined earlier, Faraday's Law gives

$$\mathcal{E}_{avg} = -\frac{\Delta\Phi_B}{\Delta t} = -\frac{(4.0\times10^{-4}\ \text{Wb})-(5.7\times10^{-4}\ \text{Wb})}{0.5\ \text{s}} = 3.4\times10^{-4}\ \text{V}$$

(b) The original magnetic flux was 5.7×10^{-4} Wb upward, and it was decreased to 4.0×10^{-4} Wb. So the change in magnetic flux is -1.7×10^{-4} Wb upward, or, equivalently, $\Delta\Phi_B = 1.7 \times 10^{-4}$ Wb downward. To oppose this change, we would need to create some magnetic flux upward. The current would be induced in the counterclockwise direction (looking down on the loop), because the Right-Hand Rule tells us that then the current would produce a magnetic field that would point up.

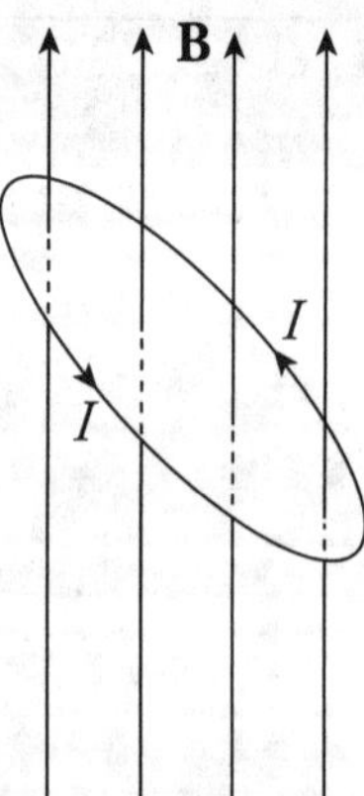

The current will flow only while the loop rotates, because emf is induced only when magnetic flux is changing. If the loop rotates 45° and then stops, the current will disappear.

Example 12 Again consider the conducting rod that's moving with constant velocity **v** along a pair of parallel conducting rails (separated by a distance ℓ), within a uniform magnetic field directed into the page, **B**:

Find the induced emf and the direction of the induced current in the rectangular circuit.

Solution. The area of the rectangular loop is ℓx, where x is the distance from the left-hand bar to the moving rod:

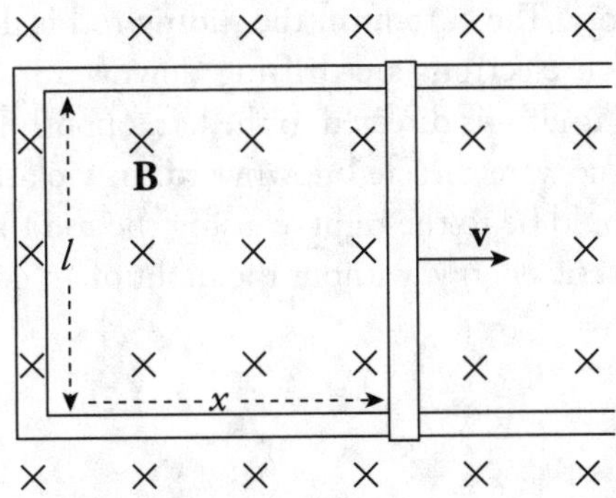

Because the area is changing, the magnetic flux through the loop is changing, which means that an emf will be induced in the loop. To calculate the induced emf, we first write $\Phi_B = BA = B\ell x$; then since $\Delta x/\Delta t = v$, we get

$$\mathcal{E}_{avg} = -\frac{\Delta \Phi_B}{\Delta t} = -\frac{\Delta(B\ell x)}{\Delta t} = -B\ell\frac{\Delta x}{\Delta t} = -B\ell v$$

We can figure out the direction of the induced current from Lenz's Law. As the rod slides to the right, the magnetic flux into the page increases. How do we oppose an increasing into-the-page flux? By producing out-of-the-page flux. In order for the induced current to generate a magnetic field that points out of the plane of the page, the current must be directed counterclockwise (according to the Right-Hand Rule).

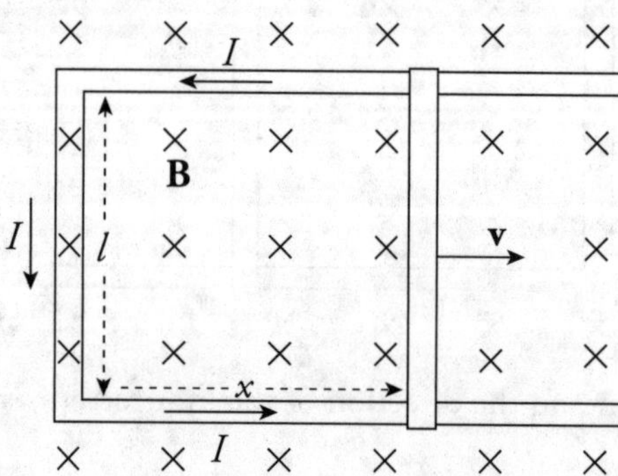

Note that the magnitude of the induced emf and the direction of the current agree with the results we derived earlier, in the section on motional emf.

This example also shows how a violation of Lenz's Law would lead directly to a violation of the Law of Conservation of Energy. The current in the sliding rod is directed upward, as given by Lenz's Law, so the conduction electrons are drifting downward. The force on these drifting electrons—and thus, the rod itself—is directed to the left, opposing the force that's pulling the rod to the right. If the current were directed downward, in violation of Lenz's Law, then the magnetic force on the rod would be to the right, causing the rod to accelerate to the right with ever-increasing speed and kinetic energy, without the input of an equal amount of energy from an external agent.

Example 13 A permanent magnet creates a magnetic field in the surrounding space. The end of the magnet at which the field lines emerge is designated the **north pole** (N), and the other end is the **south pole** (S):

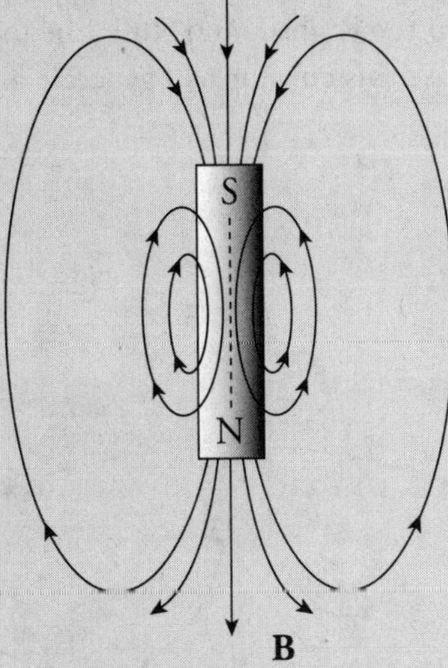

(a) The figure below shows a bar magnet moving down, through a circular loop of wire. What will be the direction of the induced current in the wire?

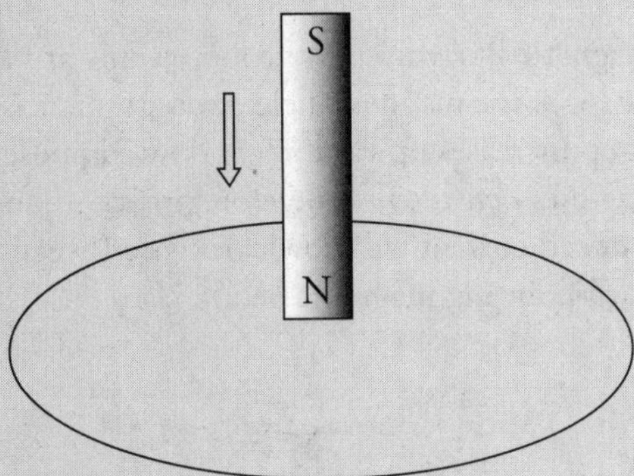

(b) What will be the direction of the induced current in the wire if the magnet is moved as shown in the following diagram?

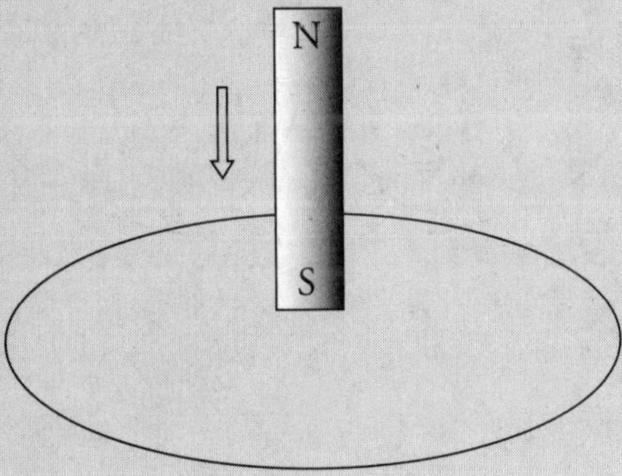

Solution.

(a) The magnetic flux down, through the loop, increases as the magnet is moved. By Lenz's Law, the induced emf will generate a current that opposes this change. How do we oppose a change of *more flux downward*? By creating flux *upward*. So, according to the Right-Hand Rule, the induced current must flow counterclockwise (because this current will generate an upward-pointing magnetic field):

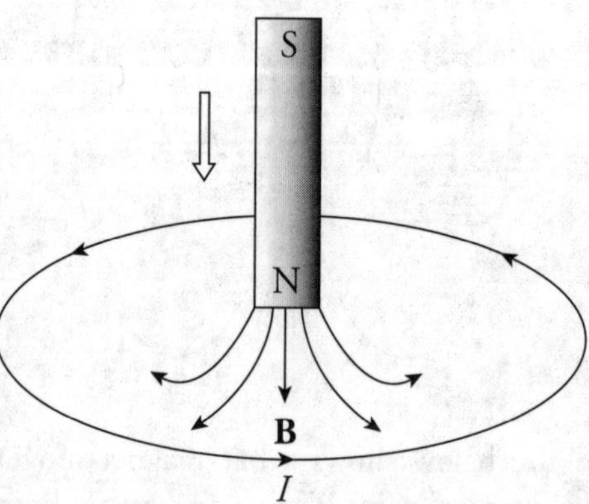

(b) In this case, the magnetic flux through the loop is upward and, as the south pole moves closer to the loop, the magnetic field strength increases so the magnetic flux through the loop increases upward. How do we oppose a change of *more flux upward*? By creating flux *downward*. Therefore, in accordance with the Right-Hand Rule, the induced current will flow clockwise (because this current will generate a downward-pointing magnetic field):

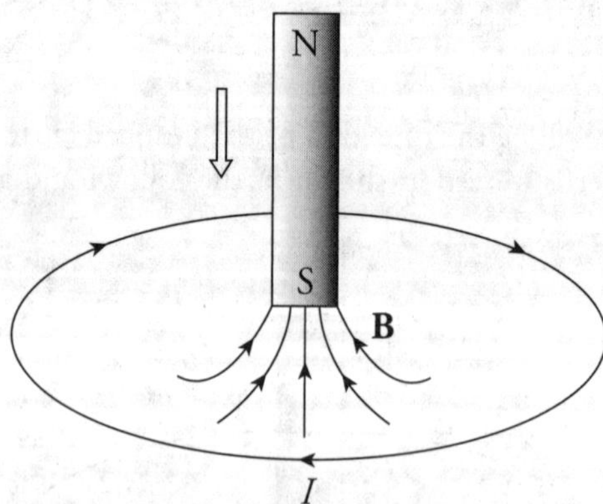

Example 14 A square loop of wire 2 cm on each side contains 5 tight turns and has a total resistance of 0.0002 Ω. It is placed 20 cm from a long, straight, current-carrying wire. If the current in the straight wire is increased at a steady rate from 20 A to 50 A in 2 s, determine the magnitude and direction of the current induced in the square loop. (Because the square loop is at such a great distance from the straight wire, assume that the magnetic field through the loop is uniform and equal to the magnetic field at its center.)

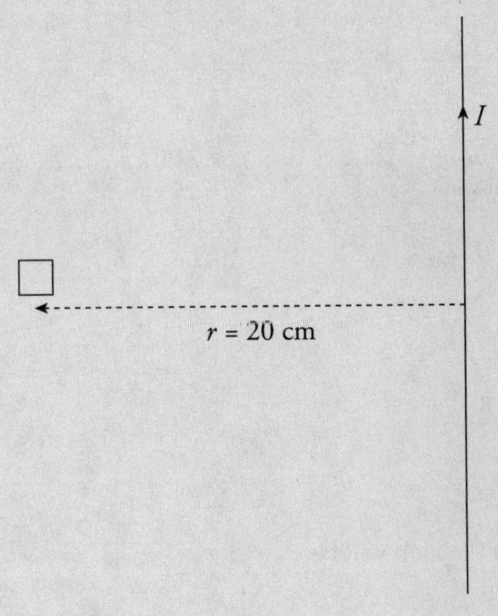

Solution. At the position of the square loop, the magnetic field due to the straight wire is directed out of the plane of the page, and its strength is given by the equation $B = (\mu_0/2\pi)(I/r)$. As the current in the straight wire increases, the magnetic flux through the turns of the square loop changes, inducing an emf and current. There are $N = 5$ turns; each loop contributes the same flux, so the total flux becomes the number of loops N times the flux of each individual loop, Φ_B. Faraday's Law becomes $\mathcal{E}_{avg} = -N(\Delta\Phi_B/\Delta t)$, and

$$\mathcal{E}_{avg} = -N\frac{\Delta\Phi_B}{\Delta t} = -N\frac{\Delta(BA)}{\Delta t} = -NA\frac{\Delta B}{\Delta t} = -NA\frac{\mu_0}{2\pi r}\frac{\Delta I}{\Delta t}$$

Substituting the given numerical values, we get

$$\begin{aligned}\mathcal{E}_{avg} &= -NA\frac{\mu_0}{2\pi r}\frac{\Delta I}{\Delta t}\\ &= -(5)(0.02\text{ m})^2\frac{4\pi\times10^{-7}\text{ T}\cdot\text{m/A}}{2\pi(0.20\text{ m})}\frac{(50\text{ A}-20\text{ A})}{2\text{ s}}\\ &= -3\times10^{-8}\text{ V}\end{aligned}$$

The magnetic flux through the loop is out of the page and increases as the current in the straight wire increases. To oppose an increasing out-of-the-page flux, the direction of the induced current should be clockwise, thereby generating an into-the-page magnetic field (and flux).

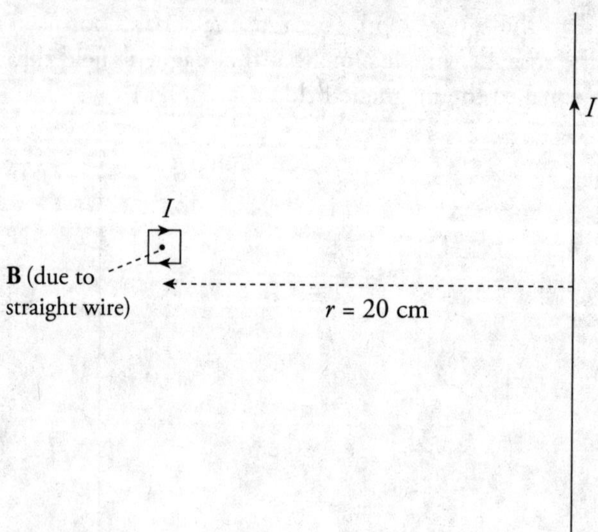

The value of the current in the loop will be

$$I = \frac{\mathcal{E}}{R} = \frac{3\times10^{-8}\ \text{V}}{0.0002\ \Omega} = 1.5\times10^{-4}\ \text{A}$$

Chapter 7 Review Questions

Answers and explanations can be found in Chapter 11.

Section I: Multiple Choice

1 Mark for Review

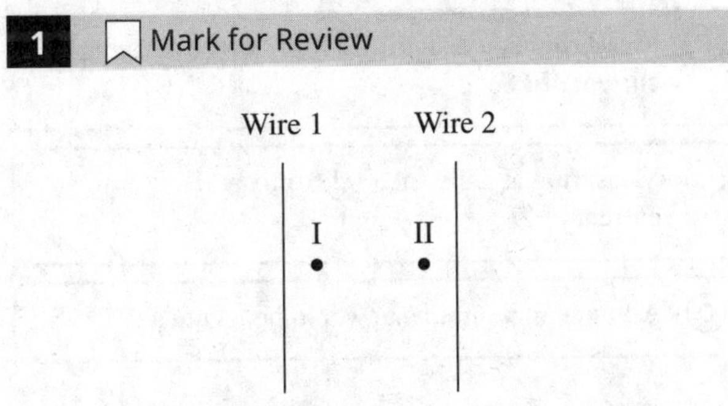

Two long parallel wires are lain as shown above. If the current through wire 1 is larger than the current through wire 2, which of the following scenarios and points could there be a magnetic field of strength 0 T?

(A) Point I when the currents flow in the same direction

(B) Point II when the currents flow in the same direction

(C) Point I when the currents flow in opposite directions

(D) Point II when the currents flow in opposite directions

2 Mark for Review

An experiment is performed with a long current-carrying wire in a region free from any other magnetic fields. The field strength is recorded at various distances from the wire. Which of the following statements about a graph of B versus I is true?

(A) The slope of the graph is directly proportional to the square of the distance the magnetic field strength was measured from the wire.

(B) The slope of the graph is directly proportional to the distance the magnetic field strength was measured from the wire.

(C) The slope of the graph is inversely proportional to the square of the distance the magnetic field strength was measured from the wire.

(D) The slope of the graph is inversely proportional to the distance the magnetic field strength was measured from the wire.

3 Mark for Review

A particle of charge $-q$ is launched to the left at speed **v** through a uniform magnetic field **B** which points up and to the left, as shown below.

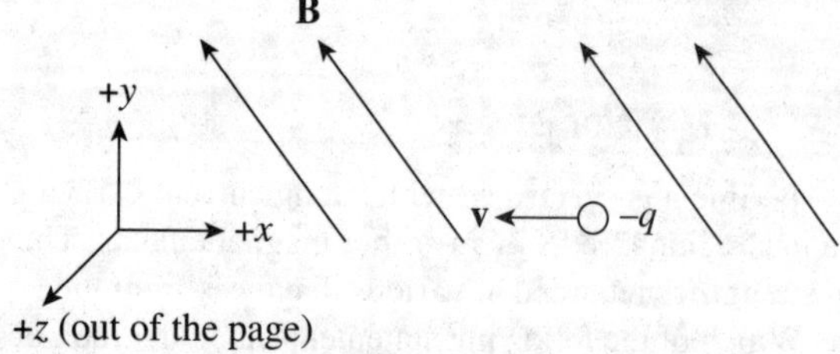

Another particle is then launch through a uniform magnetic field and experiences a magnetic force of the same magnitude and direction as the first particle. Which of the following could NOT be the conditions for the second particle?

(A) **B** is unchanged, but q is now positive and **v** is directed to the right.

(B) The charge is still negative, but **v** and **B** are rotated 90° into the plane of the paper so that **v** is directed along the $-z$-axis.

(C) The charge is still negative, but **v** and **B** are rotated 90° counterclockwise so that **v** is directed along the $-y$-axis.

(D) **v** is unchanged, but q is now positive and **B** is rotated 180° to point into the fourth quadrant.

4 Mark for Review

Which of the following situations would result in the largest measurement of magnetic field?

(A) Measuring at a distance r from a wire carrying a current of I

(B) Measuring at a distance $2r$ from a wire carrying a current of $I/2$

(C) Measuring at a distance $r/2$ from a wire carrying a current of $2I$

(D) All three measurements would be identical.

5 Mark for Review

In the figure below, what is the magnetic field at the Point P, which is midway between the two wires?

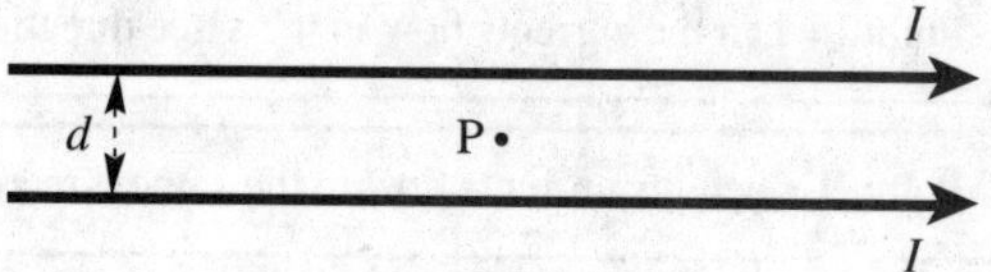

(A) $2\mu_0 I/(\pi d)$, into the plane of the page

(B) $\mu_0 I/(2\pi d)$, out of the plane of the page

(C) $\mu_0 I/(2\pi d)$, into the plane of the page

(D) Zero

6 Mark for Review

Here is a section of a wire with a current moving to the right. Where is the magnetic field strongest and pointing INTO the page?

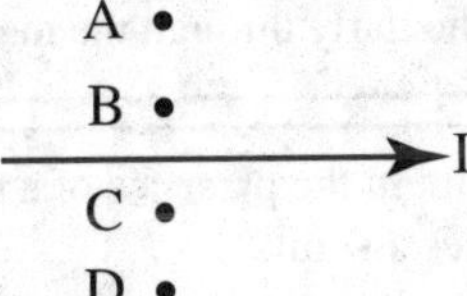

Ⓐ A

Ⓑ B

Ⓒ C

Ⓓ D

7 Mark for Review

What is the direction of force acting on the current-carrying wire as shown below?

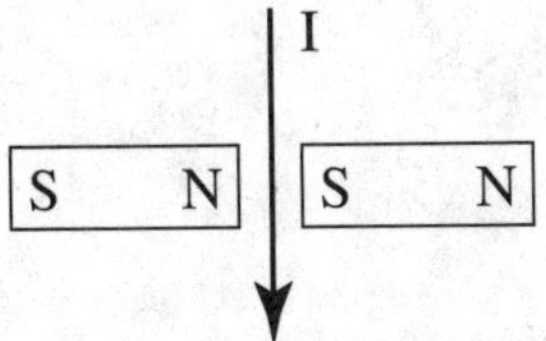

Ⓐ To the bottom of the page

Ⓑ Into the page

Ⓒ Out of the page

Ⓓ To the right of the page

8 Mark for Review

A metal rod of length L is pulled upward with constant velocity **v** through a uniform magnetic field **B** that points into the plane of the page.

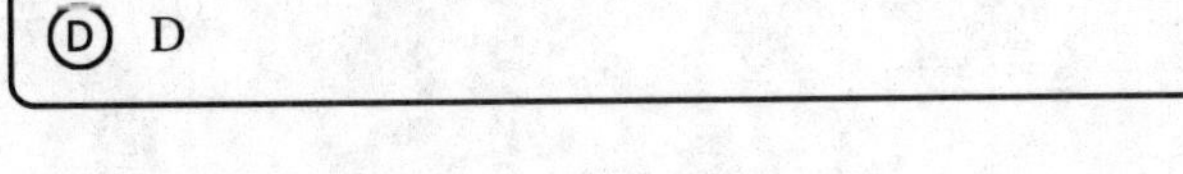

What is the potential difference between points a and b?

Ⓐ 0

Ⓑ $\frac{1}{2}vBL$, with point b at the higher potential

Ⓒ vBL, with point a at the higher potential

Ⓓ vBL, with point b at the higher potential

9 Mark for Review

A conducting rod of length 0.2 m and resistance 10 ohms between its endpoints slides without friction along a U-shaped conductor in a uniform magnetic field B of magnitude 0.5 T perpendicular to the plane of the conductor, as shown in the diagram below.

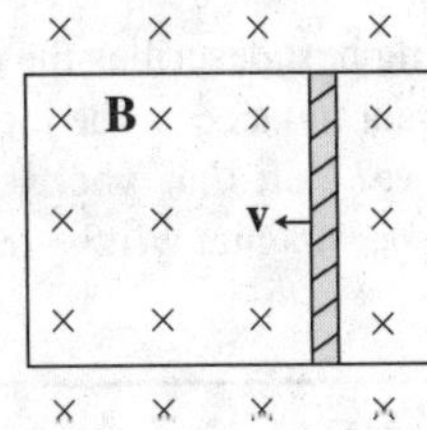

If the rod is moving with velocity **v** = 3 m/s to the left, what is the magnitude and direction of the current induced in the rod?

Ⓐ 0.03A down the rod

Ⓑ 0.03A up the rod

Ⓒ 0.3A down the rod

Ⓓ 0.3A up the rod

10 Mark for Review

As shown in the figures below, a bar magnet is moved at a constant speed through a loop of wire. Figure A shows the bar magnet when it is at a position below the loop of wire, and Figure B shows the loop of wire after the bar magnet has passed completely through the loop.

Figure A

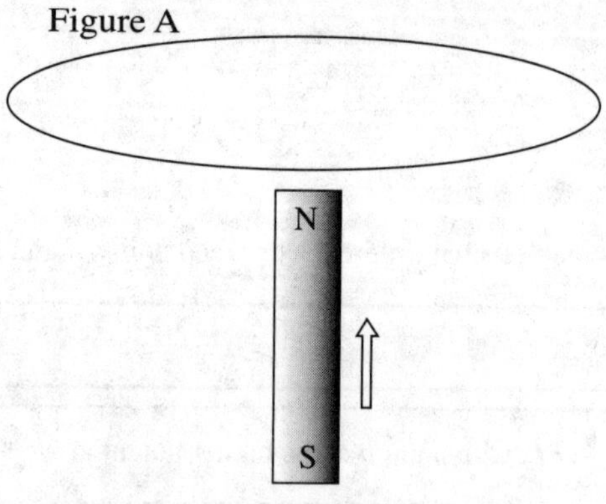

Figure B

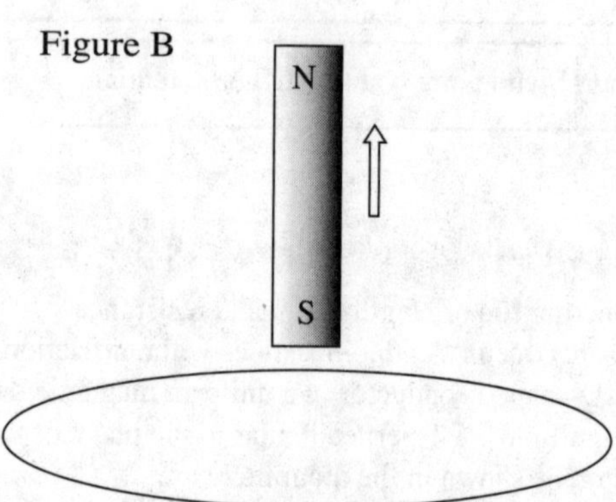

Which of the following best describes the direction or directions of the current induced in the loop when the loop is looked at from above? Note that when looking at the loop from above, the bar magnet will be moving toward the viewer.

(A) Always clockwise

(B) Always counterclockwise

(C) First clockwise, then counterclockwise

(D) First counterclockwise, then clockwise

11 Mark for Review

Which of the following statements about induced emf is true?

(A) An emf will be induced when moving a loop of wire perpendicularly through the magnetic field.

(B) Rotating a wire in the presence of a magnetic field always induces an emf.

(C) Shrinking the size of a loop of wire in constant magnetic field will induce an emf in the wire.

(D) A time varying magnetic field is required to induce an emf.

Section II: Free Response

1 Mark for Review

Figure 1 shows a simple mass spectrograph. It consists of a source of ions (charged atoms) that are accelerated (essentially from rest) by the voltage V and enter a region containing a uniform magnetic field **B**. The polarity of V may be reversed so that both positively charged ions (cations) and negatively charged ions (anions) can be accelerated. Once the ions enter the magnetic field, they follow a semicircular path and strike the front wall of the spectrograph, on which photographic plates are constructed to record the impact. Assume that the ions have mass m.

Figure 1

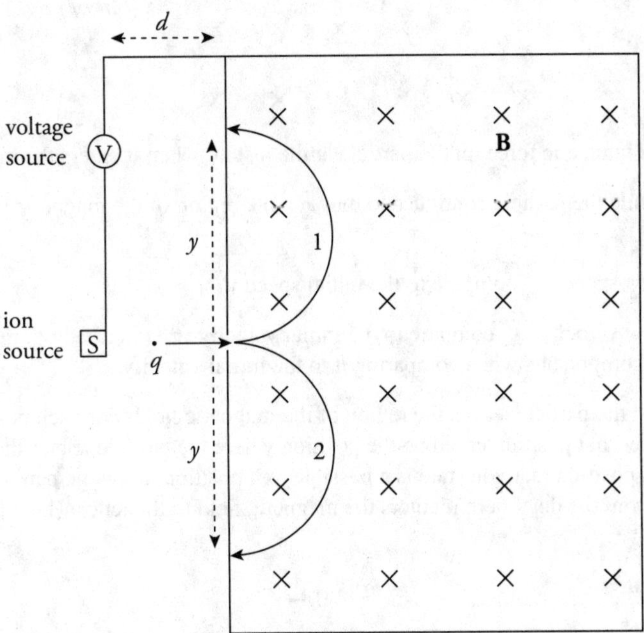

A. What is the acceleration of an ion of charge q just before it enters the magnetic field?

B. Find the speed with which an ion of charge q enters the magnetic field.

C.

i. Which semicircular path, 1 or 2, would a cation follow?

ii. Which semicircular path, 1 or 2, would an anion follow?

D. Determine the mass of a cation entering the apparatus in terms of y, q, **B**, and V.

E. Once a cation of charge q enters the magnetic field, how long does it take to strike the photographic plate?

F. What is the work done by the magnetic force in the spectrograph on a cation of charge q?

2 Mark for Review

A region of uniform magnetic field directed into the page (in the $-z$ direction) is generated. A positively charged particle traveling in the $+x$ direction at a speed v enters the region of the magnetic field, as shown in Figure 1.

Figure 1

y

x

z (out of page)

A. What direction is the magnetic force on the particle at the instant when it enters the magnetic field region?

B. Some time passes while the particle continues to move in the region of the magnetic field under the influence of the magnetic force.

i. How will the new speed v_{new} compare to the initial speed v?

ii. How will the new velocity $\overrightarrow{v_{new}}$ compare to the initial velocity $\vec{v}$? Describe both the magnitude of the velocity and its components when comparing it to the initial velocity.

C. More time passes and the particle leaves the region of the magnetic field. For each position A, B, C, and D in Figure 2, state whether that position could be the position where the particle leaves the region of the magnetic field or not. For each position that you state is a possible exit position, draw the path on the diagram below that the particle would take from the dot where it enters the magnetic field to the letter where it exits.

Figure 2

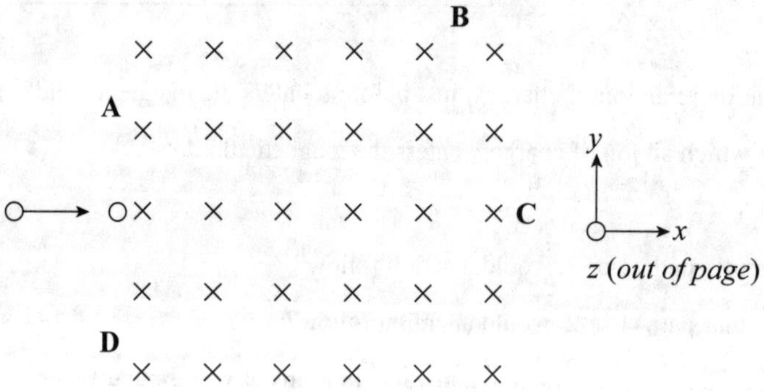

B

A

C

D

y

x

z (out of page)

3 Mark for Review

A rectangular wire is pulled through a uniform magnetic field of 2 T going into the page, as shown. The resistor has a resistance of 20 Ω.

Figure 1

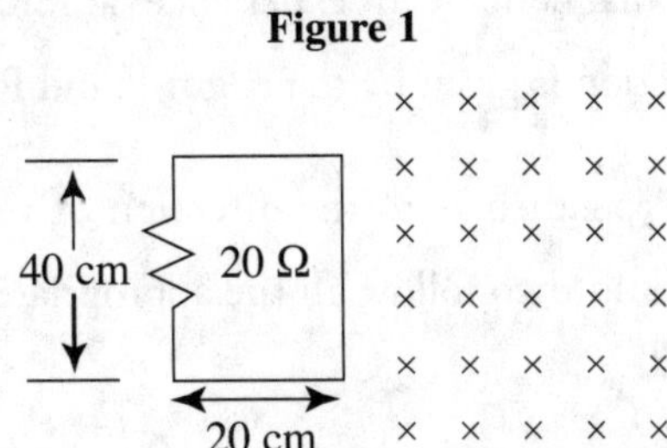

A. What is the voltage across the resistor as the wire is pulled horizontally at a velocity of 1 m/s and it just enters the field?

B. What is the current through the circuit in the above case and in what direction does it flow?

C. The region containing the magnetic field is 1 meter long in the direction the loop is being pulled. If the loop is pulled at the constant velocity of 1 m/s, describe the current in the loop from a time before the loop encounters the left edge of the region of the magnetic field until the time that the left edge of the loop completely passes out of the magnetic field region. Indicate the values and the direction of the current.

D. The loop of wire is rotated 90° clockwise so that the 20 cm side is vertical and the 40 cm side is horizontal at the top. Explain, without relying solely on equations, whether the loop would have to be pulled faster than 1 m/s, at exactly 1 m/s, or slower than 1 m/s for the current to have the same magnitude as in part B.

Chapter 7 Summary

- Charges moving through a magnetic field experience a force whose magnitude is given by $F_B = qvB\sin\theta$ and whose direction is given by the Right-Hand Rule.
- Because the force is always perpendicular to the direction of velocity, the charge may experience uniform circular motion. It would then follow all the appropriate circular motion relationships and orbit in a radius given by $r = \frac{mv}{qB}$.
- Because wires have charges moving through them, if a wire is carrying a current, it will experience a force if placed in a magnetic field. This is expressed by $F_B = BI\ell\sin\theta$.
- A current-carrying wire will produce a magnetic field whose strength is given by $B = \frac{\mu_o}{2\pi}\frac{I}{r}$, where $\mu_0 = 4\pi \times 10^{-7}$ T·m/A, I is the current through the wire, and r is the radial distance from the wire.
- An electromotive force is produced as a conducting wire's position changes or as the magnetic field changes. This idea is summarized by $\varepsilon = B\ell v$.
- Lenz's Law states that the electromagnetic force produced by Faraday's Law will act to produce a current that will create a magnetic field with direction that opposes the change in the magnetic flux.
- Faraday's Law of Induction says that if a wire is formed in a loop, an electromotive force is produced if the magnetic flux through the loop changes with time. This idea can be summarized by $|\varepsilon| = \left|\frac{\Delta\Phi_B}{\Delta t}\right|$.

Chapter 8
Geometric Optics

REFLECTION AND REFRACTION

Imagine a beam of light directed toward a smooth, transparent surface. When it hits this surface, some of its energy will be reflected off the surface and some will be transmitted into the new medium. Some of the transmitted light will be absorbed, which is the source of heating from radiation discussed in Thermodynamics, while some of the light will emerge from the new medium as a transmitted beam. We can figure out the directions of the reflected and transmitted beams by calculating the angles that the beams make with the normal to the interface. In the following figure, an incident beam strikes the boundary of another medium; it could be a beam of light in air striking a piece of glass. Notice all angles are measured from the normal.

Tricky Problems!
Remember that the normal is always perpendicular to the surface. When a question asks for the angle of incidence or reflection, it wants the angle formed with the normal, not the angle formed with the surface. Make sure you know if you're solving for the angle of incidence, the angle of reflection, or the angle of refraction.

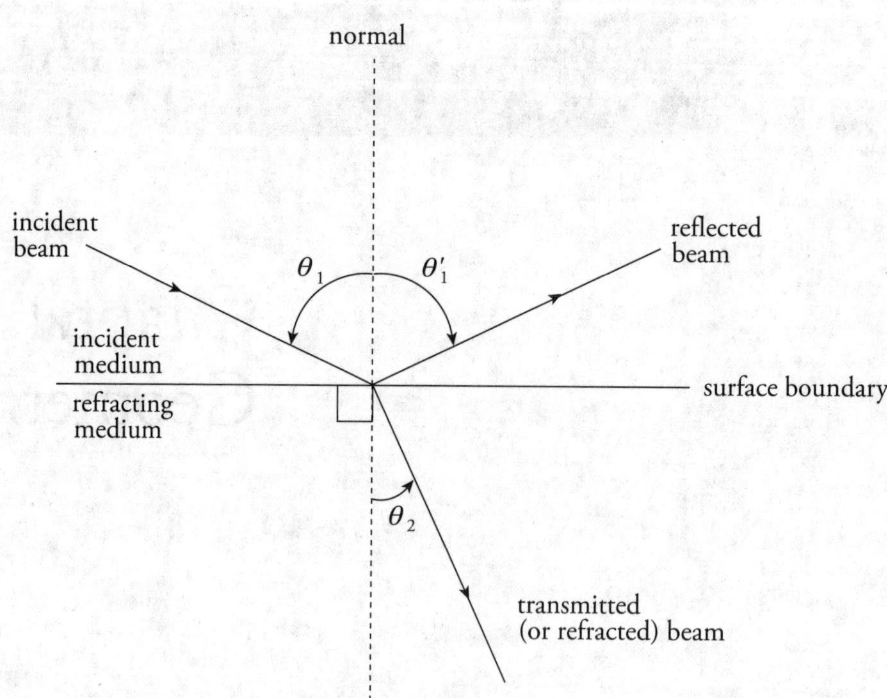

The angle that the **incident beam** makes with the normal is called the **angle of incidence**, or θ_1. The angle that the **reflected beam** makes with the normal is called the **angle of reflection**, θ'_1, and the angle that the **transmitted beam** makes with the normal is called the **angle of refraction**, θ_2. The incident, reflected, and transmitted beams of light all lie in the same plane.

The relationship between θ_1 and θ'_1 is pretty easy; it is called the **Law of Reflection**:

$$\theta_1 = \theta'_1$$

The Law of Reflection basically states the Angle of Reflection is equal to the Angle of Incidence. In order to describe how θ_1 and θ_2 are related, we first need to talk about a medium's index of refraction.

When light travels through empty space (vacuum), its speed is $c = 3.00 \times 10^8$ m/s; this is one of the fundamental constants of nature. But when light travels through a medium (such as water or glass), it's constantly being absorbed and re-emitted by the atoms that compose the material and, as a result, its apparent speed, v, is some fraction of c. The reciprocal of this fraction,

$$n = \frac{c}{v}$$

Equation Sheet

is called the medium's **index of refraction,** n. For example, since the speed of light in water is $v = 2.25 \times 10^8$ m/s, the index of refraction of water is

$$n = \frac{3.00 \times 10^8 \text{ m/s}}{2.25 \times 10^8 \text{ m/s}} = 1.33$$

Note that n has no units; it's also never less than 1, since light always travels slower in a medium than in a vacuum.

***n* of air**
The index of refraction of air is 1.00029, which is so close to 1 that the norm is to consider it to just be 1.

The equation that relates θ_1 and θ_2 involves the index of refraction of the incident medium (n_1) and the index of refraction of the refracting medium (n_2); it's called **Snell's Law**:

$$n_1 \sin \theta_1 = n_2 \sin \theta_2$$

Equation Sheet

If $n_2 > n_1$, then Snell's Law tells us that $\theta_2 < \theta_1$; that is, the beam will bend (**refract**) *toward* the normal as it enters the medium. On the other hand, if $n_2 < n_1$, then $\theta_2 > \theta_1$, and the beam will bend *away* from the normal.

Example 1 A beam of light in air is incident upon a piece of glass, striking the surface at an angle of 30°. If the index of refraction of the glass is 1.5, what are the angles of reflection and refraction?

Solution. If the light beam makes an angle of 30° with the surface, then it makes an angle of 60° with the normal; this is the angle of incidence. By the Law of Reflection, then, the angle of reflection is also 60°. We use Snell's Law to find the angle of refraction. The index of refraction of air is close to 1, so we can say that $n = 1$ for air.

$$\begin{aligned} n_1 \sin \theta_1 &= n_2 \sin \theta_2 \\ (1) \sin 60° &= 1.5 \sin \theta_2 \\ \sin \theta_2 &= 0.5774 \\ \theta_2 &= 35° \end{aligned}$$

Note that $\theta_2 < \theta_1$, as we would expect since the refracting medium (glass) has a greater index than does the incident medium (air).

Example 2 A fisherman drops a flashlight into a lake that's 10 m deep. The flashlight sinks to the bottom where its emerging light beam is upward toward the surface of the lake, making an angle (θ_1) with the normal. When leaning over the edge of the boat and placing his head just above the water, the beam seems to be coming from a depth of 9 m and a horizontal distance of 6.6 m. How far, horizontally, is the flashlight actually from the fisherman?

Solution. Take a look at the figure below.

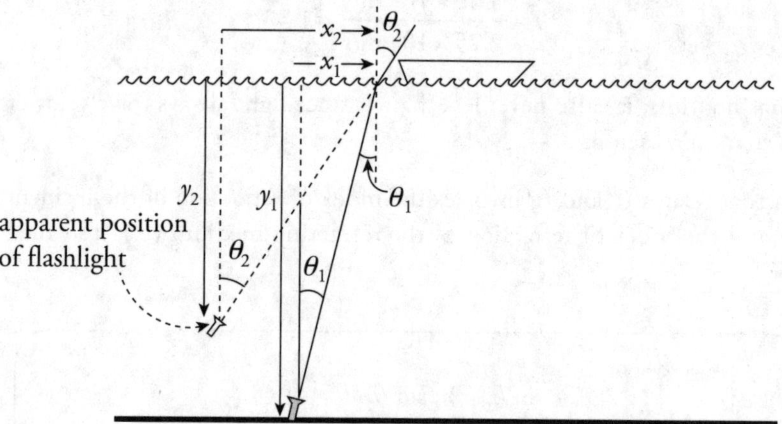

Since the refracting medium (air) has a lower index than the incident medium (water), the beam of light will bend away from the normal as it emerges from the water. As a result, the fisherman will think it's at a horizontal distance of x_2 = 6.6 m, rather than its actual horizontal distance of x_1. By simple trigonometry, we know

$$\tan\theta_1 = \frac{x_1}{y_1} \quad \text{and} \quad \tan\theta_2 = \frac{x_2}{y_2} \quad \text{or} \quad \theta_2 = \tan^{-1}\left(\frac{x_2}{y_2}\right) = \tan^{-1}\left(\frac{6.6 \text{ m}}{9 \text{ m}}\right) \approx 36.25°$$

Snell's Law tells us that $n_1 \sin\theta_1 = n_2 \sin\theta_2$, using $n_1 = n_{\text{water}} = 1.33$ and $n_2 = n_{\text{air}} = 1$

$$\theta_1 = \sin^{-1}\left(\frac{n_2}{n_1}\sin\theta_2\right) = \sin^{-1}\left(\frac{1}{1.33}\sin(36.25°)\right) \approx 26.4°$$

which means

$$x_1 = y_1 \tan(\theta_1) = (10 \text{ m})\tan(26.4°) \approx 5.0 \text{ m}$$

TOTAL INTERNAL REFLECTION

When a beam of light strikes the boundary to a medium that has a lower index of refraction, the beam bends away from the normal. As the angle of incidence increases, the angle of refraction becomes larger. At some point, when the angle of incidence reaches a **critical angle**, θ_c, the angle of refraction becomes 90°, which means the refracted beam is directed along the surface.

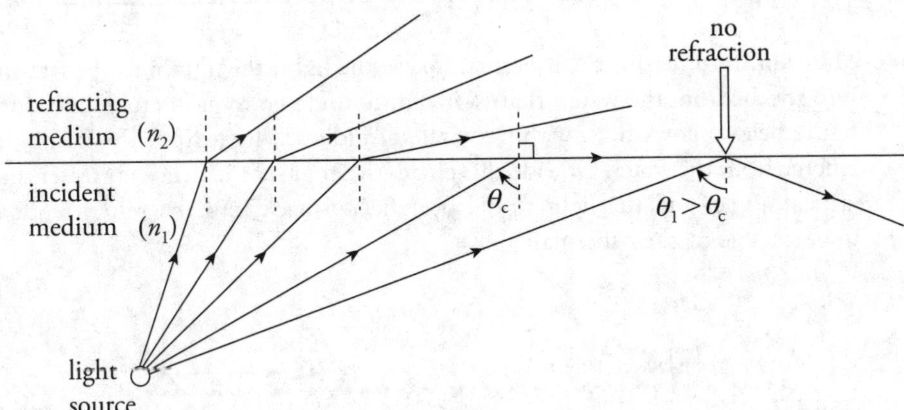

For angles of incidence that are greater than θ_c, there is *no* angle of refraction; the entire beam is reflected back into the original medium. This phenomenon is called **total internal reflection** (sometimes abbreviated **TIR**).

Total internal reflection occurs when:

1) $n_1 > n_2$

and

2) $\theta_1 > \theta_c$, where $\theta_c = \sin^{-1}(n_2/n_1)$

> **Subscripts**
> Always think of n_1 or θ_1 as the incidence ray, meaning where the light comes from. Treat n_2 or θ_2 as the resulting ray.

Notice that total internal reflection cannot occur if $n_1 < n_2$. This is because the largest output of $\sin\theta$ is 1, so the largest input of $\sin^{-1}(x)$ is 1. If $n_1 > n_2$, then total internal reflection is a possibility; it will occur if the angle of incidence is large enough; that is, if it's greater than the critical angle, θ_c.

Example 3 What is the critical angle for total internal reflection between air and water? In which of these media must light be incident for total internal reflection to occur?

Solution. First, total internal reflection can occur only when the light is incident in the medium that has the greater refractive index and strikes the boundary to a medium that has a lower index. So, in this case, total internal reflection can occur only when the light source is in the water and the light is incident upon the water/air surface. The critical angle is found as follows:

$$\sin\theta_c = \frac{n_2}{n_1} \Rightarrow \sin\theta_c = \frac{n_{\text{air}}}{n_{\text{water}}} = \frac{1}{1.33} \Rightarrow \sin\theta_c = 0.75 \Rightarrow \theta_c = 49°$$

Total internal reflection will occur if the light from the water strikes the water/air boundary at an angle of incidence greater than 49°.

> **Example 4** How close must the fisherman be to the flashlight in Example 2 in order to see the light it emits?

Remember to Check
For optics problems, you need to always remember to check if the refracted light ray undergoes total internal reflection. If it undergoes total internal reflection, there will be no refracted ray. Hence, if a trick problem asks for an angle of refraction for a situation that undergoes total internal reflection, there is no angle of refraction.

Solution. In order for the fisherman to see the light, the light must be transmitted into the air from the water; that is, it cannot undergo total internal reflection. The figure below shows that, within a circle of radius *x*, light from the flashlight will emerge from the water. Outside this circle, the angle of incidence is greater than the critical angle, and the light would be reflected back into the water, rendering it undetectable by the fisherman above.

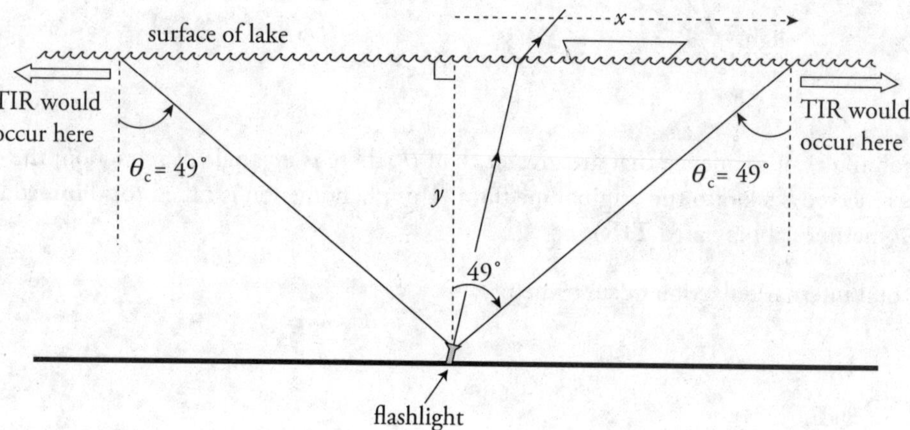

Because the critical angle for total internal reflection at a water/air interface is 49° (as we found in the preceding example), we can solve for *x*:

$$\tan 49° = \frac{x}{y} \quad \Rightarrow \quad x = y \tan 49° = (10 \text{ m}) \tan 49° = 11.5 \text{ m}$$

Example 5 The refractive index for the glass prism shown below is 1.55. In order for a beam of light to experience total internal reflection at the right-hand face of the prism, the angle θ_1 must be smaller than what value?

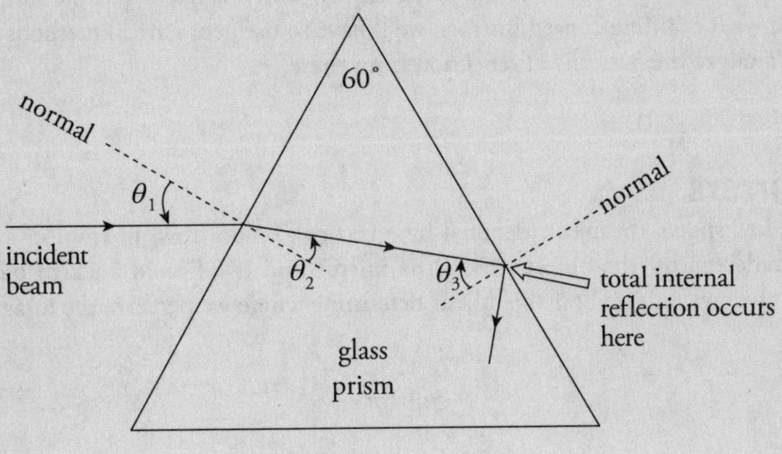

Solution. Total internal reflection will occur at the glass/air boundary if θ_3 is greater than the critical angle, θ_c, which we can calculate this way:

$$\sin \theta_c = \frac{n_{air}}{n_{glass}} = \frac{1}{1.55} \Rightarrow \theta_c = 40°$$

Because $\theta_3 = 60° - \theta_2$, total internal reflection will take place if θ_2 is smaller than 20°. Now, by Snell's Law, $\theta_2 = 20°$ if

$$\begin{aligned} n_{air} \sin \theta_1 &= n_{glass} \sin \theta_2 \\ \sin \theta_1 &= 1.55 \sin 20° \\ \theta_1 &= 32° \end{aligned}$$

Therefore, total internal reflection will occur at the right-hand face of the prism if θ_1 is smaller than 32°.

MIRRORS

A **mirror** is an optical device that forms an image by reflecting light. We've all looked into a mirror and seen images of nearby objects, and the purpose of this section is to analyze these images mathematically. We begin with a plane mirror, which is flat and is the simplest type of mirror. Then, we'll examine curved mirrors; we'll have to use geometrical methods or algebraic equations to analyze the patterns of reflection from these.

Plane Mirrors

The figure below shows an object (denoted by a vertical, bold arrow) in front of a flat mirror. Light that's reflected off the object strikes the mirror and is reflected back to our eyes. The directions of the rays reflected off the mirror determine where we perceive the image to be.

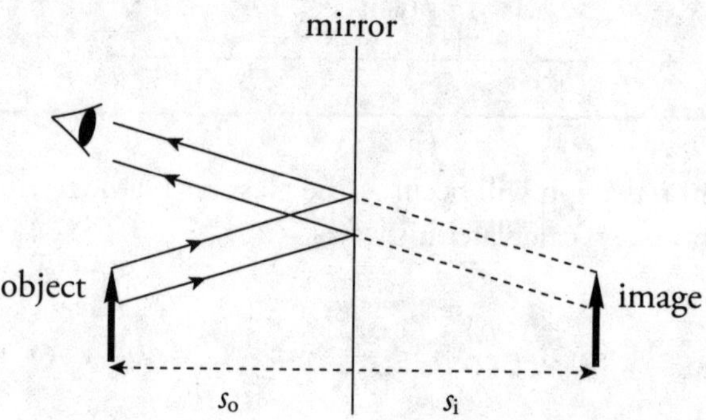

There are four questions we'll answer about the image formed by a mirror:

(1) Where is the image?

(2) Is the image real, or is it virtual?

(3) Is the image upright, or is it inverted?

(4) What is the height of the image (compared with that of the object)?

When we look at ourselves in a mirror, it seems like our image is *behind* the mirror, and, if we take a step back, our image also takes a step back. The Law of Reflection can be used to show that the image seems as far behind the mirror as the object is in front of the mirror. This answers question (1).

An image is said to be **real** if light rays actually focus at the image. A real image can be projected onto a screen. For a flat mirror, light rays bounce off the front of the mirror; so, of course, no light focuses behind it. Therefore, the images produced by a flat mirror are not real; they are **virtual**. This answers question (2).

When we look into a flat mirror, our image isn't upside down; flat mirrors produce upright images, and question (3) is answered.

Finally, the image formed by a flat mirror is neither magnified nor diminished (minified) relative to the size of the object. This answers question (4).

Spherical Mirrors

A **spherical mirror** is a mirror that's curved in such a way that its surface forms part of a sphere.

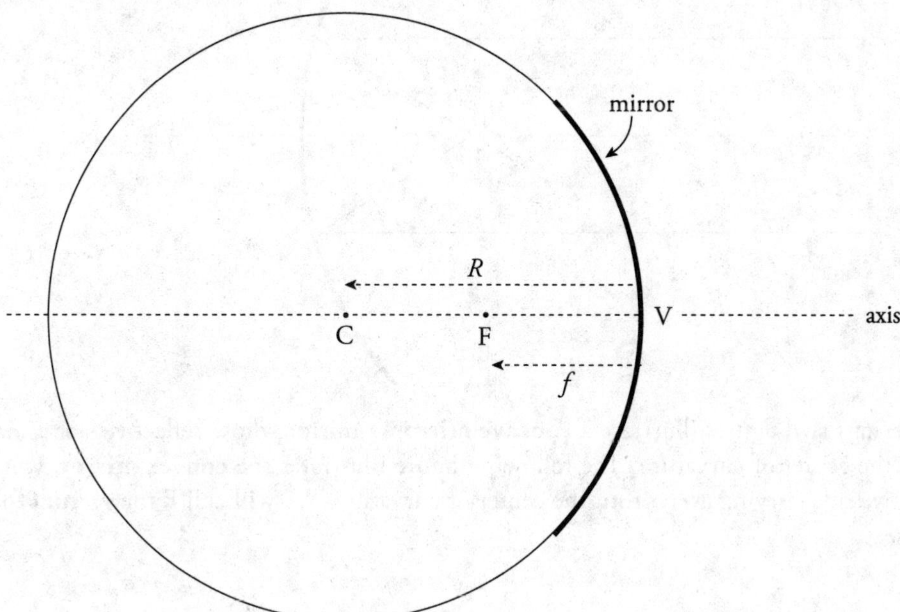

The center of this imaginary sphere is the mirror's **center of curvature**, and the radius of the sphere is called the mirror's **radius of curvature**, R. Halfway between the mirror and the center of curvature, C, is the **focus** (or **focal point**), F. The intersection of the mirror's optic **axis** (its axis of symmetry) with the mirror itself is called the **vertex**, V, and the distance from V to F is called the **focal length**, f, equal to one-half of the radius of curvature:

$$f = \frac{R}{2}$$

If the mirror had a parabolic cross-section, then any ray parallel to the axis would be reflected by the mirror through the focal point. Spherical mirrors do this for incident light rays near the axis (**paraxial rays**) because, in the region of the mirror that's close to the axis, the shapes of a parabolic mirror and a spherical mirror are nearly identical.

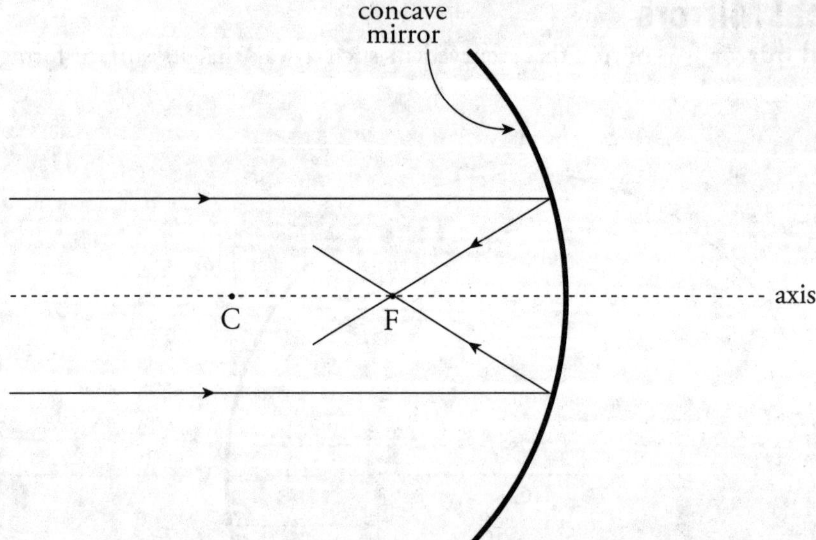

The previous two figures illustrate a **concave mirror**, a mirror whose reflective side is *caved in* toward the center of curvature. The following figure illustrates the **convex mirror**, which has a reflective side curving away from the center of curvature. We will call F the **virtual focus** in this case.

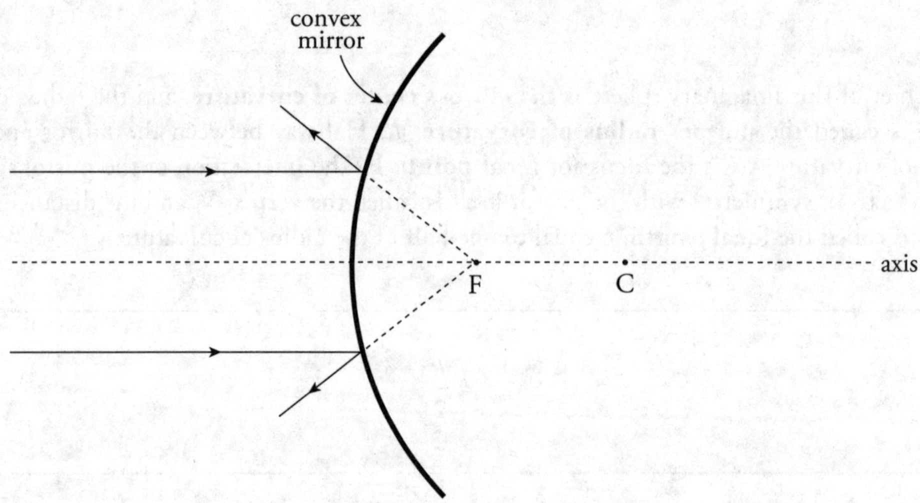

RAY TRACING FOR MIRRORS

One method of answering the four questions previously listed concerning the image formed by a mirror involves a geometric approach called **ray tracing**. Representative rays of light are sketched in a diagram that depicts the object and the mirror; the point at which the reflected rays intersect (or appear to intersect) is the location of the image.

Some rules governing rays are as follows:

Concave Mirrors

- An incident ray parallel to the axis is reflected through the focus.
- An incident ray that passes through the focus is reflected parallel to the axis.
- An incident ray that strikes the vertex is reflected at an equal angle to the axis.

Convex Mirrors

- An incident ray parallel to the axis is reflected away from the virtual focus.
- An incident ray directed toward the virtual focus is reflected parallel to the axis.
- An incident ray that strikes the vertex is reflected at an equal angle to the axis.

Example 6 The figure below shows a concave mirror and an object (the bold arrow). Use a ray diagram to locate the image of the object.

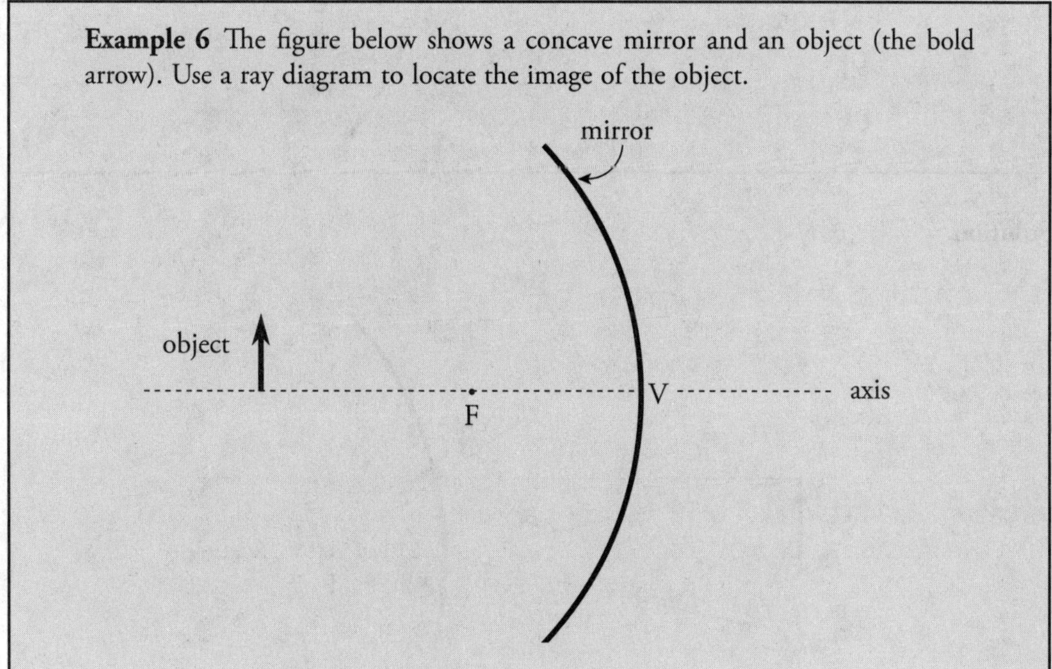

Solution. It takes only two distinct rays to locate the image:

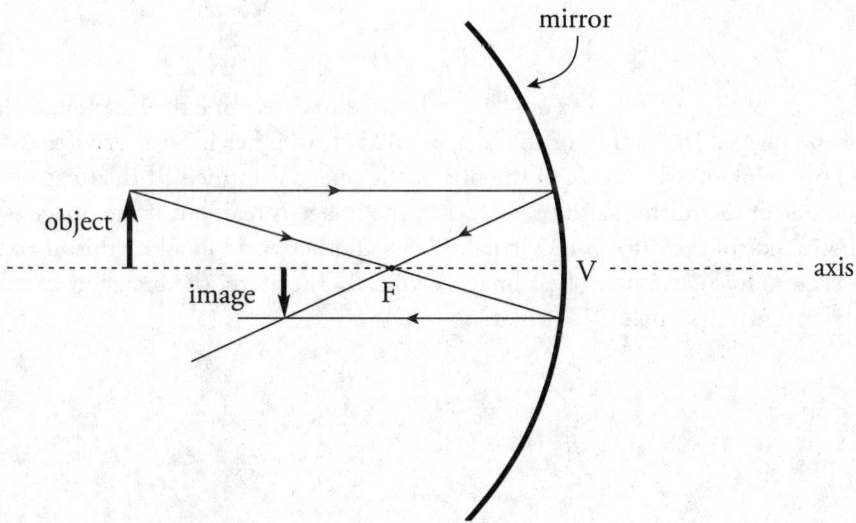

Example 7 The figure below shows a convex mirror and an object (the arrow). Use a ray diagram to locate the image of the object.

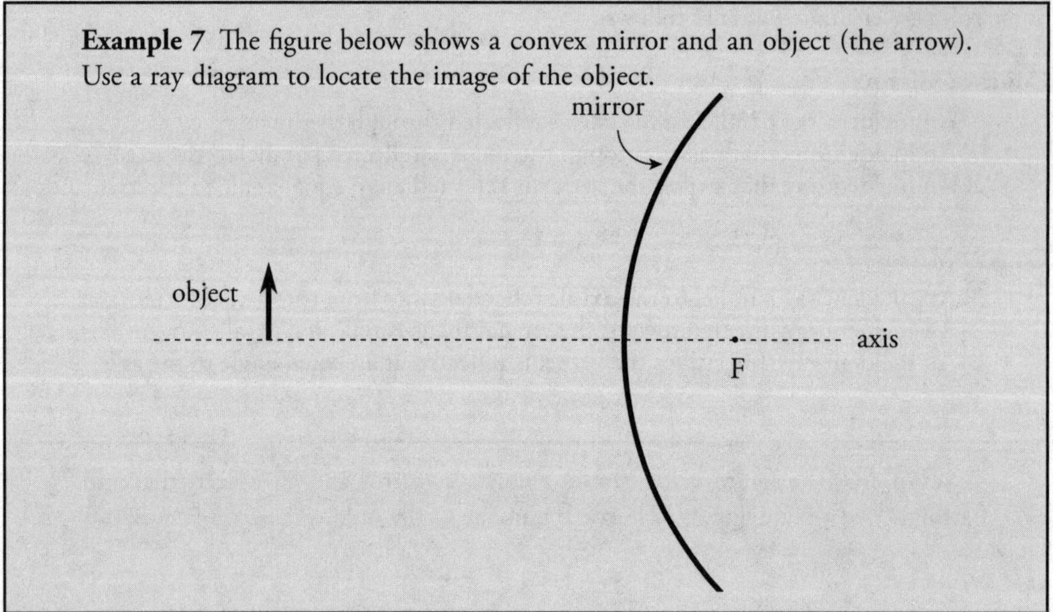

Solution.

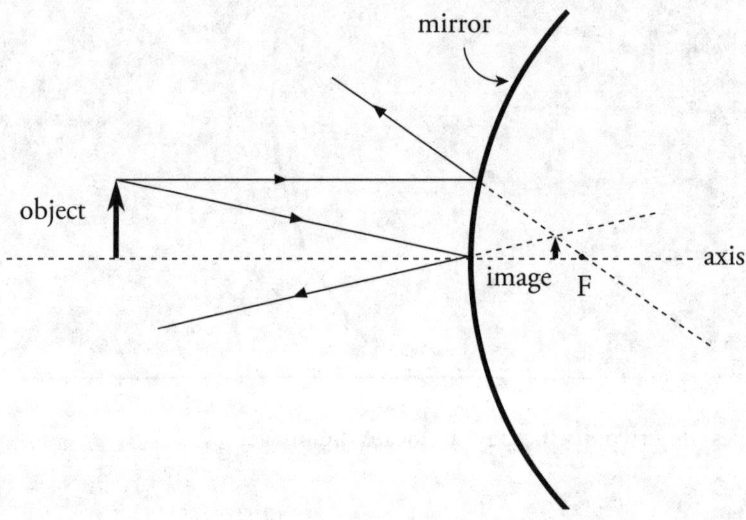

The ray diagrams of the preceding examples can be used to determine the location, orientation, and size of the image. The nature of the image—that is, whether it's real or virtual—can be determined by seeing on which side of the mirror the image is formed. If the image is formed on the same side of the mirror as the object, then the image is real, but if the image is formed on the opposite side of the mirror, it's virtual. Here's another way to look at this: if you had to trace lines back to form an image, that image is virtual. Therefore, the image in Example 6 is real, and the image in Example 7 is virtual.

Using Equations to Answer Questions About the Image

While ray diagrams can answer our questions about images completely, the fastest and easiest way to get information about an image is to use two equations and some simple conventions. The first equation, called the **mirror equation**, is

$$\frac{1}{s_o}+\frac{1}{s_i}=\frac{1}{f}$$

Equation Sheet

where s_o is the object's distance from the mirror, s_i is the image's distance from the mirror, and f is the focal length of the mirror. The value of s_o is *always* positive for a real object, but s_i can be positive or negative. The sign of s_i tells us whether the image is real or virtual: if s_i is positive, the image is real; and if s_i is negative, the image is virtual.

The second equation is called the **magnification equation**:

$$|M|=\left|\frac{h_i}{h_o}\right|=\left|\frac{s_i}{s_o}\right|$$

Equation Sheet

This gives the magnification; the height of the image, h_i, is $|M|$ times the height of the object, h_o. If M is positive, then the image is upright relative to the object; if M is negative, it's inverted relative to the object. Because s_o is always positive, we can come to two important conclusions. If s_i is positive, then M is negative, so real images are always inverted, and, if s_i is negative, then M is positive, so virtual images are always upright.

Finally, to distinguish *mathematically* between concave and convex mirrors, we always write the focal length f as a positive value for concave mirrors and a negative value for convex mirrors. With these two equations and their accompanying conventions, all four questions about an image can be answered.

Which Is the Positive Side?

We will always treat the object's position s_o as being the positive side. For mirrors, if the image is produced on the same side as the object, it is considered a positive distance. If an image is produced opposite the side of the object, it is considered a negative distance. Concave mirrors will always have a positive focal length (meaning the focal point is on the same side of the object). Convex mirrors will always have a negative focal length (meaning the focal point is on the opposite side of the object).

MIRRORS

converging—concave ⇔ f positive
diverging—convex ⇔ f negative

$$\frac{1}{s_o}+\frac{1}{s_i}=\frac{1}{f}\begin{cases} s_o \text{ always positive (real object)} \\ s_i \text{ positive} \Rightarrow \text{image is real (located on the } \textit{same} \text{ side of mirror as object)} \\ s_i \text{ negative} \Rightarrow \text{image is virtual (located on the } \textit{opposite} \text{ side of mirror from object)} \end{cases}$$

Real or Virtual?
Remember that all real images are inverted and all virtual images are upright.

$$M=\frac{h_i}{h_o}=\frac{-s_i}{s_o}\begin{cases} \text{given } h_o \text{ is positive} \\ h_i \text{ and } M \text{ positive} \Rightarrow \text{image is upright} \\ h_i \text{ and } M \text{ negative} \Rightarrow \text{image is inverted} \end{cases}$$

Example 8 An object of height 4 cm is placed 30 cm in front of a concave mirror whose focal length is 10 cm.

(a) Where's the image?
(b) Is it real or virtual?
(c) Is it upright or inverted?
(d) What's the height of the image?

Solution.

(a) With s_o = 30 cm and f = 10 cm, the mirror equation gives

$$\frac{1}{s_o}+\frac{1}{s_i}=\frac{1}{f}\Rightarrow\frac{1}{30\text{ cm}}+\frac{1}{s_i}=\frac{1}{10\text{ cm}}\Rightarrow\frac{1}{s_i}=\frac{1}{15\text{ cm}}\Rightarrow s_i=15\text{ cm}$$

The image is located 15 cm in front of the mirror.

(b) Because s_i is positive, the image is real.

(c) Real images are inverted.

(d) $$\frac{h_i}{h_o}=\frac{-s_i}{s_o}\Rightarrow h_i=\frac{-s_ih_o}{s_o}\Rightarrow h_i=\frac{(-15\text{ cm})(4\text{ cm})}{30\text{ cm}}=-2\text{ cm}$$

The –2 cm also confirms the image is inverted.

Example 9 An object of height 4 cm is placed 20 cm in front of a convex mirror whose focal length is –30 cm.

(a) Where's the image?

(b) Is it real or virtual?

(c) Is it upright or inverted?

(d) What's the height of the image?

Solution.

(a) With s_o = 20 cm and f = –30 cm, the mirror equation gives us

$$\frac{1}{s_o}+\frac{1}{s_i}=\frac{1}{f} \Rightarrow \frac{1}{20\text{ cm}}+\frac{1}{s_i}=\frac{1}{-30\text{ cm}} \Rightarrow \frac{1}{s_i}=-\frac{1}{12\text{ cm}} \Rightarrow s_i=-12\text{ cm}$$

So, the image is located 12 cm behind the mirror.

(b) Because s_i is negative, the image is virtual.

(c) Virtual images are upright.

(d) $\frac{h_i}{h_o}=\frac{-s_i}{s_o} \Rightarrow h_i=\frac{-s_i h_o}{s_o} \Rightarrow h_i=\frac{-(-12\text{ cm})(4\text{ cm})}{20\text{ cm}}=2.4\text{ cm}$

The +2.4 cm also confirms the image is upright.

What Produces What?
Only concave mirrors can produce both real images (if $s_o > f$) and virtual images (if $s_o < f$). Convex mirrors and plane mirrors can produce only virtual images.

Example 10 Show how the statements made earlier about plane mirrors can be derived from the mirror and magnification equations.

Solution. A plane mirror can be considered a spherical mirror with an infinite radius of curvature (and an infinite focal length). If $f = \infty$, then $1/f = 0$, so the mirror equation becomes

$$\frac{1}{s_o}+\frac{1}{s_i}=0 \Rightarrow s_i=-s_o$$

So, the image is as far behind the mirror as the object is in front. Also, since s_o is always positive, s_i is negative, so the image is virtual. The magnification is

$$M=-\frac{s_i}{s_o}=-\frac{-s_o}{s_o}=1$$

and the image is upright and has the same height as the object. The mirror and magnification equations confirm our description of images formed by plane mirrors.

Example 11 Show why convex mirrors can form only virtual images.

Solution. Because f is negative and s_o is positive, the mirror equation

$$\frac{1}{s_o} + \frac{1}{s_i} = \frac{1}{f}$$

immediately tells us that s_i cannot be positive (if it were, the left-hand side would be the sum of two positive numbers, while the right-hand side would be negative). Since s_i must be negative, the image must be virtual.

Example 12 An object placed 60 cm in front of a spherical mirror forms a real image at a distance of 30 cm from the mirror.

(a) Is the mirror concave or convex?

(b) What's the mirror's focal length?

(c) Is the image taller or shorter than the object?

Solution.

(a) The fact that the image is real tells us that the mirror cannot be convex, since convex mirrors form only virtual images. The mirror is concave.

(b) With s_o = 60 cm and s_i = 30 cm (s_i is positive since the image is real), the mirror equation tells us that

$$\frac{1}{s_o} + \frac{1}{s_i} = \frac{1}{f} \Rightarrow \frac{1}{60\text{ cm}} + \frac{1}{30\text{ cm}} = \frac{1}{f} \Rightarrow \frac{1}{f} = \frac{1}{20\text{ cm}} \Rightarrow f = 20\text{ cm}$$

Note that f is positive, which confirms that the mirror is concave.

(c) The magnification is

$$M = -\frac{s_i}{s_o} = -\frac{30\text{ cm}}{60\text{ cm}} = -\frac{1}{2}$$

Since the absolute value of M is less than 1, the mirror makes the object look smaller (minifies the height of the object). The image is only half as tall as the object and is inverted, since M is negative.

Example 13 A concave mirror with a focal length of 25 cm is used to create a real image that has twice the height of the object. How far is the image from the mirror?

So Many Equations!
The purpose of a mirror is to reflect an image. The purpose of a lens is to refract an image (or see through it). Fortunately, the equations for mirrors and lenses will be the same with a minor difference on the focal length.

Solution. Since h_i (the height of the image) is twice h_o (the height of the object), the value of the magnification is either +2 or –2. To figure out which, we just notice that the image is real; real images are inverted, so the magnification, M, must be negative. Therefore, $M = -2$, so

$$-\frac{s_i}{s_o} = -2 \Rightarrow s_o = \frac{1}{2}s_i$$

Substituting this into the mirror equation gives us

$$\frac{1}{s_o}+\frac{1}{s_i}=\frac{1}{f} \Rightarrow \frac{1}{\frac{1}{2}s_i}+\frac{1}{s_i}=\frac{1}{f} \Rightarrow \frac{3}{s_i}=\frac{1}{f} \Rightarrow s_i - 3f = 3(25 \text{ cm}) = 75 \text{ cm}$$

THIN LENSES

A lens is an optical device that forms an image by *refracting* light. We'll now talk about the equations and conventions that are used to analyze images formed by the two major categories of lenses: converging and diverging.

Mirrors vs. Lenses
The equations are exactly the same. The only difference is we switch the locations of the focus. A positive focal length lens will have a focus on the opposite side of the lens from the object. If the focal point is located on the same side of the object, then it is a negative focal length.

A **converging lens**—like the bi-convex one shown below—converges parallel paraxial rays of light to a focal point on the far side. (This lens is *bi-convex*; both of its faces are convex. All converging lenses have at least one convex face.) Because parallel light rays actually focus at F, this point is called a **real focus**. Its distance from the lens is the focal length, f.

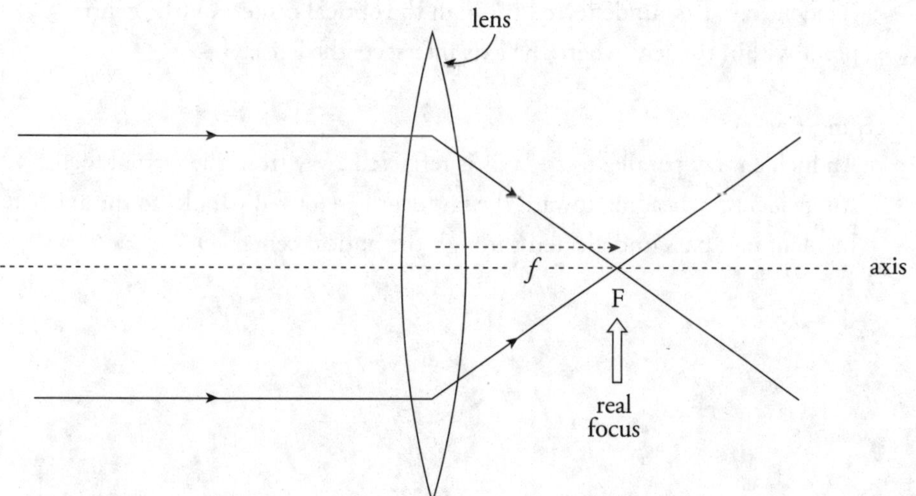

A **diverging lens**—like the *bi-concave* one shown below—causes parallel paraxial rays of light to diverge away from a **virtual focus**, F, on the same side as the incident rays. (All diverging lenses have at least one concave face.)

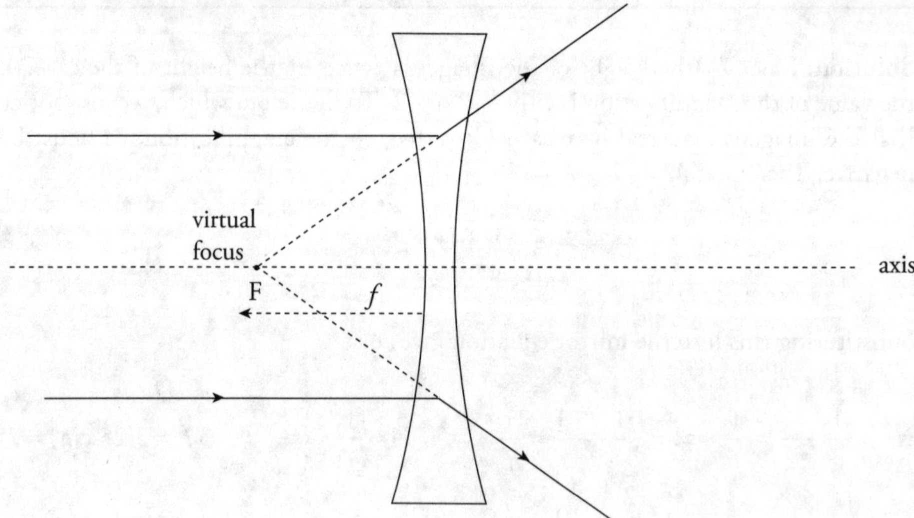

Don't Mix Them Up
Converging Lens = Bi-Convex Lens
Diverging Lens = Bi-Concave Lens

RAY TRACING FOR LENSES

Just as is the case with mirrors, representative rays of light can be sketched in a diagram along with the object and the lens; the point at which the reflected rays intersect (or appear to intersect) is the location of the image. The rules that govern these rays are as follows:

Converging Lenses

- An incident ray parallel to the axis is refracted through the real focus.
- An incident ray that passes through the focus is refracted parallel to the axis.
- Incident rays pass undeflected through the optical center, O (the central point within the lens where the axis intersects the lens).

Lenses actually have two focal points: a near focal point (on the same side of the lens as the object) and a far focal point (on the opposite side of the lens as the object).

Diverging Lenses

- An incident ray parallel to the axis is refracted away from the virtual focus.
- An incident ray heading toward the far focus is refracted parallel to the axis.
- Incident rays pass undeflected through the optical center, O.

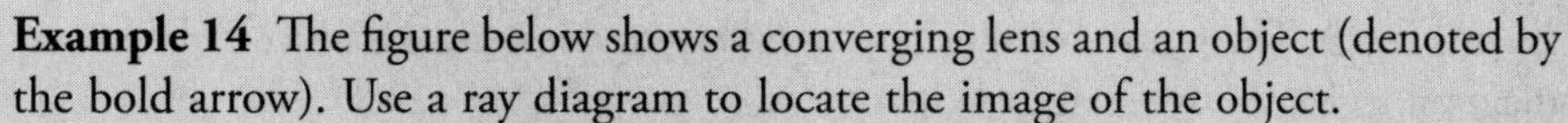

Example 14 The figure below shows a converging lens and an object (denoted by the bold arrow). Use a ray diagram to locate the image of the object.

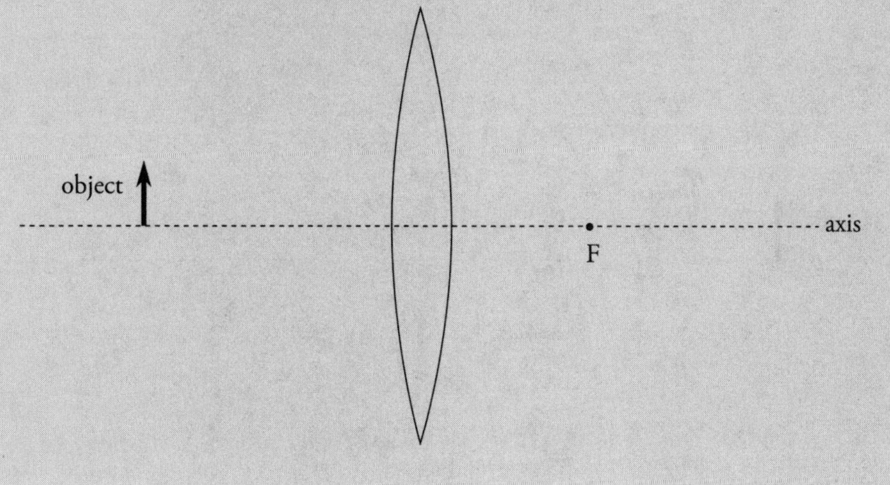

Solution.

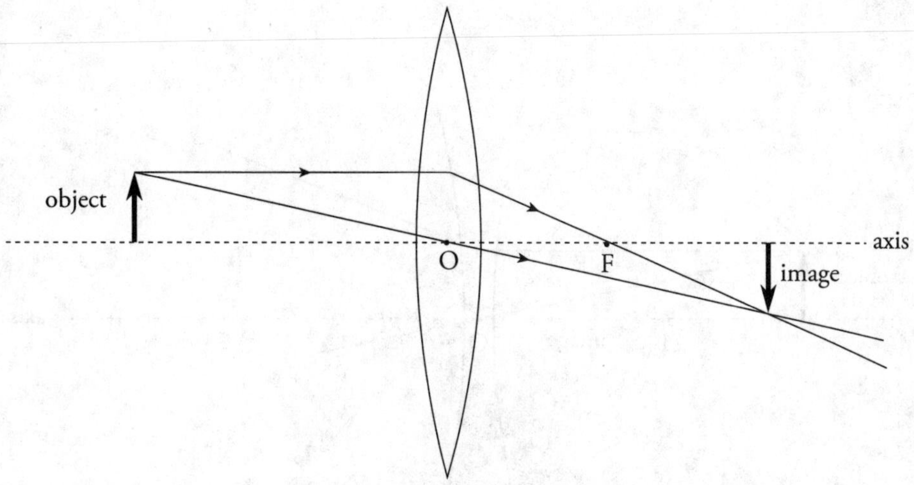

Example 15 The figure below shows a diverging lens and an object. Use a ray diagram to locate the image of the object.

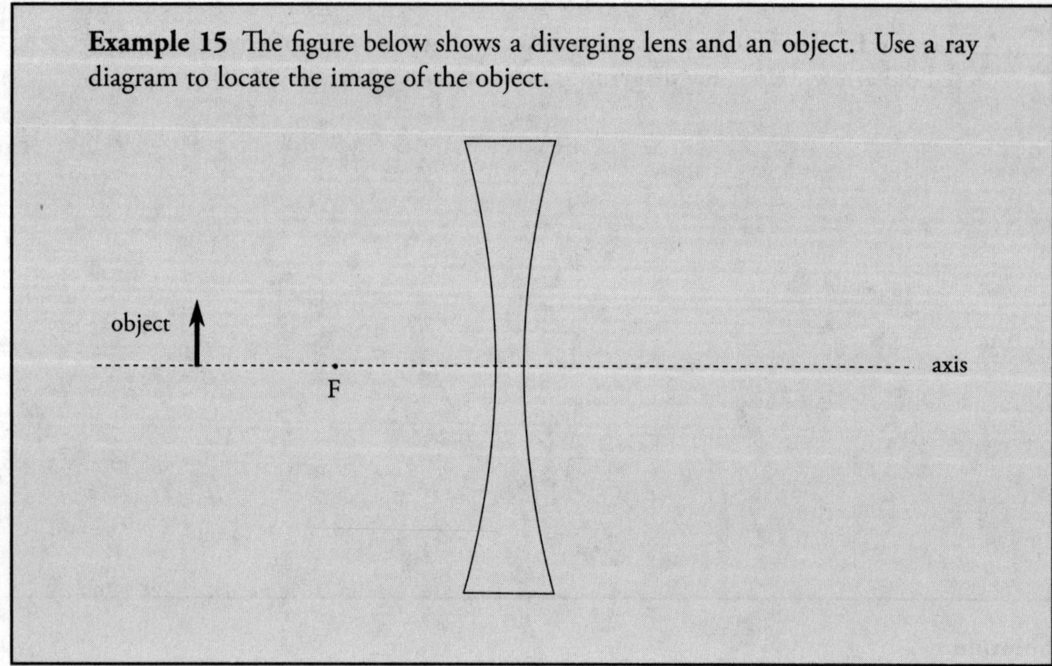

Solution.

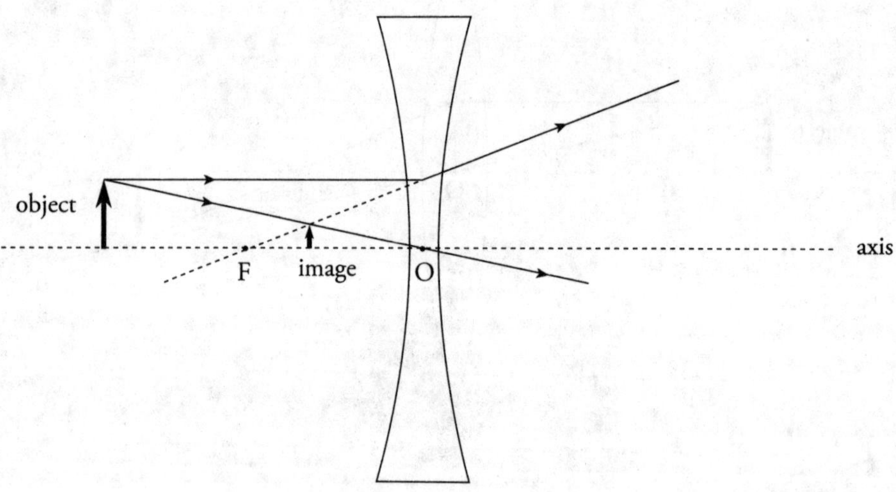

The nature of the image—that is, whether it's real or virtual—is determined by the side of the lens upon which the image is formed. If the image is formed on the side of the lens that's opposite the object, then the image is real, and if the image is formed on the same side of the lens as the object, then it's virtual. Another way of looking at this is if you had to trace lines back to form an image, that image is virtual. Therefore, the image in Example 14 is real, while the image in Example 15 is virtual.

USING EQUATIONS TO ANSWER QUESTIONS ABOUT THE IMAGE

Lenses and mirrors use the same equations, notation, and sign conventions, with the following note. Converging optical devices ($+f$) are con<u>cave</u> mirrors and con<u>vex</u> lenses. Diverging optical devices ($-f$) are con<u>vex</u> mirrors and con<u>cave</u> lenses.

LENSES

convex lens—converging ⇔ f positive
concave lens—diverging ⇔ f negative

$$\frac{1}{s_o} + \frac{1}{s_i} = \frac{1}{f}$$

- s_o always positive (real object)
- s_i positive ⇒ image is real (located on *opposite* side of lens from object)
- s_i negative ⇒ image is virtual (located on *same* side of lens as object)

$$M = \frac{h_i}{h_o} = \frac{-s_i}{s_o}$$

- given h_o is positive
- h_i and M positive ⇒ image is upright
- h_i and M negative ⇒ image is inverted

(Arrows link: real image → inverted; virtual image → upright.)

Example 16 An object of height 11 cm is placed 44 cm in front of a converging lens with a focal length of 24 cm.

(a) Where's the image?
(b) Is it real or virtual?
(c) Is it upright or inverted?
(d) What's the height of the image?

Solution.

(a) With s_o = 44 cm and f = 24 cm, the lens equation gives us

$$\frac{1}{s_o} + \frac{1}{s_i} = \frac{1}{f} \Rightarrow \frac{1}{44\text{ cm}} + \frac{1}{s_i} = \frac{1}{24\text{ cm}} \Rightarrow \frac{1}{s_i} = 0.0189\text{ cm}^{-1} \Rightarrow s_i = 53\text{ cm}$$

So, the image is located 53 cm from the lens, on the opposite side from the object.

(b) Because s_i is positive, the image is real.

(c) Real images are inverted.

(d) $$\frac{h_i}{h_o} = \frac{-s_i}{s_o} \Rightarrow h_i = \frac{-s_i h_o}{s_o} \Rightarrow h_i = \frac{(-53\text{ cm})(11\text{ cm})}{44\text{ cm}} = -13\text{ cm}$$

The negative h_i reaffirms that we have an inverted image.

Example 17 An object of height 11 cm is placed 48 cm in front of a diverging lens with a focal length of –24.5 cm.

(a) Where's the image?
(b) Is it real or virtual?
(c) Is it upright or inverted?
(d) What's the height of the image?

Solution.

(a) With s_o = 48 cm and f = –24.5 cm, the lens equation gives us

$$\frac{1}{s_o}+\frac{1}{s_i}=\frac{1}{f} \Rightarrow \frac{1}{48\text{ cm}}+\frac{1}{s_i}=\frac{1}{-24.5\text{ cm}} \Rightarrow \frac{1}{s_i}=-0.0616\text{ cm}^{-1} \Rightarrow s_i=-16\text{ cm}$$

The image is 16 cm from the lens, on the same side as the object.

(b) Because s_i is negative, the image is virtual.

(c) Virtual images are upright.

(d) $$\frac{h_i}{h_o}=\frac{-s_i}{s_o} \Rightarrow h_i=\frac{-s_i h_o}{s_o} \Rightarrow h_i=\frac{-(-16\text{ cm})(11\text{ cm})}{48\text{ cm}}=3.7\text{ cm}$$

The positive h_i reaffirms that we have an upright image.

What Produces What?
Only converging lenses can produce both real images (if $s_o > f$) and virtual images (if $s_o < f$). Diverging lenses can produce only virtual images.

Chapter 8 Review Questions

Answers and explanations can be found in Chapter 11.

Section I: Multiple Choice

1 Mark for Review

A beam of light in air is incident upon the smooth surface of a piece of flint glass, as shown, and refracts into the glass:

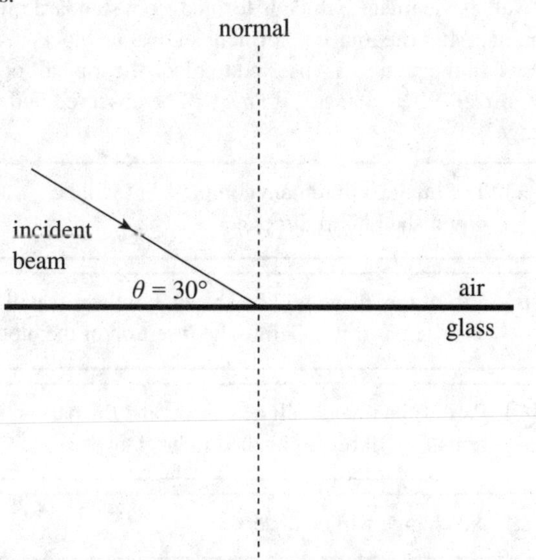

As the angle from the surface, θ, for this beam is increased toward $\theta = 90°$, what observation is made of the refracted ray?

(A) The angle of the refracted ray from the surface will also increase, but it will always be smaller than θ.

(B) The angle of the refracted ray from the surface will also increase, but it will always be larger than θ.

(C) The angle of the refracted ray will also increase, but at some point, total internal reflection begins to occur.

(D) The angle of the refracted ray from the surface will decrease.

2 Mark for Review

A convex lens constructed of glass makes a real image of an object. When the object is located d_o in front of the lens, the image appears at a distance of d_i behind the lens. If the object remains at its position, d_o, and the lens is replaced with a convex lens of identical shape but with a higher index of refraction, which of the following is possible?

(A) The image is still at d_i and is still real.

(B) The image is at a position closer to the lens than d_i and is real.

(C) The image is at a position farther from the lens than d_i and is real.

(D) The image becomes virtual.

3 Mark for Review

A beam of light traveling in Medium 1 strikes the interface to another transparent medium with a lower index of refraction, Medium 2. If the intensity of light is measured to be less in Medium 2 than in Medium 1, what can be concluded?

(A) The decrease in intensity was caused by the change in the speed of light in the different media.

(B) Total internal reflection occurred at the interface.

(C) The angle the light travels relative to the normal in Medium 2 will be the same as the angle the light had traveled relative to the normal in Medium 1.

(D) Some part of the light was reflected and/or absorbed at the interface.

4 Mark for Review

If a clear liquid has a refractive index of 1.45 and a transparent solid has an index of 2.90, then, for total internal reflection to occur at the interface between these two media, which of the following must be true?

	incident beam originates in	at an angle of incidence greater than
(A)	The solid	30°
(B)	The liquid	30°
(C)	The liquid	60°

(D) Total internal reflection cannot occur.

5 Mark for Review

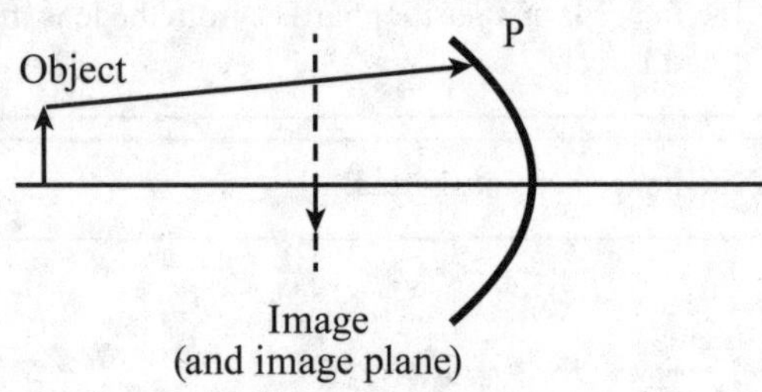

The above picture shows a converging mirror and an object, as well as the image, which is formed from standard ray tracing. Also shown, a dotted vertical line indicates the image plane for this configuration. Also shown is a ray that travels from the top of the image to the point P on the curved mirror. The point P lies above the line which is parallel to the optic axis from the tip of the image. Where will the reflected ray from point P intersect the image plane?

(A) On the optic axis

(B) Below the optic axis but above the tip of the image

(C) At the tip of the image

(D) Farther from the optic axis than the tip of the image

6 Mark for Review

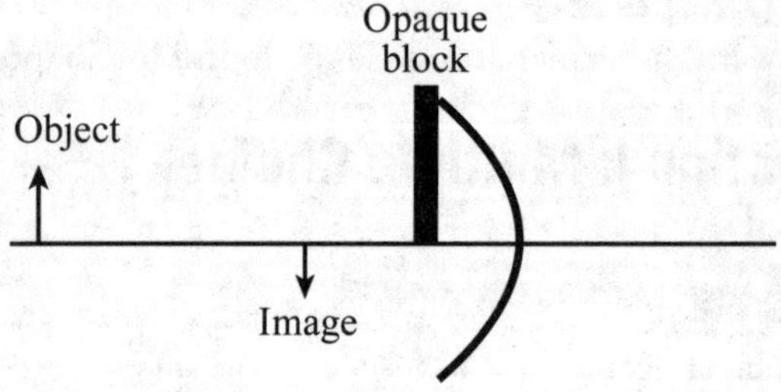

The above picture shows a converging mirror and an object, as well as the image, which is formed from standard ray tracing. After the image is formed, an opaque block, also shown in the picture, is inserted to block the top half of the mirror. What changes, if any, will be observed in the image?

(A) The image will remain complete but will be diminished in brightness.

(B) Part of the image will be absent, but the rest will be as bright as it was before the insertion of the block.

(C) Part of the image will be absent, and the part which remains will be diminished in brightness.

(D) No change will be observed.

7 Mark for Review

You are tasked with creating a real image using a diverging lens as your imaging system. Which of the following criteria is true about both the image and the object?

(A) A real image can be created only if the object is farther away from the lens than the focal length.

(B) A real image can be created only if the object is closer to the lens than the focal length.

(C) A real image can be created regardless of whether the object is farther away from the lens or closer to the lens than the focal length.

(D) A real image cannot be created using only a diverging lens.

Section II: Free Response

1 Mark for Review

A beam of light shines on a screen with only air between the light source and the screen. The spot where the light strikes the screen is marked and called point P. Then, as Figure 1 shows, a piece of glass with thickness T is placed in the path of the beam. The glass is surrounded on top and bottom by air. The normal to the glass slab at the point where the incident beam strikes the glass is drawn as a dashed line.

Figure 1

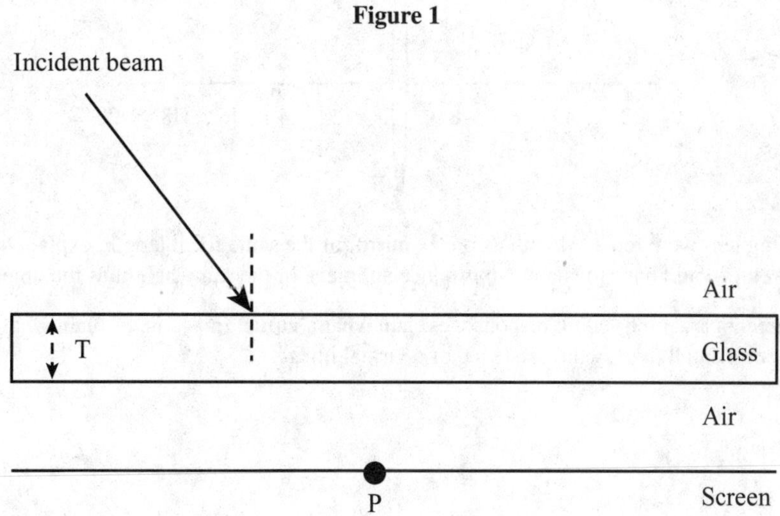

A. Two students are discussing what observations will be made in this new arrangement.

Student 1: The spot on the screen will still be at point P. According to Snell's Law, the light will bend at the first interface and then unbend at the second interface and travel parallel to its original path. This will cause the spot on the screen to be at the same location.

Student 2: The beam after the glass cannot be parallel to the beam before the glass. Snell's Law is $n_{in}\sin(\theta_{in}) = n_{out}\sin(\theta_{out})$, and because θ_{out} is different from θ_{in} at the first interface, the angles have to be different at the second interface as well.

i. Which parts, if any, of Student 1's reasoning are correct?

ii. Which parts, if any, of Student 2's reasoning are correct?

iii. Which parts, if any, of Student 1's reasoning are incorrect?

iv. Which parts, if any, of Student 2's reasoning are incorrect?

B. Use ray tracing to draw the path of the beam from where it is incident on the glass to where it strikes the screen. Use appropriate labels to indicate which angles are the same and which are different.

C. In a paragraph-length explanation without relying solely on equations, explain how your answer to part B would be different if the area between the screen and the glass were filled with water (whose index of refraction is between that of air and that of glass).

2 Mark for Review

An experiment is set up with an object, a single converging lens, and a screen to make an image appear on the screen. The lens has a focal length of 3 cm. The arrangement is configured so that the image which appears on the screen is the same size as the object. The lens is placed at a position labeled 10 cm on the optic axis.

A. Draw a diagram of the experimental setup along the optic axis below. Justify your answer.

Figure 1

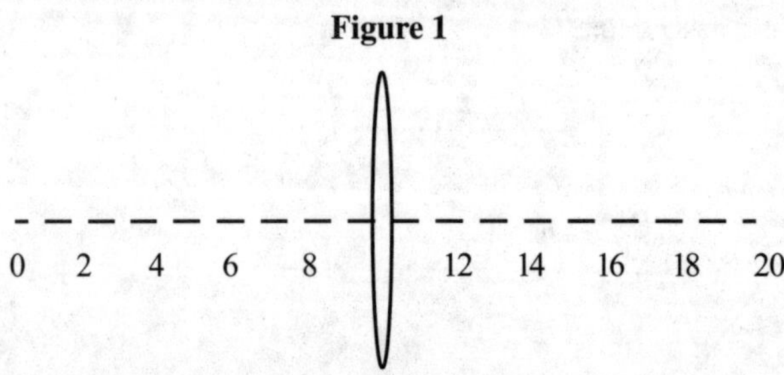

B. If the converging lens were replaced with a convex mirror of the same focal length, explain how (if at all) your experimental setup would have to change to produce an image of the same height as the object.

C. In a clear, coherent paragraph-length response, explain what a virtual image is. Explain what steps would have to be taken to experimentally verify the existence of a virtual image.

Chapter 8 Summary

- Geometric optics always measures angles from the normal line. Topics include the fact that the angle of incidence is equal to the angle of reflection and that these angles lie in the same plane.
- Snell's Law states when light enters a new medium, it changes speed and may change direction. This is stated in the formula

  $$n_1 \sin\theta_1 = n_2 \sin\theta_2$$

  where n is the index of refraction, which is a ratio of the speed of light in a vacuum to the speed of light in the substance ($n = \frac{c}{v}$). The index of refraction is always greater than 1 and has no units. When going from a higher index of refraction to a lower index of refraction, light may experience total internal reflection if the angle of incidence is larger than the critical angle ($\sin\theta_c = \frac{n_2}{n_1}$).
- For both thin lenses and curved mirrors, you can use the mirror equation: $\frac{1}{s_o} + \frac{1}{s_i} = \frac{1}{f}$.

  Note that f is positive for a convex lens or a concave mirror, f is negative for a concave lens or a convex mirror, and s_o is always positive.
- For a lens, if s_i is negative, the lens image is virtual and located on the same side of the lens as the object. If s_i is positive, the lens image is real and located on the opposite side of the lens from the object.
- For a mirror, if s_i is negative, the image is virtual and located on the opposite side of the mirror from the object. If s_i is positive, the image is real and located on the same side of the mirror as the object.
- The magnification equation is given by $|M| = \left|\frac{h_i}{h_o}\right| = \left|\frac{s_i}{s_o}\right|$.

- A **diverging optical device** could be either a convex mirror or a concave lens. The thing these two have in common is that no matter where the object is located, they form images that are virtual and upright. The sign of h_i will always be positive and the sign of s_i will always be negative. The sign of the magnification will be positive and its value will be less than one. The image will always be located between the focal length and the optical device.
- A **converging optical device** could be either a concave mirror or a convex lens. One thing these two have in common is that when s_o is *outside* the focal length, they form images that are real and inverted. The sign of h_i will be negative and the sign of s_i will be positive. The sign of the magnification will be negative and its absolute value can be any number greater than zero. Another thing they have in common is that when s_o is *inside* the focal length, they form images that are virtual and upright. The sign of h_i will be positive and the sign of s_i will be negative. The sign of the magnification will be positive and its value will be greater than one.

Chapter 9
Mechanical Waves, Sound, and Physical Optics

TRANSVERSE TRAVELING WAVES

Let's take a look at a long rope. Someone standing near the system would see peaks and valleys actually moving along the rope, in what's called a **traveling wave.**

Transverse Wave
A transverse wave travels (propagates) in a direction perpendicular to the direction in which the medium is vibrating. Basically, the wave oscillates perpendicular to its direction of travel.

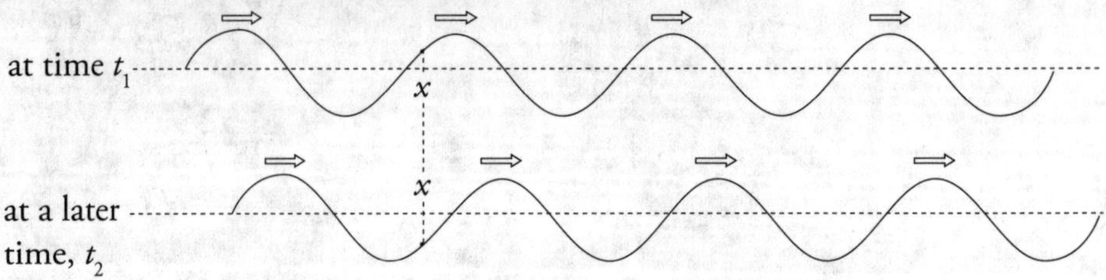

What features of the wave can we see in this point of view? Well, we can see the points at which the rope has its maximum vertical displacement above the horizontal; these points are called **crests.** The points at which the rope has its maximum vertical displacement below the horizontal are called **troughs**. These crests and troughs repeat themselves at regular intervals along the rope, and the distance between two adjacent crests (or two adjacent troughs) is the length of one wave, and is called the **wavelength** (λ, *lambda*). Also, the maximum displacement from the horizontal equilibrium position of the rope is also measurable; this is known as the **amplitude** (*A*) of the wave.

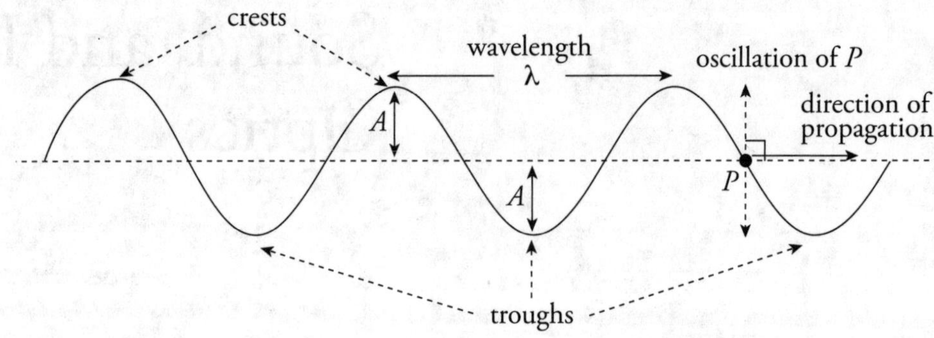

Future Physicist?
If you're planning to pursue physics beyond this course, pay particular attention to this chapter. Waves are a topic that will come up in one way or another for just about every physics course you take. Forming a strong foundation now will serve you well.

Consider the point P on the rope, as shown in the figure. As the wave **propagates** (or travels) to the right, point P oscillates up and down. Since the direction in which the rope oscillates (vertically) is perpendicular to the direction in which the wave propagates (horizontally), this wave is transverse.

The time it takes for one complete vertical oscillation of a point on the rope is called the period, *T*, of the wave, and the number of cycles it completes in one second is called its frequency, *f*. The period and frequency are established by the source of the wave and, of course, $T = 1/f$.

In addition to the amplitude, period, frequency, and wavelength, another important characteristic of a traveling wave is its speed, v. Consider point P in the second figure on the previous page moving from its crest position, down to its trough position, and then back up to the crest position. The wave took a time period of T to move a distance of one wavelength, signified by λ. Therefore, the equation *distance* = *rate* × *time* becomes

$$\lambda = vT$$

$$\lambda = v\left(\frac{1}{f}\right)$$

$$\lambda = v / f$$

The simple equation $\lambda = v/f$ shows how the wave speed, wavelength, and frequency are interconnected. It's the most basic equation in wave theory.

Example 1 A traveling wave on a rope has a frequency of 2.5 Hz. If the speed of the wave is 1.5 m/s, what are its period and wavelength?

Solution. The period is the reciprocal of the frequency:

$$T = \frac{1}{f} = \frac{1}{2.5\text{ Hz}} = 0.4\text{ s}$$

and the wavelength can be found from the equation $\lambda = v/f$:

$$\lambda = \frac{v}{f} = \frac{1.5\text{ m/s}}{2.5\text{ Hz}} = 0.6\text{ m}$$

Example 2 The period of a traveling wave is 0.5 s, its amplitude is 10 cm, and its wavelength is 0.4 m. What are its frequency and wave speed?

Solution. The frequency is the reciprocal of the period: $f = 1/T = 1/(0.5\text{ s}) = 2\text{ Hz}$. The wave speed can be found from the equation $\lambda = v/f$:

$$v = \lambda f = (0.4\text{ m})(2\text{ Hz}) = 0.8\text{ m/s}$$

Note that the frequency, period, wavelength, and wave speed have nothing to do with the amplitude.

WAVE SPEED ON A STRETCHED STRING

We can also derive an equation for the speed of a transverse wave on a stretched string or rope. Let the mass of the string be m and its length be L. If the tension in the string is F_T, then the speed of a traveling transverse wave on this string is given by

$$v = \sqrt{\frac{F_T}{m/L}}$$

Note that v depends only on the physical characteristics of the string: its tension and linear density. So, because $\lambda = v/f$ for a given stretched string, varying f will create different waves that have different wavelengths, but v will not vary.

Big Wave Rules

Notice that the equation for the wave speed on a stretched rope shows that v does not depend on f (or λ). While this may seem to contradict the first equation, $\lambda = v/f$, it really doesn't. The speed of the wave depends on the characteristics of the rope: how tense it is and what it's made of. We can wiggle the end at any frequency we want, and the speed of the wave we create will be a constant. However, because $\lambda = v/f$ must always be true, a higher f will mean a shorter λ, and a lower f will mean a longer λ. Thus, changing f doesn't change v: it changes λ. This brings up our first big wave rule:

> Wave Rule #1: The speed of a wave is determined by the type of wave and the characteristics of the medium, not by the frequency.

Rule #1 deals with a wave in a single medium. What travels to you faster, a yell or a whisper? Neither does; they both travel to you at the speed of sound. In a single medium, the speed of a wave is constant and is represented by an inverse relationship between wavelength and frequency: $\lambda_1 f_1 = \lambda_2 f_2$.

Notice that two different types of waves can move at different speeds through the same medium. For example, sound and light move through air at very different speeds.

Our second wave rule addresses what happens when a wave passes from one medium into another. Because wave speed is determined by the characteristics of the medium, a change in the medium causes a change in wave speed, but the frequency won't change.

> Wave Rule #2: When a wave passes into another medium, its speed changes, but its frequency does not.

Example 3 A horizontal rope with linear mass density μ = 0.5 kg/m sustains a tension of 60 N. The non-attached end is oscillated vertically with a frequency of 4 Hz.

(a) What are the speed and wavelength of the resulting wave?

(b) How would you answer these questions if f were increased to 5 Hz?

Solution.

(a) Wave speed is established by the physical characteristics of the rope:

$$v = \sqrt{\frac{F_T}{\mu}} = \sqrt{\frac{60 \text{ N}}{0.5 \text{ kg/m}}} = 11 \text{ m/s}$$

With v, we can find the wavelength: $\lambda = v/f$ = (11 m/s)/(4 Hz) = 2.8 m.

(b) If f were increased to 5 Hz, then v would not change, but λ would; the new wavelength would be

$$\lambda' = v/f' = (11 \text{ m/s})/(5 \text{ Hz}) = 2.2 \text{ m}$$

Example 4 Two ropes of unequal linear densities are connected, and a wave is created in the rope on the left, which propagates to the right, toward the interface with the heavier rope.

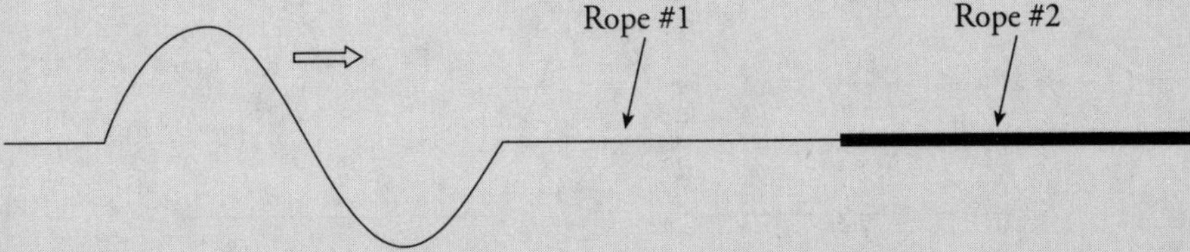

When a wave strikes the boundary to a new medium (in this case, the heavier rope), some of the wave's energy is reflected and some is transmitted. The frequency of the transmitted wave is the same, but the speed and wavelength are not. How do the speed and wavelength of the transmitted wave compare to the speed and wavelength of the incident wave?

Solution. Since the wave enters a new medium, it will have a new wave speed. Because Rope #2 has a greater linear mass density than Rope #1, and because v is inversely proportional to the square root of the linear mass density, the speed of the wave in Rope #2 will be less than the speed of the wave in Rope #1. Since $\lambda = v/f$ must always be satisfied and f does not change, the fact that v changes means that λ must change too. In particular, since v decreases upon entering Rope #2, so will λ.

SUPERPOSITION OF WAVES

When two or more waves meet, the displacement at any point of the medium is equal to the algebraic sum of the displacements due to the individual waves: This is **superposition**. The figure below shows two wave pulses traveling toward each other along a stretched string. Note that when they meet and overlap (**interfere**), the displacement of the string is equal to the sum of the individual displacements, but after they pass, the wave pulses continue, unchanged by their meeting.

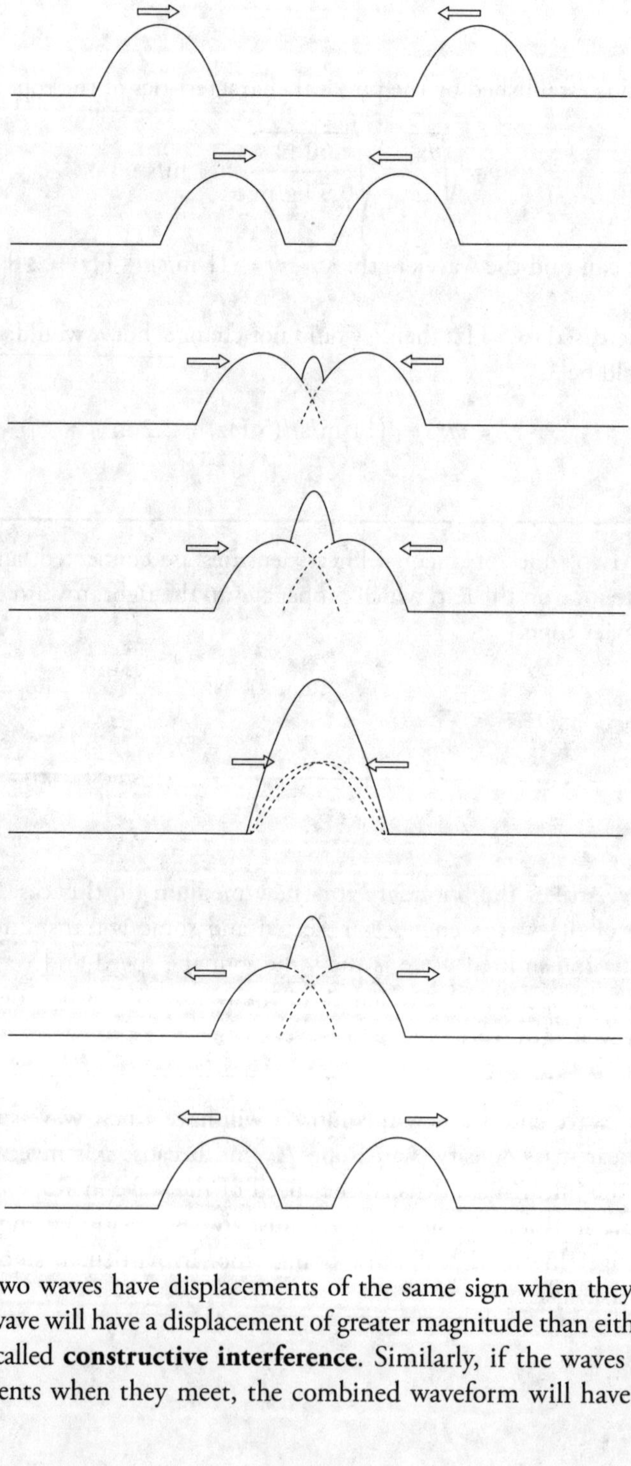

Constructive = Add

Destructive = Subtract

If the two waves have displacements of the same sign when they overlap, the combined wave will have a displacement of greater magnitude than either individual wave; this is called **constructive interference**. Similarly, if the waves have opposite displacements when they meet, the combined waveform will have a displacement of smaller magnitude than either individual wave; this is called **destructive interference**. If the waves travel in the same direction, the amplitude of the combined wave depends on the relative phase of the two waves. If the waves are exactly **in phase**—that is, if crest meets crest and trough meets trough—then the waves will constructively interfere completely, and the amplitude of the combined wave will be the sum of the individual amplitudes. However, if the waves are exactly **out of phase**—that is, if crest meets trough and trough meets crest—then they will destructively interfere completely, and the amplitude of the combined wave will be the difference between the individual amplitudes. In general, the waves will be somewhere in between exactly in phase and exactly out of phase.

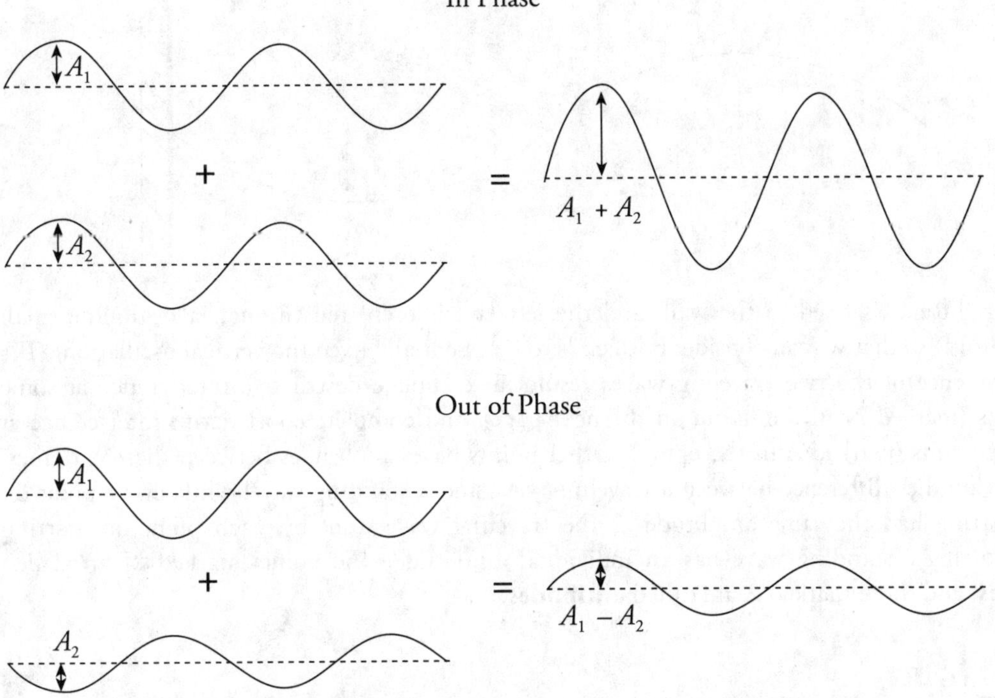

> **Example 5** Two waves, one with an amplitude of 8 cm and the other with an amplitude of 3 cm, travel in the same direction on a single string and overlap. What are the maximum and minimum possible amplitudes of the string while these waves overlap?

Solution. The maximum amplitude occurs when the waves are exactly in phase; the amplitude of the combined waveform will be 8 cm + 3 cm = 11 cm. The minimum amplitude occurs when the waves are exactly out of phase; the amplitude of the combined waveform will then be 8 cm – 3 cm = 5 cm. Without more information about the relative phase of the two waves, all we can say is that the amplitude will be at least 5 cm and no greater than 11 cm.

STANDING WAVES

When our prototype traveling wave on a string strikes the wall, the wave will reflect and travel back toward us. The string now supports two traveling waves: the wave we generated at our end, which travels toward the wall, and the reflected wave. What we actually see on the string is the superposition of these two oppositely directed traveling waves, which have the same frequency, amplitude, and wavelength. If the length of the string is just right, the resulting pattern will oscillate vertically and remain fixed. The crests and troughs no longer travel down the length of the string. This is a **standing wave**.

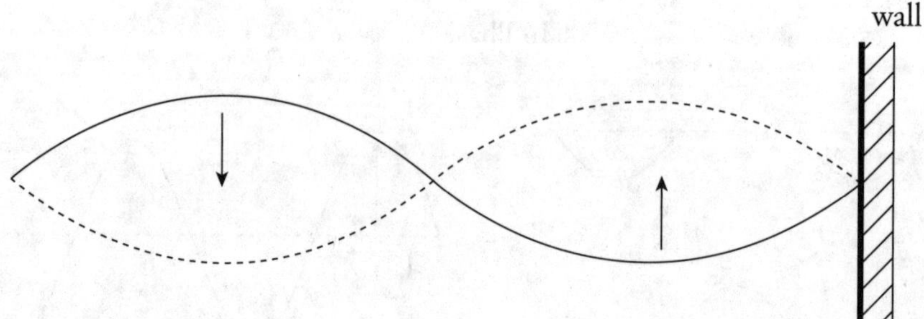

The right end is fixed to the wall, and the left end is oscillated through a negligibly small amplitude so that we can consider both ends to be essentially fixed (no vertical oscillation). The interference of the two traveling waves results in complete destructive interference at some points (marked N in the figure on the next page), and complete constructive interference at other points (marked A in the figure). Other points have amplitudes between these extremes. Note another difference between a traveling wave and a standing wave: while every point on the string had the same amplitude as the traveling wave went by, each point on a string supporting a standing wave has an individual amplitude. The points marked N are called **nodes**, and those marked A are called **antinodes**.

Tip
You can remember nodes as areas of "**no d**isplac**e**ment." Antinodes are the opposite of that.

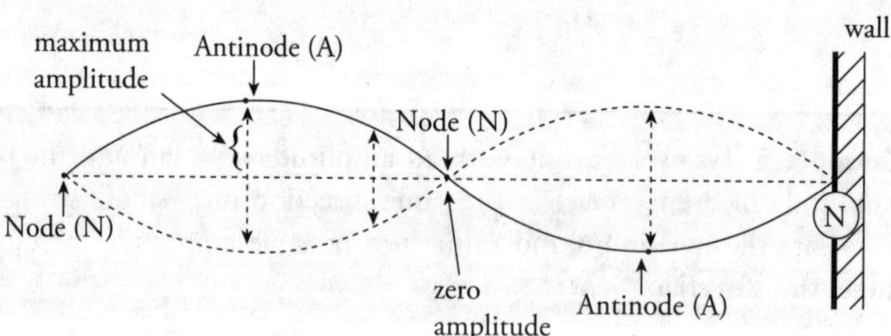

Nodes and antinodes always alternate, they're equally spaced, and the distance between two successive nodes (or antinodes) is equal to $\frac{1}{2}\lambda$. This information can be used to determine how standing waves can be generated. The following figures show the three simplest standing waves that our string can support. The first standing wave has one antinode, the second has two, and the third has three. The length of the string in all three diagrams is L.

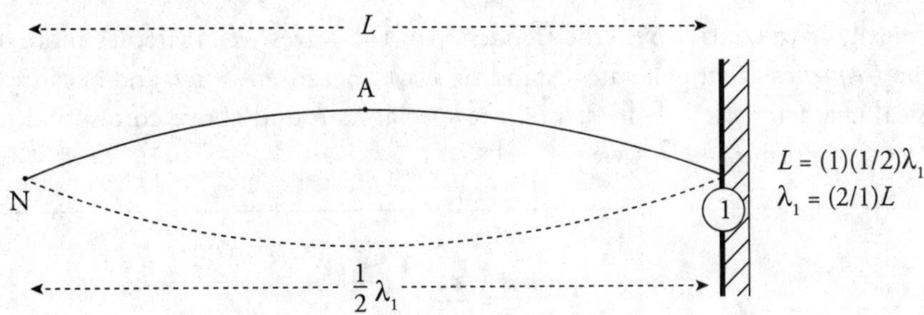

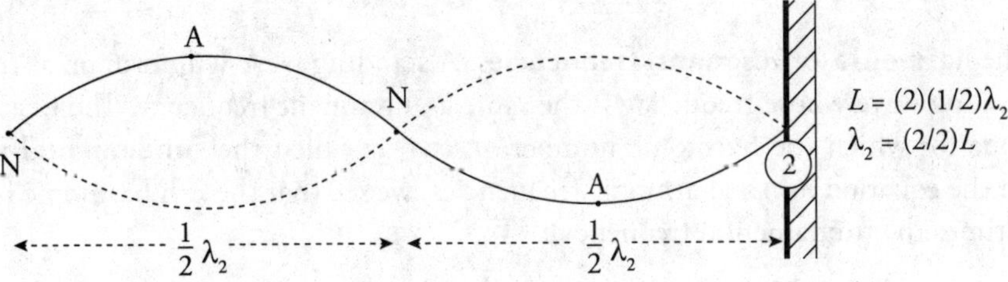

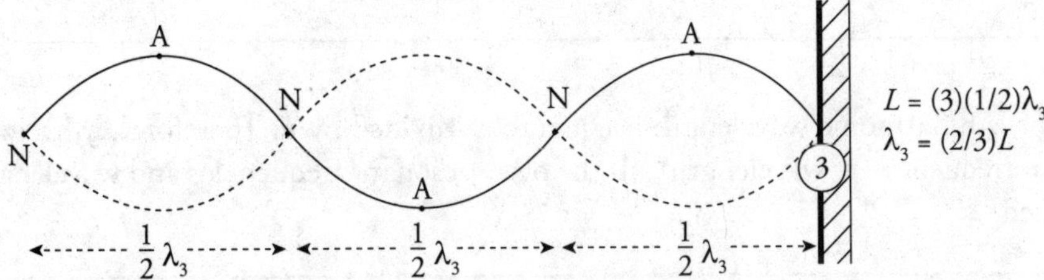

For the first standing wave, notice that L is equal to $1(\frac{1}{2}\lambda)$. For the second standing wave, L is equal to $2(\frac{1}{2}\lambda)$, and for the third, $L = 3(\frac{1}{2}\lambda)$. A pattern is established: a standing wave can form only when the length of the string is a multiple of $\frac{1}{2}\lambda$:

$$L = n(\frac{1}{2}\lambda)$$

Solving this for the wavelength, we get

$$\lambda_n = \frac{2L}{n}$$

These are called the **harmonic** (or **resonant**) **wavelengths**, and the integer n is known as the **harmonic number**.

Since we typically have control over the frequency of the waves we create, it's more helpful to figure out the *frequencies* that generate a standing wave. Because $\lambda = v/f$, and because v is fixed by the physical characteristics of the string, the special λs found above correspond to equally special frequencies. From $f_n = v/\lambda_n$, we get

$$f_n = \frac{nv}{2L}$$

These are the **harmonic** (or **resonant**) **frequencies**. A standing wave will form on a string if we create a traveling wave whose frequency is the same as a resonant frequency. The first standing wave, the one for which the harmonic number, n, is 1, is called the **fundamental standing wave**. From the equation for the harmonic frequencies, we see that the nth harmonic frequency is simply n times the fundamental frequency:

$$f_n = nf_1$$

Similarly, the nth harmonic wavelength is equal to λ_1 divided by n. Therefore, by knowing the fundamental frequency (or wavelength), all the other resonant frequencies and wavelengths can be determined.

> **Example 6** A string of length 12 m that's fixed at both ends supports a standing wave with a total of 5 nodes. What are the harmonic number and wavelength of this standing wave?

Solution. First, draw a picture.

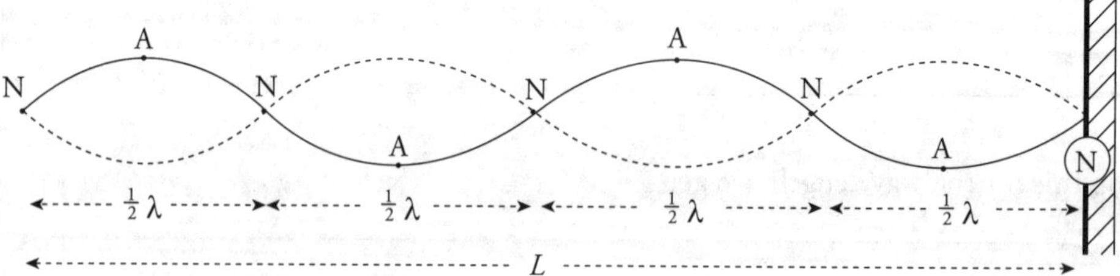

This shows that the length of the string is equal to $4(\frac{1}{2}\lambda)$, so

$$L = 4(\frac{1}{2}\lambda) \quad \Rightarrow \quad \lambda = \frac{2L}{4}$$

This is the fourth-harmonic standing wave, with wavelength λ_4 (because the expression above matches $\lambda_n = 2L/n$ for $n = 4$). Since $L = 12$ m, the wavelength is

$$\lambda_4 = \frac{2(12 \text{ m})}{4} = 6 \text{ m}$$

Example 7 A string of length 10 m and mass 300 g is fixed at both ends, and the tension in the string is 40 N. What is the frequency of the standing wave for which the distance between a node and the closest antinode is 1 m?

Solution. Because the distance between two successive nodes (or successive antinodes) is equal to $\frac{1}{2}\lambda$, the distance between a node and the closest antinode is half this, or $\frac{1}{4}\lambda$. Therefore, $\frac{1}{4}\lambda = 1$ m, so $\lambda = 4$ m. Since the harmonic wavelengths are given by the equation $\lambda_n = 2L/n$, we can find that

$$4 \text{ m} = \frac{2(10 \text{ m})}{n} \quad \Rightarrow \quad n = 5$$

The frequency of the fifth harmonic is

$$f_5 = \frac{5v}{2L} = \frac{5}{2L}\sqrt{\frac{F_T}{\mu}} = \frac{5}{2L}\sqrt{\frac{F_T}{m/L}} = \frac{5}{2(10 \text{ m})}\sqrt{\frac{40 \text{ N}}{(0.3 \text{ kg}/10 \text{ m})}} = 9.1 \text{ Hz}$$

If you attached a rope or string to a ring that could slide up and down a pole without friction, you would make a rope that is fixed at one end but free at another end. This would create nodes at the closed end and antinodes at the open end. Below are some possible examples of this.

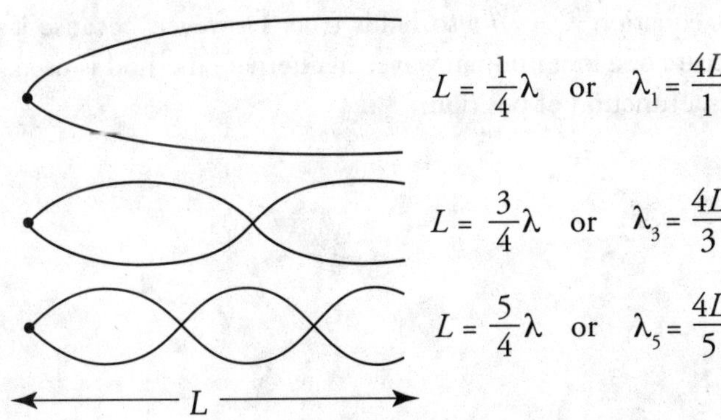

If one end of the rope is free to move, standing waves can form for wavelengths of $\lambda_n = \frac{4L}{n}$ or by frequencies of $f_n = \frac{nv}{4L}$, where L is the length of the rope, v is the speed of the wave, and n must be an odd integer.

SOUND WAVES

Speed of Sound
At sea level, approximately 340 m/s.

Sound waves are produced by the vibration of an object, such as your vocal cords, a plucked string, or a jackhammer. The vibrations cause pressure variations in the conducting medium (which can be gas, liquid, or solid), and if the frequency is between 20 Hz and 20,000 Hz, the vibrations may be detected by human ears and perceived as sound. The variations in the conducting medium can be positions at which the molecules of the medium are bunched together (where the pressure is above normal), which are called **compressions**, and positions where the pressure is below normal, called **rarefactions**. In the figure below, a vibrating diaphragm sets up a sound wave in an air-filled tube. Each dot represents many, many air molecules:

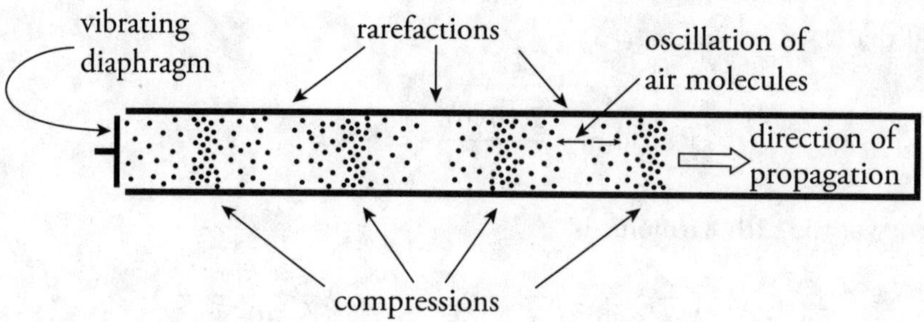

Longitudinal Waves
A longitudinal wave travels and oscillates in the same direction.

An important difference between sound waves and the waves we've been studying on stretched strings is that the molecules of the medium transmitting a sound wave move *parallel* to the direction of wave propagation, rather than perpendicular to it. For this reason, sound waves are said to be **longitudinal**. Despite this difference, all of the basic characteristics of a wave—amplitude, wavelength, period, frequency—apply to sound waves as they did for waves on a string. Furthermore, the all-important equation $\lambda = v/f$ also holds true. However, because it's very difficult to draw a picture of a longitudinal wave, an alternate method is used: we can graph the pressure as a function of position:

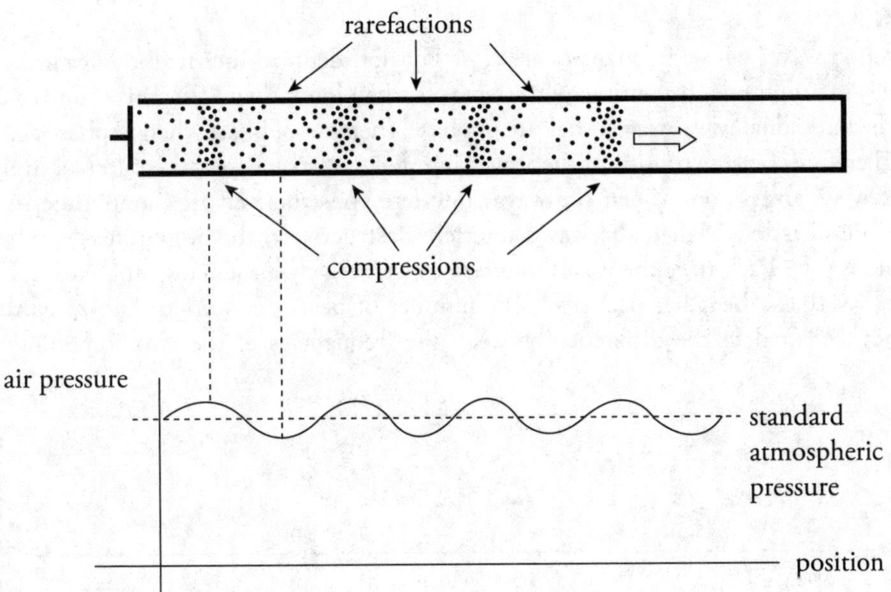

The speed of a sound wave depends on the medium through which it travels. At room temperature (approximately 20°C) and normal atmospheric pressure, sound travels at 343 m/s. This value increases as air warms.

Example 8 A sound wave with a frequency of 300 Hz travels through the air.

(a) What is its wavelength?

(b) If its frequency increased to 600 Hz, what are the wave's speed and wavelength?

Solution.

(a) Using v = 343 m/s for the speed of sound through air, we find that

$$\lambda = \frac{v}{f} = \frac{343 \text{ m/s}}{300 \text{ Hz}} = 1.14 \text{ m}$$

(b) Unless the ambient pressure or the temperature of the air changed, the speed of sound would not change. Wave speed depends on the characteristics of the medium, not on the frequency, so v would still be 343 m/s. However, a change in frequency would cause a change in wavelength. Since f increased by a factor of 2, λ would decrease by a factor of 2, to $\frac{1}{2}$(1.14 m) = 0.57 m.

Example 9 A sound wave traveling through water has a frequency of 500 Hz and a wavelength of 3 m. How fast does sound travel through water?

Solution. $v = \lambda f$ = (3 m)(500 Hz) = 1500 m/s.

Beats

If two sound waves whose frequencies are close but not identical interfere, the resulting sound modulates in amplitude, becoming loud, then soft, then loud, then soft. This is due to the fact that as the individual waves travel, they are in phase, then out of phase, then in phase again, and so on. Therefore, by superposition, the waves interfere constructively, then destructively, then constructively, and so on. When the waves interfere constructively, the amplitude increases, and the sound is loud; when the waves interfere destructively, the amplitude decreases, and the sound is soft. Each time the waves interfere constructively, producing an increase in sound level, we say that a **beat** has occurred. The number of beats per second, known as the **beat frequency**, is equal to the difference between the frequencies of the two combining sound waves:

$$f_{\text{beat}} = |f_1 - f_2|$$

If frequencies f_1 and f_2 match, then the combined waveform doesn't waver in amplitude, and no beats are heard.

Example 10 A piano tuner uses a tuning fork to adjust the key that plays the A note above middle C (whose frequency should be 440 Hz). The tuning fork emits a perfect 440 Hz tone. When the tuning fork and the piano key are struck, beats of frequency 3 Hz are heard.

(a) What is the frequency of the piano key?

(b) If it's known that the piano key's frequency is too high, should the piano tuner tighten or loosen the wire inside the piano in order to tune it?

Solution.

(a) Since $f_{\text{beat}} = 3$ Hz, the tuning fork and the piano string are off by 3 Hz. Since the fork emits a tone of 440 Hz, the piano string must emit a tone of either 437 Hz or 443 Hz. Without more information, we can't determine which.

(b) If we know that the frequency of the tone emitted by the out-of-tune string is too high (that is, it's 443 Hz), we need to find a way to lower the frequency. Remember that the resonant frequencies for a stretched string fixed at both ends are given by the equation $f = nv/2L$, and that $v = \sqrt{F_T/\mu}$. Since f is too high, v must be too high. To lower v, we must reduce F_T. The piano tuner should loosen the string and listen for beats again, adjusting the string until the beats disappear.

RESONANCE FOR SOUND WAVES

Just as standing waves can be set up on a vibrating string, standing *sound* waves can be established within an enclosure. In the figure below, a vibrating source at one end of an air-filled tube produces sound waves that travel the length of the tube.

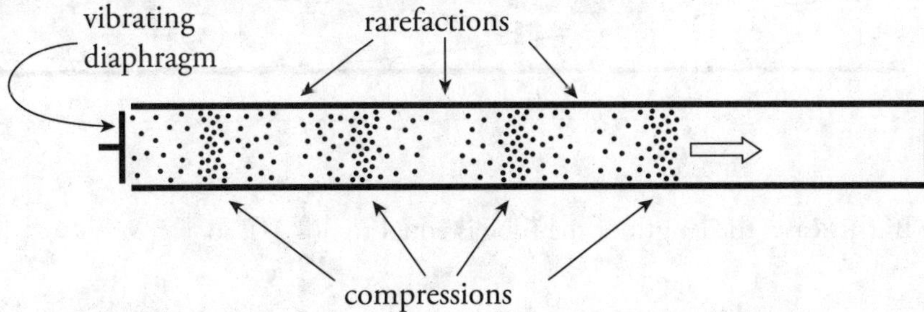

These waves reflect off the far end, and the superposition of the forward and reflected waves can produce a standing wave pattern if the length of the tube and the frequency of the waves are related in a certain way.

Notice that air molecules at the far end of the tube can't oscillate horizontally because they're up against a wall. So, the far end of the tube is a displacement node. But the other end of the tube (where the vibrating source is located) is a displacement antinode. A standing wave with one antinode (A) and one node position (N) can be depicted as follows:

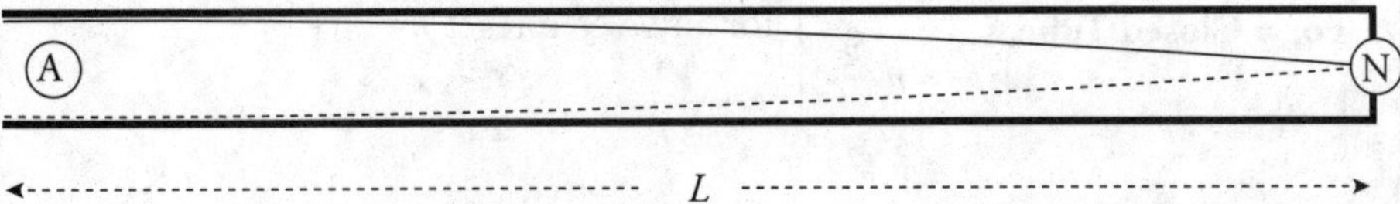

Although sound waves in air are longitudinal, here we'll show the wave as transverse so that it's easier to determine the wavelength. Since the distance between an antinode and an adjacent node is always $\frac{1}{4}$ of the wavelength, the length of the tube, L, in the figure above is $\frac{1}{4}$ the wavelength. This is the longest standing wavelength that can fit in the tube, so it corresponds to the lowest standing wave frequency, the fundamental:

$$L = \frac{\lambda_1}{4} \Rightarrow \lambda_1 = 4L \Rightarrow f_1 = \frac{v}{\lambda_1} = \frac{v}{4L}$$

Our condition for resonance was a node at the closed end and an antinode at the open end. Therefore, the next higher-frequency standing wave that can be supported in this tube must have two antinodes and two nodes:

Make sure you're comfortable with drawing these standing wave diagrams in addition to understanding the equations. Free-response questions often require you to produce images.

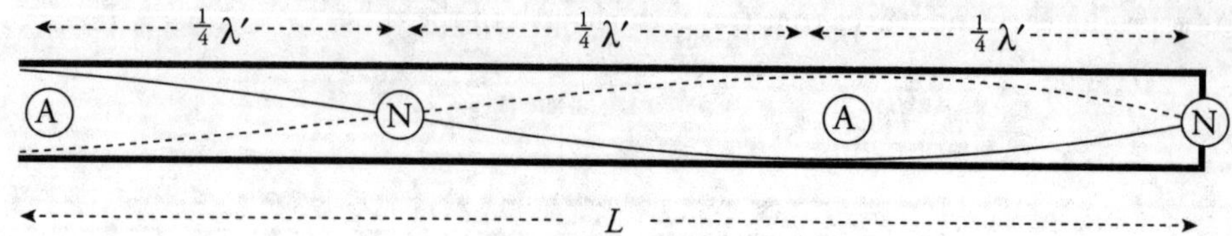

In this case, the length of the tube is equal to $3(\frac{1}{4}\lambda')$, so

$$L = \frac{3\lambda'}{4} \Rightarrow \lambda' = \frac{4L}{3} \Rightarrow f' = \frac{v}{\lambda'} = \frac{3v}{4L}$$

Here's the pattern: standing sound waves can be established in a tube that's closed at one end if the tube's length is equal to an *odd* multiple of $\frac{1}{4}\lambda$. The resonant wavelengths and frequencies are given by the following equations:

For a Closed Tube, $\left.\begin{aligned}\lambda_n &= \frac{4L}{n} \\ f_n &= n\frac{v}{4L}\end{aligned}\right\}$ for any *odd* integer n

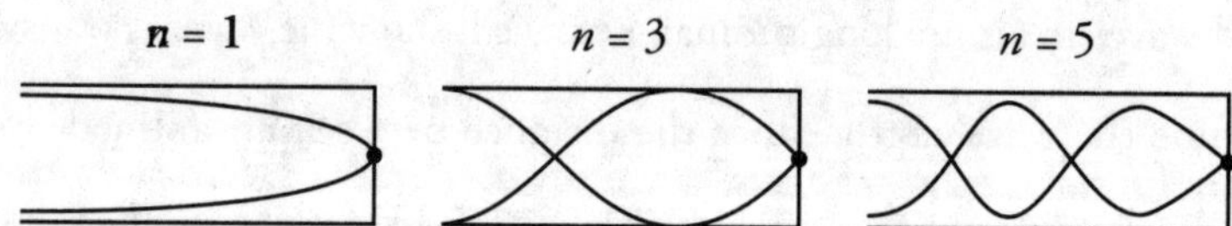

If the far end of the tube is not sealed, standing waves can still be established in the tube, because sound waves can be reflected from the open air. A closed end is a displacement node, but an open end is a displacement antinode. In this case, then, the standing waves will have two displacement antinodes (at the ends of the tube), and the resonant wavelengths and frequencies will be given by

For an Open Tube, $\left.\begin{aligned} \lambda_n &= \frac{2L}{n} \\ f_n &= n\frac{v}{2L} \end{aligned}\right\}$ for any integer n

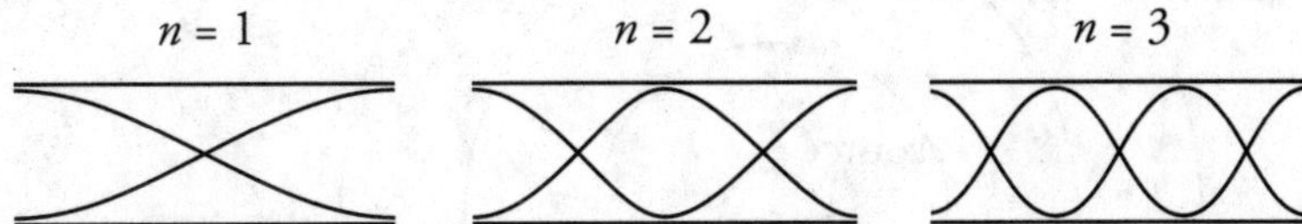

Note that, while an open-end tube can support any harmonic, a closed-end tube can support only odd harmonics.

Example 11 A closed-end tube resonates at a fundamental frequency of 440.0 Hz. The air in the tube is at a temperature of 20°C, and it conducts sound at a speed of 343 m/s.

(a) What is the length of the tube?

(b) What is the next higher harmonic frequency?

(c) Answer the questions posed in (a) and (b) assuming that the tube is open at its far end.

Solution.

(a) For a closed-end tube, the harmonic frequencies obey the equation $f_n = nv/(4L)$. The fundamental corresponds to $n = 1$, so

$$f_1 = \frac{v}{4L} \quad \Rightarrow \quad L = \frac{v}{4f_1} = \frac{343 \text{ m/s}}{4(440.0 \text{ Hz})} = 0.195 \text{ m} = 19.5 \text{ cm}$$

(b) Since a closed-end tube can support only *odd* harmonics, the next higher harmonic frequency (the first **overtone**) is the *third* harmonic, f_3, which is

$$3f_1 = 3(440.0 \text{ Hz}) = 1320 \text{ Hz}$$

(c) For an open-end tube, the harmonic frequencies obey the equation $f_n = nv/(2L)$. The fundamental corresponds to $n = 1$, so

$$f_1 = \frac{v}{2L'} \quad \Rightarrow \quad L' = \frac{v}{2f_1} = \frac{343 \text{ m/s}}{2(440.0 \text{ Hz})} = 0.390 \text{ m} = 39.0 \text{ cm}$$

And, since an open-end tube can support any harmonic, the first overtone would be the second harmonic, $f_2 = 2f_1 = 2(440.0 \text{ Hz}) = 880.0$ Hz.

THE DOPPLER EFFECT

When a source of sound waves and a detector are not in relative motion, the frequency that the source emits matches the frequency that the detector receives.

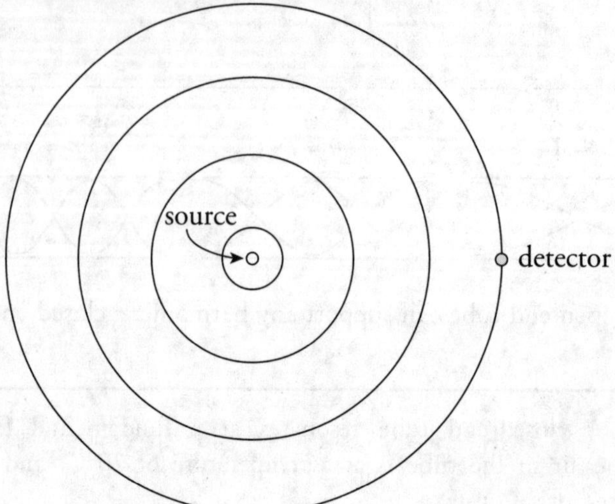

However, if there *is* relative motion between the source and the detector, then the waves that the detector receives are different in frequency (and wavelength). For example, if the detector moves toward the source, then the detector intercepts the waves at a rate higher than the one at which they were emitted; the detector hears a higher frequency than the source emitted. In the same way, if the source moves toward the detector, the wavefronts pile up, and this results in the detector receiving waves with shorter wavelengths and higher frequencies:

You're almost done with this chapter! If you haven't taken a break in a while, consider putting this book aside to take a walk, get some fresh air, or unwind by listening to music or reading a book (not this one!). Mental breaks are essential to test prep!

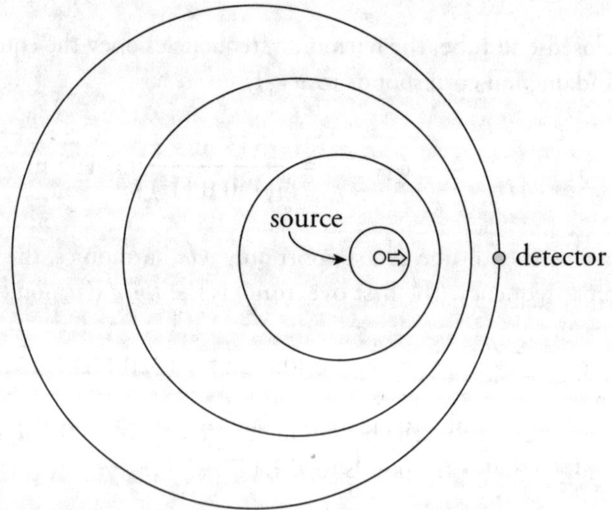

Conversely, if the detector is moving away from the source or if the source is moving away from the detector,

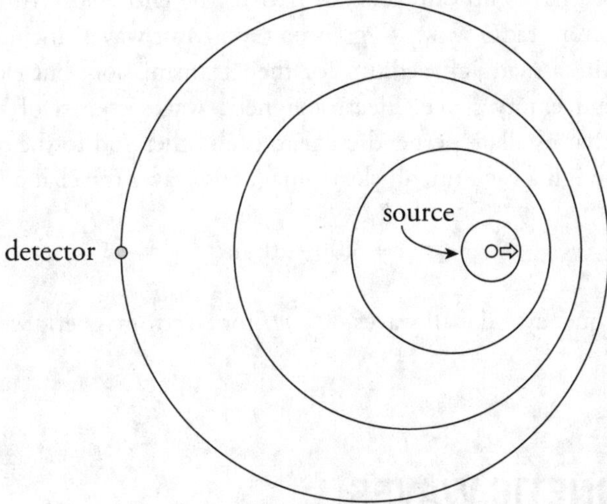

then the detected waves have a longer wavelength and a lower frequency than they had when they were emitted by the source. The shift in frequency and wavelength that occurs when the source and detector are in relative motion is known as the **Doppler effect**. In general, relative motion *toward* each other results in a higher perceived frequency, and relative motion *away* from each other results in a lower perceived frequency.

While equations for solving the changes due to Doppler effect exist, you don't need to know them for the AP Exam. Only a qualitative understanding is required.

Example 12 A ambulance blares a siren at a constant frequency, f_0, describe how the frequency detected would differ from f_0 for

(a) A person standing by the road before the ambulance passes them.

(b) A person standing by the road after the ambulance passes them.

(c) A car behind the ambulance going faster than the ambulance and accelerating.

Solution. The frequency detected is higher if the detector and source have relative motion toward each other, and lower if the detector and source have relative motion away from each other, so

(a) The ambulance is approaching the person by the road, so they would detect a frequency higher than f_0.

(b) The ambulance is now leaving the person by the road, so they would detect a frequency lower than f_0.

(c) The car is catching up to the ambulance, so they would detect a frequency higher than f_0, and the detected frequency would increase even more as the car sped up.

PHYSICAL OPTICS

Light (or visible light) makes up only a small part of the entire spectrum of electromagnetic waves, which ranges from radio waves to gamma rays. Most waves, including those we've discussed thus far, require a material medium for their transmission, but electromagnetic waves can propagate through empty space. Electromagnetic waves consist of time-varying electric and magnetic fields that oscillate perpendicular to each other and to the direction of propagation of the wave. Through a vacuum, all electromagnetic waves travel at a fixed speed:

$$c = 3.00 \cdot 10^8 \text{ m/s}$$

regardless of their frequency. Like all waves, $\lambda = v/f$ for electromagnetic waves.

ELECTROMAGNETIC WAVES

We saw previously that if we oscillate one end of a long rope, we generate a wave that travels down the rope and has the same frequency as that of the oscillation.

You can think of an electromagnetic wave in a similar way: an oscillating electric charge generates an electromagnetic (EM) wave, which is composed of oscillating electric and magnetic fields. These fields oscillate with the same frequency at which the electric charge that created the wave oscillated. The fields oscillate in phase with each other, perpendicular to each other and the direction of propagation. For this reason, electromagnetic waves are transverse waves. The direction in which the wave's electric field oscillates is called the direction of polarization of the wave.

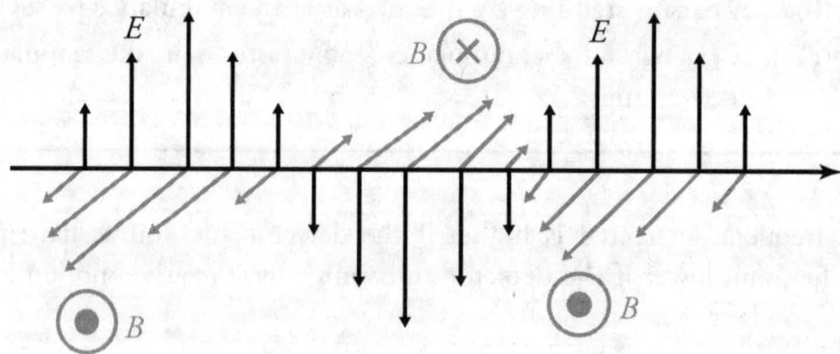

THE ELECTROMAGNETIC SPECTRUM

Electromagnetic waves can be categorized by their frequency (or wavelength); the full range of waves is called the **electromagnetic** (or **EM**) **spectrum**. Types of waves include **radio waves**, **microwaves**, **infrared**, **visible light**, **ultraviolet**, **X-rays**, and **γ-rays** (**gamma rays**) and, although they've been delineated in the spectrum below, there's no universal agreement on all the boundaries, so many of these bands overlap. You should be familiar with the names of the major categories, and, in particular, memorize the order of the colors within the visible spectrum (which, as you can see, accounts for only a tiny sliver of the full EM spectrum). In order of increasing wave frequency, the colors are red, orange, yellow, green, blue, and violet, which is commonly remembered as ROYGBV ("roy-gee-biv"). The wavelengths of the colors in the visible spectrum are usually expressed in nanometers (nm). For example, electromagnetic waves whose wavelengths are between 577 nm and 597 nm are seen as yellow light.

Electromagnetic Wave Speed
All electromagnetic waves, regardless of frequency, travel through a vacuum at this speed. The most important equation for waves, $\lambda = v/f$, is also true for electromagnetic waves. For EM waves traveling through a vacuum, $v = c$, so the equation becomes $\lambda f = c$.

Visible Spectrum
White light is not colorless. White light is actually composed of a combination of all of the colors of the visible spectrum.

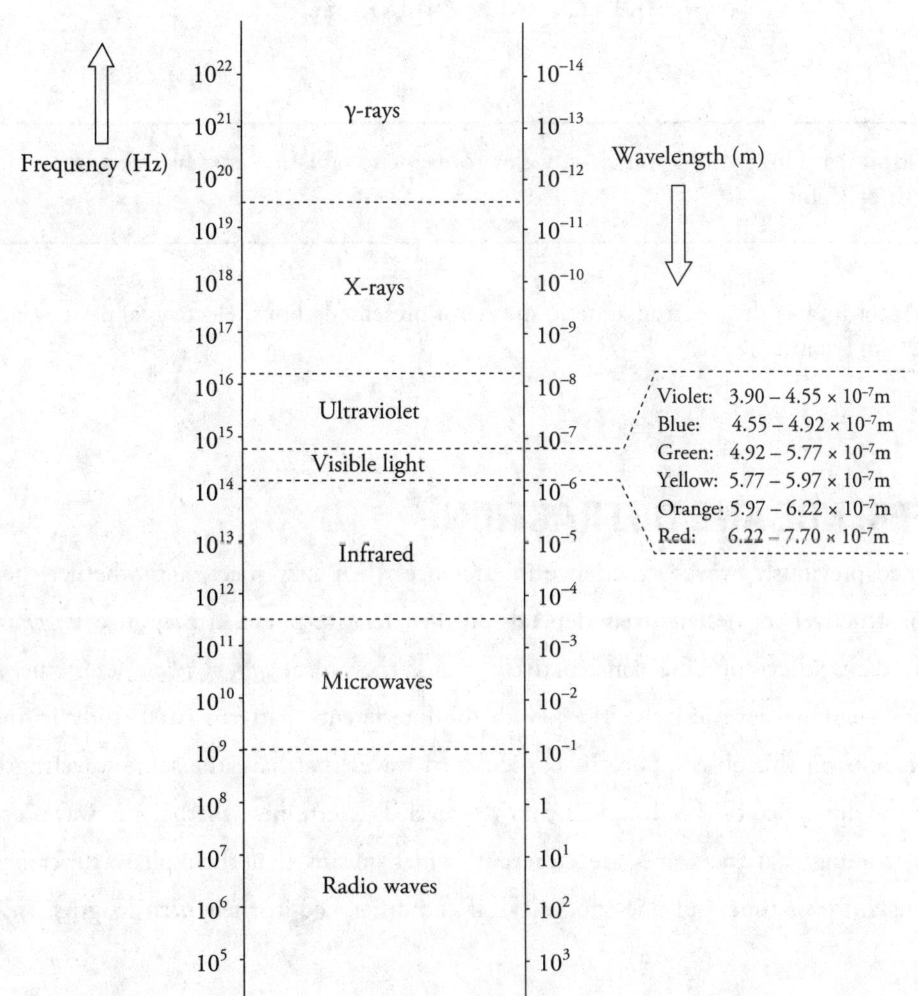

Example 13 What's the frequency range for green light?

Solution. According to the spectrum, light is green if its wavelength is between 4.92×10^{-7} m and 5.77×10^{-7} m. Using the equation $\lambda = v/f$, we find that the upper end of this wavelength range corresponds to a frequency of

$$f_1 = \frac{v}{\lambda_1} = \frac{3.00 \times 10^8 \text{ m/s}}{5.77 \times 10^{-7} \text{ m}} = 5.20 \times 10^{14} \text{ Hz}$$

while the lower end corresponds to

$$f_2 = \frac{v}{\lambda_2} = \frac{3.00 \times 10^8 \text{ m/s}}{4.92 \times 10^{-7} \text{ m}} = 6.10 \times 10^{14} \text{ Hz}$$

So, the frequency range for green light is

$$5.20 \times 10^{14} \text{ Hz} \le f_{\text{green}} \le 6.10 \times 10^{14} \text{ Hz}$$

Example 14 How would you classify electromagnetic radiation that has a wavelength of 1 cm?

Solution. According to the electromagnetic spectrum presented above, electromagnetic waves with $\lambda = 10^{-2}$ m are microwaves.

INTERFERENCE AND DIFFRACTION

As we learned previously, waves experience interference when they meet, and whether they interfere constructively or destructively depends on their relative phase. If they meet *in phase* (crest meets crest), they combine constructively, but if they meet *out of phase* (crest meets trough), they combine destructively. The key to the interference patterns we'll study in the next section rests on this observation. In particular, if waves that have the same wavelength meet, then the difference in the distances they've traveled determines whether the waves are in phase. Assuming that the waves are **coherent** (which means that their phase difference remains constant over time and does not vary), if the difference in their path lengths, ΔD, is a whole number of wavelengths—0, $\pm\lambda$, $\pm 2\lambda$, etc.—they'll arrive in phase at the meeting point. On the other hand, if this difference is a whole number plus one-half a wavelength—$\pm\frac{1}{2}\lambda$, $\pm(1+\frac{1}{2})\lambda$, $\pm(2+\frac{1}{2})\lambda$, etc.—then they'll arrive exactly out of phase. That is,

$$\left.\begin{array}{ll} \text{constructive interference:} & \Delta D = m\lambda \\ \text{destructive interference:} & \Delta D = (m+\frac{1}{2})\lambda \end{array}\right\} \text{where } m = 0, 1, 2\ldots$$

Young's Double-Slit Interference Experiment

The following figure shows incident light on a barrier that contains two narrow slits (perpendicular to the plane of the page), separated by a distance *d*. On the right is a screen whose distance from the barrier, *L*, is much greater than *d*. The question is, what will we see on the screen? You might expect that we'll just see two bright narrow strips of light, directly opposite the slits in the barrier. As reasonable as this may sound, it doesn't take into account the wave nature of light.

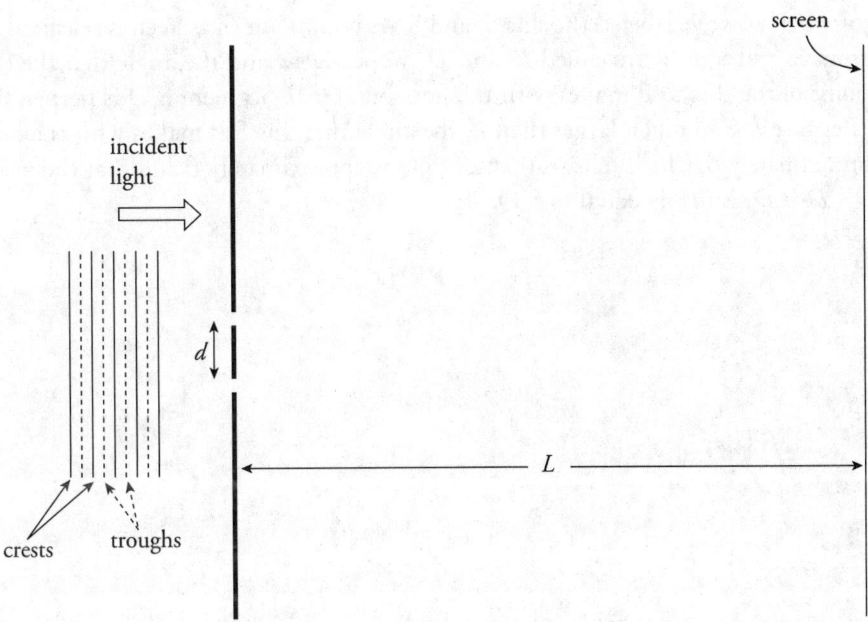

When a wave encounters an aperture whose width is comparable to its wavelength, the wave will fan out after it passes through. The alteration in the straight-line propagation of a wave when it encounters a barrier is called **diffraction**. In the setup above, the waves will diffract through the slits, and spread out and interfere as they travel toward the screen.

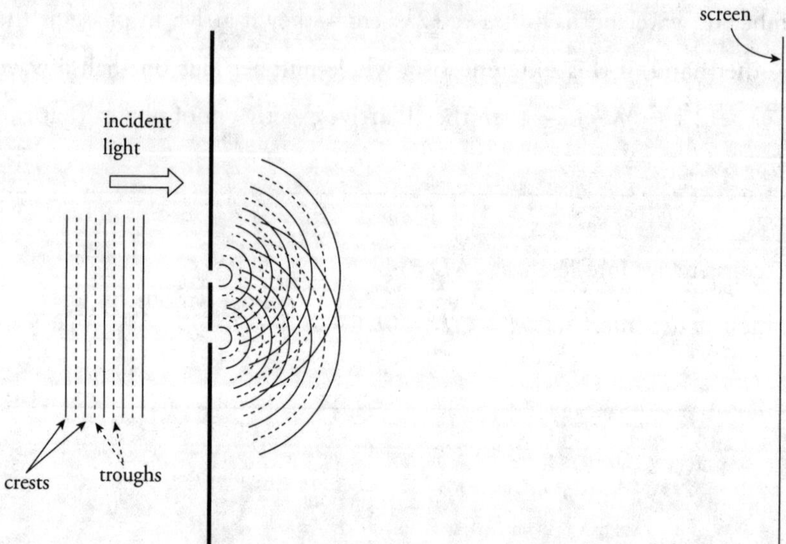

You can clearly see points of interference in the figure above. Look for solid lines intersecting other solid lines—these are points of constructive interference. Look for solid lines intersecting dashed lines—these are points of destructive interference.

The screen will show the results of this interference: there will be bright bands (bright **fringes**) centered at those points at which the waves interfere constructively, alternating with dark fringes, where the waves interfere destructively. Let's determine the locations of these fringes.

In the figure below, we've labeled the slits S_1 and S_2. A point P on the screen is selected, the path lengths from S_1 and S_2 to P are called D_1 and D_2, respectively, and the angle that the line from the midpoint of the slits to P makes with the horizontal is θ. Segment S_1Q is perpendicular to line S_2P. Because L is so much larger than d, the angle that line S_2P makes with the horizontal is also approximately θ, which tells us that $\angle S_2S_1Q$ is approximately θ and that the path difference, $\Delta D = D_2 - D_1$, is nearly equal to S_2Q.

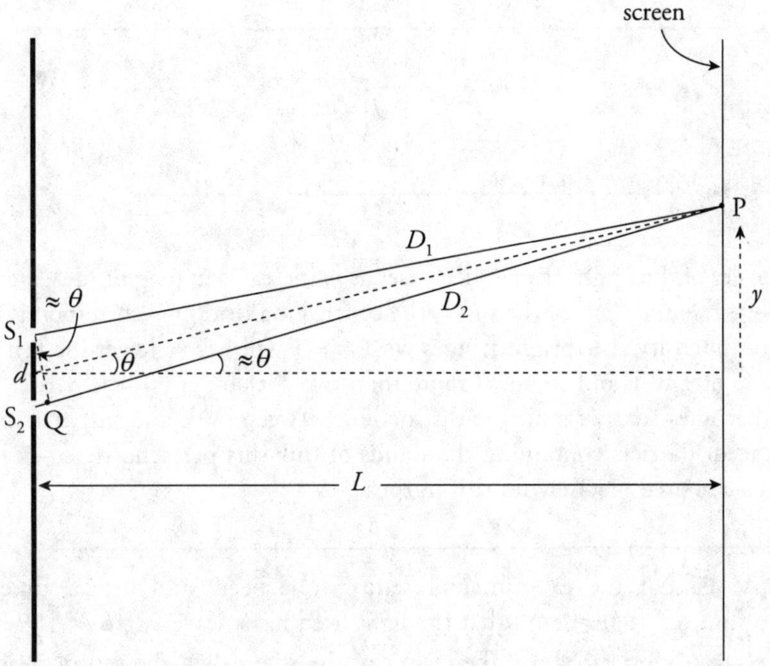

From what we learned earlier about constructive interference, there will be an intensity maximum when the path difference is some multiple of the wavelength, λ. Because the path difference is $S_2Q = d\sin\theta$, we can write:

constructive interference
(intensity maximum
bright fringe on screen): $\Delta D = m\lambda = d\sin\theta$ where $m = 0, 1, 2...$

Equation Sheet

Similarly, there will be an intensity maximum when the path difference is half a wavelength off, so

destructive interference
(intensity minimum
dark fringe on screen): $\Delta D = (m + \frac{1}{2})\lambda = d\sin\theta$ where $m = 0, 1, 2...$

Equation Sheet

To locate the positions of the bright fringes on the screen, we use the fact that $\tan\theta = \frac{y}{L}$. If θ is small, then $\tan\theta \approx \sin\theta$, so we can write $\sin\theta = \frac{y}{L}$ (we can tell this from the figure). Since $d\sin\theta = m\lambda$ for bright fringes, we get

Equation Sheet

$$d\left(\frac{y_{max}}{L}\right) = m\lambda$$

Also, the intensity of the bright fringes decreases as m increases in magnitude. The bright fringe directly opposite the midpoint of the slits—the **central maximum**—corresponds to $m = 0$ and has the greatest intensity. The bright fringes with $m = 1$ will have a lower intensity, those with $m = 2$ will be fainter still, and so on. If more than two slits are cut in the barrier, the interference pattern becomes sharper, and the distinction between dark and bright fringes becomes more pronounced. Barriers containing thousands of tiny slits per centimeter—called **diffraction gratings**—are used precisely for this purpose.

Example 15 For the experimental setup we've been studying, assume that $d = 1.5$ mm, $L = 6.0$ m, and that the light used has a wavelength of 589 nm.

(a) How far above the center of the screen will the second brightest maximum appear?

(b) How far below the center of the screen is the third dark fringe?

(c) What would happen to the interference pattern if the slits were moved closer together?

Solution.

(a) The central maximum corresponds to $m = 0$ ($y = 0$). The first maximum above the central one is labeled y_1 (since $m = 1$). The other bright fringes on the screen are labeled accordingly.

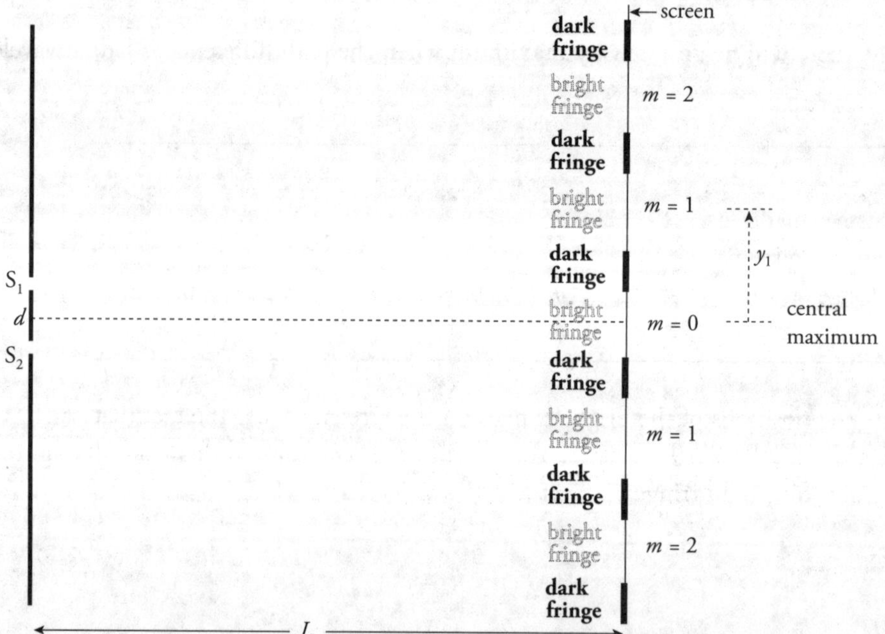

The value of y_1 is

$$y_1 = \frac{1 \cdot \lambda L}{d} = \frac{(589 \times 10^{-9}\ \text{m})(6.0\ \text{m})}{1.5 \times 10^{-3}\ \text{m}} = 2.4 \times 10^{-3}\ \text{m} = 2.4\ \text{mm}$$

(b) The first dark fringe occurs when the path difference is 0.5λ, the second dark fringe occurs when the path difference is 1.5λ, and the third dark fringe occurs when the path difference is 2.5λ, so

$$y_{\substack{\text{3rd minimum} \\ \text{below central max}}} = \frac{(2+\frac{1}{2})\lambda L}{d} = \frac{(2+\frac{1}{2})(589 \times 10^{-9}\ \text{m})(6.0\ \text{m})}{1.5 \times 10^{-3}\ \text{m}} = 5.9 \times 10^{-3}\text{m} = 5.9\ \text{mm}$$

(c) Since $y = m\lambda L/d$, a decrease in d would cause an increase in y_{max}. That is, the fringes would become larger; the interference pattern would be more spread out.

Single-Slit Diffraction

A diffraction pattern will also form on the screen if the barrier contains only one slit, as long as the size of the slit is roughly comparable to the wavelength. This occurs because, as the light passes through the slit, it can be treated as multiple light rays spread along the slit. The path difference for each ray can be found using a similar method to what we saw in the double-slit case:

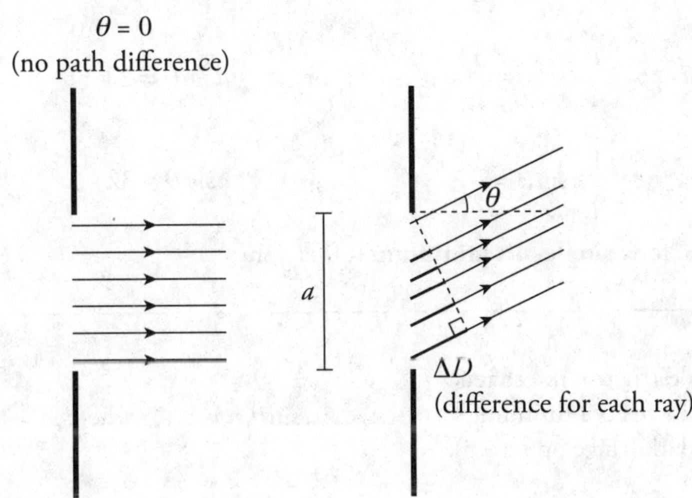

In the case of $\theta = 0$, there is no path difference between the rays and there will be a very bright central maximum from constructive interference. When going at an angle, the path difference between the ray at the top and the bottom will, by the same argument from the double-slit case, be $a\sin\theta$. However, because each ray will have a different path difference, it becomes harder to see when they will all combine constructively or destructively; and, indeed, for most angles it will be a combination of constructive and destructive interference and it will show a dim light.

There is a case where the rays will combine to get near-perfect destructive interference. To see how, we can pair the rays off so that the vertical distance between them is always $a/2$:

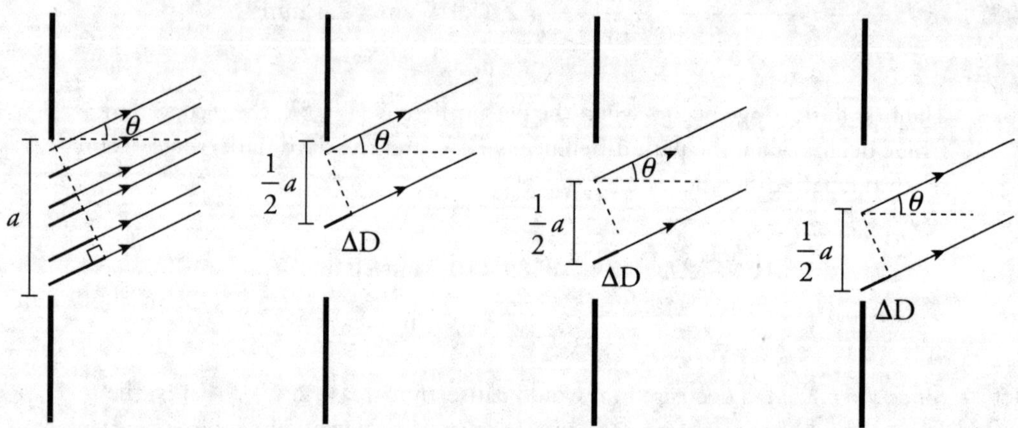

Now the path difference, ΔD, is always the same for each pair, and it will be $\frac{1}{2}a\sin\theta$. When this path difference is exactly half a wavelength, the pairs combine destructively and we get a dark fringe on the screen. This is the first angle where we get a minimum. We can repeat this process by pairing up rays with a vertical distance of $a/4$, $a/6$, $a/8$, etc. (Note: It needs to be even to avoid having constructive interference.) When the path distance between these pairs is half a wavelength, there will be a minimum in the intensity:

First minimum: $\frac{1}{2}a\sin\theta = \frac{1}{2}\lambda$ or $a\sin\theta = \lambda$

Second minimum: $\frac{1}{4}a\sin\theta = \frac{1}{2}\lambda$ or $a\sin\theta = 2\lambda$

Third minimum: $\frac{1}{6}a\sin\theta = \frac{1}{2}\lambda$ or $a\sin\theta = 3\lambda$

The general pattern for a **single-slit minimum** is therefore

Equation Sheet

destructive interference (intensity minimum dark fringe on screen): $a\sin\theta = m\lambda$ where m = 1, 2, 3...

For a circular pinhole aperture, the diffraction pattern will consist of a central, bright, circular disk surrounded by rings of decreasing intensity. The equation above still gives the angles of the dark rings.

Example 16 A single-slit experiment is performed by shining a light through a pinhole with diameter of 2 μm. The screen is placed 5 m from the aperture, and the central maximum shows up as a bright disk with diameter of 2.5 m on the screen. What is the wavelength of the light?

Solution.

The center of the bright disk will correspond to when the angle, θ, is zero. The distance to the first dark fringe on the screen, y_1, is the distance to the edge of the disk and so will be 1.25 m, the radius of the disk. This occurs when $m = 1$.

$$a \sin\theta = \lambda$$

For small angles, we can use the approximation of $\sin\theta \approx \tan\theta = \frac{y_1}{L}$, so

$$\lambda = a\left(\frac{y_1}{L}\right) = (2\times10^{-6}\text{ m})\frac{1.25\text{ m}}{5\text{ m}} = 5\times10^{-7}\text{ m} = 500\text{ nm}$$

This can be used in general, where the distance to the mth dark fringe on the screen is given by

$$a\left(\frac{y_{\min}}{L}\right) = m\lambda$$

Equation Sheet

Chapter 9 Review Questions

Answers and explanations can be found in Chapter 11.

Section I: Multiple Choice

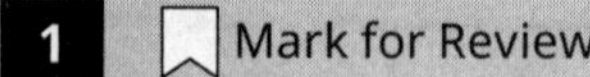

What is the wavelength of a 5 Hz wave that travels with a speed of 10 m/s?

(A) 0.25 m

(B) 0.5 m

(C) 2 m

(D) 50 m

2 Mark for Review

A rope of length 5 m is stretched to a tension of 80 N. If its mass is 1 kg, at what speed would a 10 Hz transverse wave travel down the string?

(A) 2 m/s

(B) 5 m/s

(C) 20 m/s

(D) 200 m/s

3 Mark for Review

A transverse wave on a long horizontal rope with a wavelength of 8 m travels at 2 m/s. At $t = 0$, a particular point on the rope has a vertical displacement of $+A$, where A is the amplitude of the wave. At what time will the vertical displacement of this same point on the rope be $-A$?

(A) $t = \frac{1}{4}$ s

(B) $t = \frac{1}{2}$ s

(C) $t = 2$ s

(D) $t = 4$ s

4 Mark for Review

A string, fixed at both ends, supports a standing wave with a total of 4 nodes. If the length of the string is 6 m, what is the wavelength of the wave?

(A) 0.67 m

(B) 1.2 m

(C) 3 m

(D) 4 m

5 Mark for Review

A string, fixed at both ends, has a length of 6 m and supports a standing wave with a total of 4 nodes. If a transverse wave can travel at 40 m/s down the rope, what is the frequency of this standing wave?

(A) 6.7 Hz

(B) 10 Hz

(C) 20 Hz

(D) 26.7 Hz

6 Mark for Review

A sound wave travels through a metal rod with wavelength λ and frequency *f*. Which of the following best describes the wave when it passes into the surrounding air?

	Wavelength	Frequency
(A)	Less than λ	Equal to f
(B)	Less than λ	Less than f
(C)	Greater than λ	Equal to f
(D)	Greater than λ	Less than f

7 Mark for Review

In the figure below, two speakers, S_1 and S_2, emit sound waves of wavelength 2 m, in phase with each other.

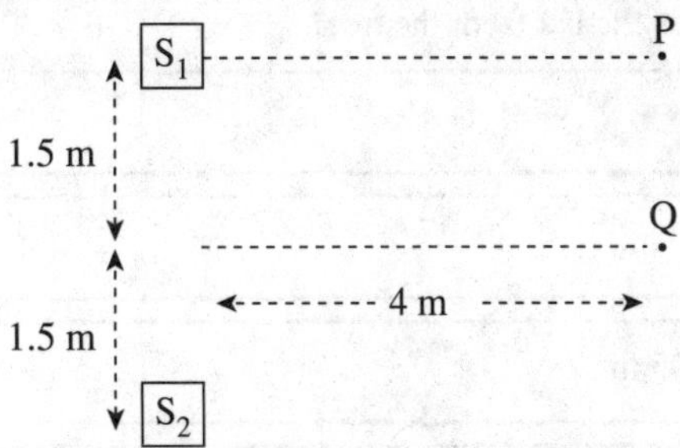

Let A_P be the amplitude of the resulting wave at Point P, and A_Q the amplitude of the resultant wave at Point Q. How does A_P compare to A_Q?

(A) $A_P < A_Q$

(B) $A_P = A_Q$

(C) $A_P > A_Q$

(D) $A_P < 0, A_Q > 0$

8 Mark for Review

An organ pipe that's closed at one end has a length of 17 cm. If the speed of sound through the air inside is 340 m/s, what is the pipe's fundamental frequency?

(A) 250 Hz

(B) 500 Hz

(C) 1000 Hz

(D) 1500 Hz

9 Mark for Review

A bat emits a 40 kHz "chirp" with a wavelength of 8.75 mm toward a tree and receives an echo 0.4 s later. How far is the bat from the tree?

(A) 35 m

(B) 70 m

(C) 105 m

(D) 140 m

10 Mark for Review

An X-ray with a frequency of 1.0×10^{18} Hz is an example of an electromagnetic wave. The equation for the electric field of such an X-ray could be given by $E = A\cos(\omega t)$ with

(A) A measured in N/C and $\omega = 1.0 \times 10^{18}$ Hz

(B) A measured in N · s/C and $\omega = 1.0 \times 10^{18}$ Hz

(C) A measured in N/C and $\omega = 2.0\pi \times 10^{18}$ Hz

(D) A measured in N · s/C and $\omega = 2.0\pi \times 10^{18}$ Hz

11 Mark for Review

In Young's double-slit interference experiment, what is the difference in path length of the light waves from the two slits to the center of the first bright fringe above the central maximum?

(A) $\frac{1}{4}\lambda$

(B) $\frac{1}{2}\lambda$

(C) λ

(D) $\frac{3}{2}\lambda$

12 Mark for Review

Three double-slit experiments are conducted and an interference pattern is observed in each. The second experiment has an identical setup to the original except a shorter wavelength of light is used. The third experiment has an identical setup to the original except the slits are closer together. What is true about the observed interference patterns for the experiments?

(A) Both the second and third experiments have closer fringes than the original.

(B) The second experiment has closer fringes than the original and the third has farther fringes.

(C) The second experiment has farther fringes than the original and the third has closer fringes.

(D) Both the second and third experiments have farther fringes than the original.

Section II: Free Response

1 Mark for Review

A rope is stretched between two vertical supports. The points where it's attached (P and Q) are fixed. The linear density of the rope is 0.4 kg/m, and the speed of a transverse wave on the rope is 12 m/s.

Figure 1

A. What's the tension in the rope?

B. With what frequency must the rope vibrate to create a traveling wave with a wavelength of 2 m?

The rope can support standing waves of lengths 4 m and 3.2 m, whose harmonic numbers are consecutive integers.

C. Find

i. the length of the rope

ii. the mass of the rope

D. What is the harmonic number of the 4 m standing wave?

E. On the diagram above, draw a sketch of the 4 m standing wave, labeling the nodes and antinodes.

2 Mark for Review

A group of students performs a set of physics experiments using two tuning forks, one with a frequency of 400 Hz and the other 440 Hz.

A. What is the observed beat frequency when the two tuning forks are struck?

B. Describe the changes to the frequency, wavelength, and speed of the sound waves from the tuning forks as they travel from air into water.

C. One student strikes a tuning fork and then throws it straight up to a student on the second floor of a building. What happens to the frequency of the sound that the student on the second floor hears as the tuning fork travels upward?

3 Mark for Review

Two trials of a double-slit interference experiment are set up as follows. The slit separation is $d = 0.50$ mm, and the distance to the screen, L, is 4.0 m.

Figure 1

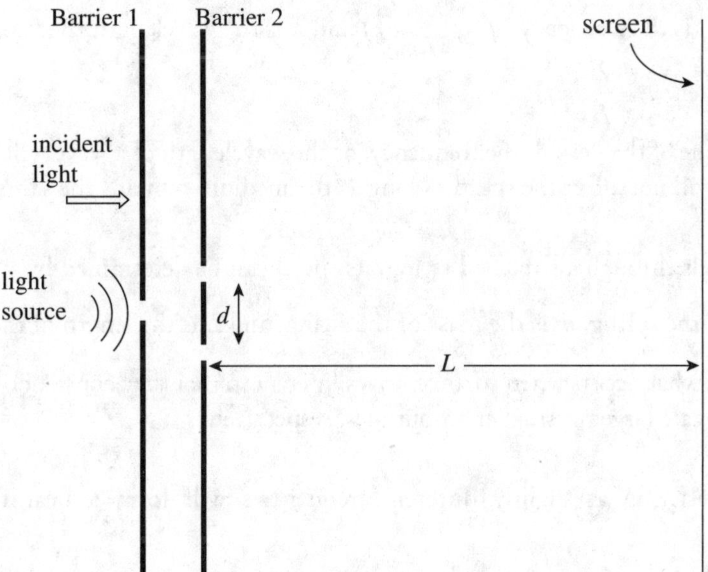

A. What is the purpose of the first (single-slit) barrier? Why not use two light sources, one at each slit at the second barrier? Explain briefly.

In the first trial, white light is used.

B. What is the vertical separation on the screen (in mm) between the first-order maxima for red light ($\lambda = 750$ nm) and violet light ($\lambda = 400$ nm)?

C. Locate the nearest point to the central maximum where an intensity maximum for violet light ($\lambda = 400$ nm) coincides with an intensity maximum for orange-yellow light ($\lambda = 600$ nm).

In the second trial, the entire region between the double-slit barrier and the screen is filled with a large slab of glass of refractive index $n = 1.5$, and monochromatic green light ($\lambda = 500$ nm in air) is used.

D. What is the separation between adjacent bright fringes on the screen?

Chapter 9 Summary

- The speed of a wave depends on the medium through which it travels. Speed can be determined by $v = f\lambda$, where f is the frequency ($f = \frac{\#\,cycles}{time}$) and λ is the wavelength. Because $f = \frac{1}{T}$, the speed can also be written $v = \frac{\lambda}{T}$.
- Changing any one of the period, the frequency, or the wavelength of a wave will affect the other two quantities, but will not affect the speed as long as the medium remains the same.
- If the wave travels through a stretched string, its speed can be determined by $v = \sqrt{\frac{F_T}{m/L}}$, where F_T is the tension in the spring, m is the mass of the string, and L is the length of the string.
- Superposition is what occurs when parts of waves interact so that they constructively or destructively interfere (e.g., create larger or smaller amplitudes, respectively).
- Two waves interfering at slightly different frequencies will form a beat frequency given by $f_{beat} = |f_1 - f_2|$.
- All electromagnetic waves travel at the speed of light ($c = 3.00 \times 10^8$ m/s). They obey the wave equation $\lambda = v/f$.

- When monochromatic coherent light goes through double slits (or a diffraction grating):
  - the path difference is given by $\Delta D = d\sin\theta$
  - there will be constructive interference when the path difference is $\Delta D = m\lambda$
  - there will be destructive interference when the path difference is $\Delta D = (m + \frac{1}{2})\lambda$
  - for small angles, we can say $d\left(\frac{y_{max}}{L}\right) = m\lambda$, where y_{max} is the distance to the mth maximum, or bright fringe, from the central maximum and shows where constructive interference occurs.
- When monochromatic coherent light goes through a single slight or circular aperture:
  - there will be destructive interference when the path difference between pairs of rays is half a wavelength, which occurs when $a\sin\theta = m\lambda$
  - for small angles, we can say $a\left(\frac{y_{min}}{L}\right) = m\lambda$, where y_{min} is the distance to the mth minimum, or dark fringe, from the central maximum and shows where destructive interference occurs.

Chapter 10
Modern Physics

INTRODUCTION

The subject matter of the previous chapters was developed in the 17th, 18th, and 19th centuries, but as we delve into the physics of the very small, we enter the 20th century. Max Planck first proposed the idea of light being emitted as individual packets of constant energy, called **quanta**. This is where the name quantum mechanics comes from. The particle nature of light was pioneered by Einstein in 1905, when he showed light transferred energy like a particle, and Arthur Compton in 1923, when he showed light has momentum and can undergo elastic collisions.

BLACKBODY RADIATION

Planck first proposed the idea that energy is quantized to explain blackbody radiation. A blackbody is an idealized object that absorbs all incident radiation. Blackbody radiation describes the electromagnetic waves that the blackbody radiates. Blackbodies emit a continuous spectrum of electromagnetic radiation in which the intensity varies for different wavelengths. The wavelength that is emitted with the highest intensity, λ_{max}, depends on the temperature, T, of the blackbody, which is described by **Wien's Law**,

Equation Sheet

$$\lambda_{max} = \frac{b}{T}$$

in which b is a proportionality constant. As the temperature of a blackbody increases, the maximum wavelength of its emission decreases. In addition, the total power, P, of the emitted radiation increases with temperature as well, as described by the **Stefan-Boltzmann Law**,

Equation Sheet

$$P = A\sigma T^4$$

in which A is the surface area of the object, and σ is the Stefan-Boltzmann constant.

Although a blackbody is an idealized object, we can observe this general trend in objects around us. For example, consider the heating element of an electric stove. When it is cold, we do not see it emitting any light because its λ_{max} is in the infrared. As it heats up, it starts to glow with a dark red color. As it gets hotter, the color turns reddish orange. As the temperature increases, the peak wavelength of light we see decreases from infrared to red to reddish orange.

Although it's not obvious from this conceptual description, it was unclear why blackbodies behaved in this way based on classical physics. Planck was only able to explain the behavior of blackbodies when he considered that energy was quantized.

PHOTONS AND THE PHOTOELECTRIC EFFECT

Planck realized that electromagnetic energy is quantized. It was Einstein who characterized how this occurs. A quantum of electromagnetic energy is known as a **photon**. Light behaves like a stream of photons, and this is illustrated by the **photoelectric effect**.

When a piece of metal is illuminated by electromagnetic radiation (specifically visible light, ultraviolet light, or X-rays), the energy absorbed by electrons near the surface of the metal can liberate them from their bound state, and these electrons can fly off. The released elecrrons are known as **photoelectrons**. In this case, the classical, wave-only theory of light would predict three results:

(1) There would be a significant time delay between the moment of illumination and the ejection of photoelectrons. This is because the metal would have to heat up to the point that the thermal energy was enough to overcome the binding energy.

(2) Increasing the intensity of the light would cause the electrons to leave the metal surface with greater kinetic energy. This is because the energy carried by a wave is related to its intensity.

(3) Photoelectrons would be emitted regardless of the frequency of the incident energy, as long as the intensity was high enough.

Surprisingly, none of these predictions was observed. Photoelectrons were ejected within just a few billionths of a second after illumination, disproving prediction (1). Secondly, increasing the intensity of the light did not cause photoelectrons to leave the metal surface with greater kinetic energy. Although more electrons were ejected as the intensity was increased, there was a maximum photoelectron kinetic energy; prediction (2) was false. And, for each metal, there was a certain **threshold frequency**, f_0: if light of frequency lower than f_0 were used to illuminate the metal surface, *no* photoelectrons were ejected, regardless of how intense the incident radiation was; prediction (3) was also false. Clearly, something was wrong with the wave-only theory of light.

Einstein explained these observations by borrowing from Planck's idea that light came in individual quanta of energy, which were later called photons by Gilbert Lewis. The energy of a photon is proportional to the frequency of the wave,

$$E = hf$$

Equation Sheet

where h is **Planck's constant** (about $6.63 \cdot 10^{-34}$ J·s). A certain amount of energy had to be imparted to an electron on the metal surface in order to liberate it; this was known as the metal's **work function**, or ϕ. If an electron absorbed a photon whose energy E was greater than ϕ, it would leave the metal with a maximum kinetic energy equal to $E - \phi$. This process could occur *very* quickly, which accounts for the rapidity with which photoelectrons are produced after illumination. This explains why prediction (1) was incorrect.

In this view, increasing the intensity (and therefore the energy) just means increasing the number of photons and results in the ejection of more photoelectrons—but since the energy of each incident photon is fixed by the equation $E = hf$, the value of K_{max} will still be $E - \phi$. This can be expressed as

Equation Sheet

$$K_{max} = hf - \phi$$

This accounts for the observation that disproved prediction (2).

Finally, if the incoming photon's energy was less than the work function (or, $E = hf < \phi$), the photon energy would not be enough to liberate electrons. Blasting the metal surface with more photons (that is, increasing the intensity of the incident beam) would also do nothing; none of the photons would have enough energy to eject electrons. This accounts for the observation of a threshold frequency, which we now know is ϕ/h. This can be expressed as

$$f_o = \phi/h$$

and is why prediction (3) was incorrect. Before we get to some examples, it's worthwhile to introduce a new unit of energy. The SI unit for energy is the joule, but it's too large to be convenient in the domains we're studying now. We'll use a much smaller unit, the **electronvolt** (abbreviated **eV**). The eV is equal to the energy gained (or lost) by an electron accelerated through a potential difference of one volt. Using the equation $\Delta U_E = qV$, we find that

$$1 \text{ eV} = (1 \text{ e})(1 \text{ V}) = (1.6 \times 10^{-19} \text{ C})(1 \text{ V}) = 1.6 \times 10^{-19} \text{ J}$$

In terms of electronvolts, the value of Planck's constant is 4.14×10^{-15} eV·s.

Example 1 The work function, ϕ, for aluminum is 4.08 eV.

(a) What is the threshold frequency required to produce photoelectrons from aluminum?

(b) Classify the electromagnetic radiation that can produce photoelectrons.

(c) If light of frequency $f = 4.00 \times 10^{15}$ Hz is used to illuminate a piece of aluminum,

(i) what is K_{max}, the maximum kinetic energy of ejected photoelectrons?

(ii) what's the maximum speed of the photoelectrons? (Electron mass = 9.11×10^{-31} kg)

(d) If the light described in part (b) were increased by a factor of 2 in intensity, what would happen to the value of K_{max}?

Solution.

(a) We know from the statement of the question that, in order for a photon to be successful in liberating an electron from the surface of the aluminum, its energy cannot be less than 4.08 eV. Therefore, the minimum frequency of the incident light—the threshold frequency—must be

$$f_0 = \frac{\phi}{h} = \frac{4.08 \text{ eV}}{4.14 \times 10^{-15} \text{ eV} \cdot \text{s}} = 9.86 \times 10^{14} \text{ Hz}$$

(b) Based on the electromagnetic spectrum given in the previous chapter, the electromagnetic radiation used to produce photoelectrons from aluminum must be at least in the ultraviolet region of the EM spectrum.

(c) (i) The maximum kinetic energy of photoelectrons is found from the equation

$$\begin{aligned} K_{\text{max}} &= hf - \phi \\ &= (4.14 \times 10^{-15} \text{ eV} \cdot \text{s})(4.00 \times 10^{15} \text{ Hz}) - (4.08 \text{ eV}) \\ &= 12.5 \text{ eV} \end{aligned}$$

(ii) Using the above result and $K = \frac{1}{2}mv^2$, we can find v_{max}:

$$\begin{aligned} v_{\text{max}} &= \sqrt{\frac{2}{m_e} K_{\text{max}}} \\ &= \sqrt{\frac{2}{9.11 \times 10^{-31} \text{ kg}} \left(12.5 \text{ eV} \cdot \frac{1.6 \times 10^{-19} \text{ J}}{1 \text{ eV}} \right)} \\ &= 2.1 \times 10^{6} \text{ m/s} \end{aligned}$$

(d) Nothing will happen. Increasing the intensity of the illuminating radiation will cause more photons to impinge on the metal surface, thereby ejecting more photoelectrons, but their maximum kinetic energy will remain the same. The only way to increase K_{max} would be to increase the frequency of the incident energy.

COMPTON SCATTERING

Compton provided further evidence that light behaves like a particle. Like the photoelectric effect, Compton scattering involves light interacting with electrons. Unlike the photoelectric effect, the photon is not absorbed. Instead, the light is scattered. In other words, an incident photon, with an incident wavelength λ_0, interacts with an electron and moves off with a longer wavelength λ' at an angle θ relative to its original direction.

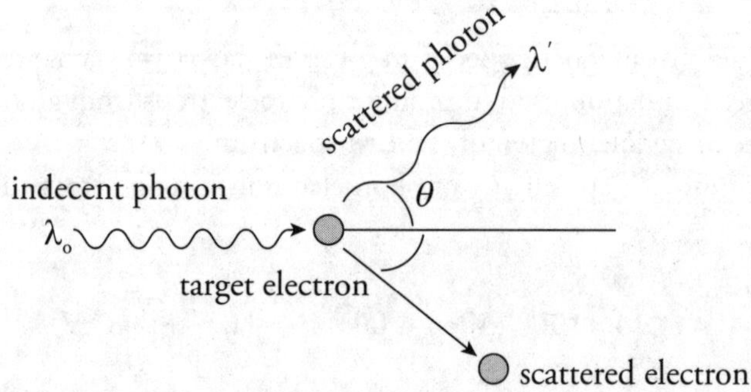

The resulting Compton shift, $\Delta\lambda$, depends only on the scattered angle:

Equation Sheet

$$\Delta\lambda = \lambda' - \lambda_0 = \frac{h}{m_e c}(1 - \cos\theta)$$

This equation follows from applying conservation of momentum and energy to the collision of the photon and electron, just like it would for the elastic collision of any two objects. This result provides another example of how light can behave like a particle.

Example 2 A coherent light source with a wavelength of 4×10^{-12} m is shone through a thin sheet of graphite.

(a) At what scattering angle would you expect to detect light with a wavelength of 5×10^{-12}?

(b) How much energy would the electron that the light scattered off of at this angle receive?

Solution.

(a) From the Compton shift equation,

$$\lambda' - \lambda_0 = \frac{h}{m_e c}(1 - \cos\theta)$$

so

$$\begin{aligned}\theta &= \cos^{-1}\left(1 - \frac{m_e c}{h}(\lambda' - \lambda_0)\right)\\ &= \cos^{-1}\left(1 - \frac{(9.11\times10^{-31}\ \text{kg})(3\times10^{8}\ \text{m/s})}{6.63\times10^{-34}\ \text{J}\cdot\text{s}}(5\times10^{-12}\ \text{m} - 4\times10^{-12}\ \text{m})\right)\\ &\approx 54^\circ\end{aligned}$$

(b) The energy given to the electron must be the energy lost by the light during the scattering. Because the energy of a photon is given by $E = hf$ and, from the wave equation, $\lambda = c/f$, we have

$$\begin{aligned}\Delta E = E_f - E_i = hf_f - hf_i &= \frac{hc}{\lambda'} - \frac{hc}{\lambda_0}\\ &= hc\left(\frac{1}{\lambda'} - \frac{1}{\lambda_0}\right)\\ &= (6.63\times10^{-34}\ \text{J}\cdot\text{s})(3\times10^{8}\ \text{m/s})\left(\frac{1}{5\times10^{-12}\ \text{m}} - \frac{1}{4\times10^{-12}\ \text{m}}\right)\\ &= -9.9\times10^{-15}\ \text{J}\end{aligned}$$

so the photon loses 9.9×10^{-15} J during scattering which is how much energy the electron will gain.

THE BOHR MODEL OF THE ATOM

In the years immediately following Rutherford's announcement of his nuclear model of the atom, a young physicist, Niels Bohr, would add an important piece to the atomic puzzle. Rutherford told us where the positive charge of the atom was located; Bohr would tell us about the electrons.

Chemistry
Modern physics begins to explore atoms, which bridges into chemistry. Fortunately, you will not need to know much about chemistry for the AP Physics 2 Exam—just a basic understanding of some topics.

For 50 years, it had been known that atoms in a gas discharge tube emit and absorb light only at specific wavelengths. The light from a glowing gas, passed through a prism to disperse the beam into its component wavelengths, produces patterns of sharp lines called **atomic spectra**. The visible wavelengths that appear in the emission spectrum of hydrogen had been summarized by the *Balmer formula*:

$$\frac{1}{\lambda_n} = R\left(\frac{1}{2^2} - \frac{1}{n^2}\right)$$

where R is the *Rydberg constant* (about 1.1×10^7 m^{-1}). The formula worked—that is, it fit the observational data—but it had no theoretical basis. So, the question was, *why* do atoms emit (or absorb) radiation only at certain discrete wavelengths?

Bohr's model of the atom explains the spectroscopists' observations. Using the simplest atom, hydrogen (with only one electron), Bohr postulated that the electron would orbit the nucleus only at certain discrete radii. When the electron is in one of these special orbits, it does not radiate away energy (as the classical theory would predict). However, if the electron absorbs a certain amount of energy, it is **excited** to a higher orbit, one with a greater radius. After a short time in this excited state, it returns to a lower orbit, emitting a photon in the process. Since each allowed orbit—or **energy level**—has a specific radius (and corresponding energy), the photons emitted in each jump have only specific wavelengths.

When an excited electron drops from energy level $n = j$ to a lower one, $n = i$, the transition causes a photon of energy to be emitted, and the energy of the photon is the difference between the two energy levels. When an electron absorbs a photon and transitions from a lower energy level, $n = i$, to a higher energy level, $n = j$, the energy of the absorbed photon is the difference between the two energy levels.

$$E_{\text{emitted or absorbed photon}} = |\Delta E| = E_j - E_i$$

The wavelength of this photon is

Speed of Light
Photons travel at the speed of light.

$$\lambda = \frac{c}{f} = \frac{c}{|E_{\text{photon}}|/h} = \frac{hc}{|E_j - E_i|}$$

Example 3 The first five energy levels of an atom are shown in the diagram below:

−3 eV	$n = 5$
−4 eV	$n = 4$
−7 eV	$n = 3$
−15 eV	$n = 2$
−62 eV	$n = 1$ ground state

(a) If the atom begins in the $n = 3$ level, what photon energies could be emitted as it returns to its ground state?

(b) What could happen if this atom, while in its ground state, were bombarded with a photon of energy 10 eV?

Solution.

(a) If the atom is in the $n = 3$ level, it could return to ground state by a transition from $3 \to 1$, or from $3 \to 2$ and then $2 \to 1$. The energy emitted in each of these transitions is simply the difference between the energies of the corresponding levels:

$$E_{3\to1} = E_3 - E_1 = (-7 \text{ eV}) - (-62 \text{ eV}) = 55 \text{ eV}$$

$$E_{3\to2} = E_3 - E_2 = (-7 \text{ eV}) - (-15 \text{ eV}) = 8 \text{ eV}$$

$$E_{2\to1} = E_2 - E_1 = (-15 \text{ eV}) - (-62 \text{ eV}) = 47 \text{ eV}$$

(b) The ground state has an energy level of –62 eV, and since there is no level with an energy of –52 eV, the atom could not absorb a 10 eV photon. In its ground state, this atom would be transparent to light of energy 10 eV.

Example 4 The ground state of hydrogen is –13.6 eV. The first excited state is –3.4 eV. The second excited state is –1.5 eV. The third excited state is –0.85 eV.

(a) How much energy must a ground state electron in a hydrogen atom absorb to be excited to the $n = 4$ energy level?

(b) With the electron in the $n = 4$ level, what wavelengths are possible for the photon emitted when the electron drops to a lower energy level? In what regions of the EM spectrum do these photons lie?

Solution.

(a) The ground-state energy level ($n = 1$) for hydrogen is –13.6 eV, and

$E_4 = -0.85$ eV.

Therefore, in order for an electron to make the transition from E_1 to E_4, it must absorb energy in the amount $E_4 - E_1 = (-0.85 \text{ eV}) - (-13.6 \text{ eV}) = 12.8$ eV.

(b) An electron in the $n = 4$ energy level can make several different transitions: it can drop to $n = 3$, $n = 2$, or all the way down to the ground state, $n = 1$.

The following diagram shows the electron dropping from $n = 4$ to $n = 3$:

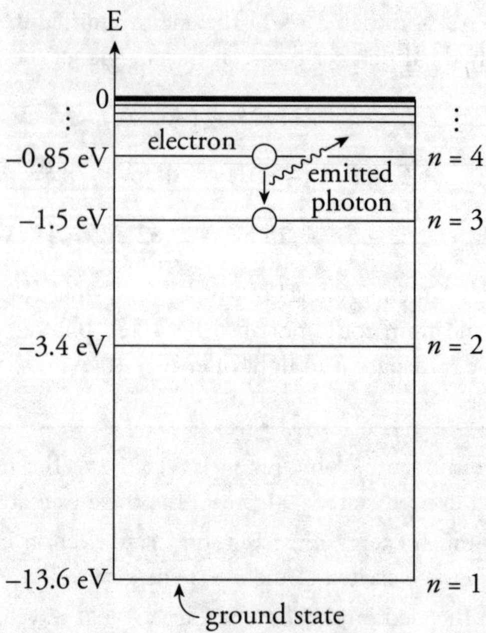

There are three possible values for the energy of the emitted photon, $E_{4\to 3}$, $E_{4\to 2}$, or $E_{4\to 1}$:

$$E_{4\to 3} = E_4 - E_3 = (-0.85\text{ eV}) - (-1.5\text{ eV}) = 0.65\text{ eV}$$
$$E_{4\to 2} = E_4 - E_2 = (-0.85\text{ eV}) - (-3.4\text{ eV}) = 2.55\text{ eV}$$
$$E_{4\to 1} = E_4 - E_1 = (-0.85\text{ eV}) - (-13.6\text{ eV}) = 12.8\text{ eV}$$

From the equation $E = hf = hc/\lambda$, we get $\lambda = hc/E$, so

$$\lambda_{4\to 3} = \frac{hc}{E_{4\to 3}} = \frac{(4.14\times 10^{-15}\text{ eV}\cdot\text{s})(3.00\times 10^{8}\text{ m/s})}{0.65\text{ eV}} = 1910\text{ nm}$$
$$\lambda_{4\to 2} = \frac{hc}{E_{4\to 2}} = \frac{(4.14\times 10^{-15}\text{ eV}\cdot\text{s})(3.00\times 10^{8}\text{ m/s})}{2.55\text{ eV}} = 487\text{ nm}$$
$$\lambda_{4\to 1} = \frac{hc}{E_{4\to 1}} = \frac{(4.14\times 10^{-15}\text{ eV}\cdot\text{s})(3.00\times 10^{8}\text{ m/s})}{12.8\text{ eV}} = 97\text{ nm}$$

Note that $\lambda_{4\to 2}$ is in the visible spectrum; this wavelength corresponds to the color blue-green; $\lambda_{4\to 1}$ is an ultraviolet wavelength, and $\lambda_{4\to 3}$ is infrared.

WAVE-PARTICLE DUALITY

Light and other electromagnetic waves exhibit wave-like characteristics through interference and diffraction. However, as we saw in the photoelectric effect, light also behaves as if its energy were granular, composed of particles. This is **wave-particle duality**: electromagnetic radiation propagates like a wave but exchanges energy like a particle.

Since an electromagnetic wave can behave like a particle, can a particle of matter behave like a wave? In 1924, the French physicist Louis de Broglie proposed that the answer is "yes." His conjecture, which has since been supported by experiment, is that a particle of mass *m* and speed *v*—and thus, linear momentum *p*—has an associated wavelength, which is called its **de Broglie wavelength**:

$$\lambda = \frac{h}{p}$$

Equation Sheet

Particles in motion can display wave characteristics and behave as if they had a wavelength.

Since the value of *h* is so small, ordinary macroscopic objects do not display wave-like behavior. For example, a baseball (mass = 0.15 kg) thrown at a speed of 40 m/s has a de Broglie wavelength of

$$\lambda = \frac{h}{p} = \frac{h}{mv} = \frac{6.63\times10^{-34}\text{ J}\cdot\text{s}}{(0.15\text{ kg})(40\text{ m/s})} = 1.1\times10^{-34}\text{ m}$$

This is much too small to measure. However, with subatomic particles, the wave nature is clearly evident. The 1937 Nobel Prize in physics was awarded for experiments by C. J. Davisson and G. P. Thomson that revealed that a stream of electrons exhibited diffraction patterns when scattered by crystals—a behavior that's characteristic of waves.

Example 5 Electrons in a diffraction experiment are accelerated through a potential difference of 200 V. What is the de Broglie wavelength of these electrons?

Solution. By definition, the kinetic energy of these electrons is 200 eV. Since the relationship between linear momentum and kinetic energy is $p = \sqrt{2mK}$,

$$\lambda = \frac{h}{p} = \frac{h}{\sqrt{2mK}} = \frac{6.63\times10^{-34}\text{ J}\cdot\text{s}}{\sqrt{2(9.11\times10^{-31}\text{ kg})\ 200\text{ eV}\cdot\dfrac{1.6\times10^{-19}\text{ J}}{1\text{ eV}}}} = 8.7\times10^{-11}\text{ m} = 0.087\text{ nm}$$

This wavelength is characteristic of X-rays.

Interpreting Wave-Function Graphs
Wave-function graphs will often have values that are positive as well as negative. When interpreting the graph, only the absolute value is important in ranking relative likelihoods of experimental outcomes (technically the likelihood is proportional to Ψ^2, but the absolute value will give correct interpretations and is easier to read from a graph).

THE WAVE FUNCTION

There are several major differences between classical physics and modern physics. One of the most important is that the mathematics that underpin modern physics (which are too complicated for treatment here) no longer specify a definite position at a definite time for a particle such as an electron. Instead, the mathematics only give the *probability* that a particle will be measured to be at a particular position when the position is measured. That probability is related to a new physical parameter called the *wave function*, Ψ.

In one interpretation of that mathematics, the act of measuring the position changes the wave function, Ψ, so that it has a single position with a probability of 100%, and that location corresponds to the result of the measurement (this is referred to as "collapsing the wave-function"). However, before the position is measured (or after a measurement has taken place and some time has passed so the position is no longer certain), there are a range of probable locations that can be represented as a graph of Ψ versus x. To interpret such a graph, the position with the largest absolute value is the most probable location for the result of a measurement. Any position where the graph has a height of 0, it is certain that the position will not be observed.

NUCLEAR PHYSICS

The nucleus of the atom is composed of particles called **protons** and **neutrons**, which are collectively called **nucleons**. The number of protons in a given nucleus is called the atom's **atomic number**, and is denoted Z, and the number of neutrons (the **neutron number**) is denoted N. The total number of nucleons, $Z + N$, is called the **mass number** (or **nucleon number**), and is denoted A. The number of protons in the nucleus of an atom defines the element. For example, the element chlorine (abbreviated Cl) is characterized by the fact that the nucleus of every chlorine atom contains 17 protons, so the atomic number of chlorine is 17; but, different chlorine atoms may contain different numbers of neutrons. In fact, about three-fourths of all naturally occurring chlorine atoms have 18 neutrons in their nuclei (mass number = 35), and most of the remaining one-fourth contain 20 neutrons (mass number = 37).

Nuclei that contain the same numbers of protons but different numbers of neutrons are called **isotopes**. The notation for a **nuclide**—the term for a nucleus with specific numbers of protons and neutrons—is to write Z and A before the chemical symbol of the element:

$$^{A}_{Z}\mathrm{X}$$

The isotopes of chlorine mentioned above would be written as follows:

$$^{35}_{17}\mathrm{Cl} \quad \text{and} \quad ^{37}_{17}\mathrm{Cl}$$

Example 6 How many protons and neutrons are contained in the nuclide ${}^{63}_{29}\text{Cu}$?

Solution. The subscript (the atomic number, Z) gives the number of protons, which is 29. The superscript (the mass number, A) gives the total number of nucleons. Since $A = 63 = Z + N$, we find that $N - 63 - 29 - 34$.

Example 7 The element neon (abbreviated Ne, atomic number 10) has several isotopes. The most abundant isotope contains 10 neutrons, and two others contain 11 and 12. Write symbols for these three nuclides.

Solution. The mass numbers of these isotopes are 10 + 10 = 20, 10 + 11 = 21, and 10 + 12 = 22. So, we'd write them as follows:

$${}^{20}_{10}\text{Ne}, \quad {}^{21}_{10}\text{Ne}, \quad \text{and} \quad {}^{22}_{10}\text{Ne}$$

Another common notation—which we also use—is to write the mass number after the name of the element. These three isotopes of neon would be written as neon-20, neon-21, and neon-22.

The Nuclear Force

Why wouldn't any nucleus that has more than one proton be unstable? After all, protons are positively charged and would therefore experience a repulsive Coulomb force from each other. Why don't these nuclei explode? And what holds neutrons—which have no electric charge—in the nucleus? These issues are resolved by the presence of another fundamental force, the **strong nuclear force**, which binds neutrons and protons together to form nuclei. Although the strength of the Coulomb force can be expressed by a simple mathematical formula (it's inversely proportional to the square of their separation), the nuclear force is much more complicated; no simple formula can be written for the strength of the nuclear force.

Binding Energy

The masses of the proton and neutron are listed below.

$$\text{proton: } m_p = 1.6726 \times 10^{-27} \text{ kg}$$

$$\text{neutron: } m_n = 1.6749 \times 10^{-27} \text{ kg}$$

Because these masses are so tiny, a much smaller mass unit is used. With the most abundant isotope of carbon (carbon-12) as a reference, the **atomic mass unit** (abbreviated **amu** or simply **u**) is defined as 1/12 the mass of a ${}^{12}\text{C}$ atom. The conversion between kg and u is 1 u = 1.6605 × 10^{-27} kg. In terms of atomic mass units,

$$m_p = 1.00728 \text{ u}$$

$$m_n = 1.00867 \text{ u}$$

Now consider the **deuteron**, the nucleus of **deuterium**, an isotope of hydrogen that contains 1 proton and 1 neutron. The mass of a deuteron is 2.01356 u, which is a little *less* than the sum of the individual masses of the proton and neutron. The difference between the mass of any bound nucleus and the sum of the masses of its constituent nucleons is called the **mass defect**, Δm. In the case of the deuteron (symbolized **d**), the mass defect is

$$\begin{aligned}\Delta m &= (m_p + m_n) - m_d \\ &= (1.00728 \text{ u} + 1.00867 \text{ u}) - (2.01356 \text{ u}) \\ &= 0.00239 \text{ u}\end{aligned}$$

What happened to this missing mass? It was converted to energy when the deuteron was formed. In 1905, Einstein gave us the famous equation:

Equation Sheet

$$E = mc^2$$

which tells us how much energy mass contains. Because of this, the mass-difference gives us the amount of energy needed to break the deuteron into a separate proton and neutron. Since this tells us how strongly the nucleus is bound, it is called the **binding energy** of the nucleus.

$$E_B = (\Delta m)c^2$$

Using $E = mc^2$, the energy equivalent of 1 atomic mass unit is

$$\begin{aligned}E &= (1.6605 \times 10^{-27} \text{ kg})(2.9979 \times 10^8 \text{ m/s})^2 \\ &= 1.4924 \times 10^{-10} \text{ J} \\ &= 1.4924 \times 10^{-10} \text{ J} \cdot \frac{1 \text{ eV}}{1.6022 \times 10^{-19} \text{ J}} \\ &= 9.31 \times 10^8 \text{ eV} \\ &= 931 \text{ MeV}\end{aligned}$$

In terms of electronvolts, then, the binding energy of the deuteron is

$$E_B \text{ (deuteron)} = 0.00239 \text{ u} \times \frac{931 \text{ MeV}}{1 \text{ u}} = 2.23 \text{ MeV}$$

Since the deuteron contains 2 nucleons, the **binding-energy-per-nucleon** is

$$\frac{2.23 \text{ MeV}}{2 \text{ nucleons}} = 1.12 \text{ MeV/nucleon}$$

This is the lowest value of all nuclides. The highest, 8.8 MeV/nucleon, is for an isotope of nickel, ^{62}Ni. Typically, when nuclei smaller than nickel are fused to form a single nucleus, the binding energy per nucleon increases, which tells us that energy is released in the process. On the other hand, when nuclei *larger* than nickel are *split*, binding energy per nucleon again increases, releasing energy. This is the basis of nuclear fission.

Example 8 What is the maximum wavelength of EM radiation that could be used to photodisintegrate a deuteron?

Solution. The binding energy of the deuteron is 2.23 MeV, so a photon would need to have at least this much energy to break the deuteron into a proton and neutron. Since $E = hf$ and $f = c/\lambda$,

$$E = \frac{hc}{\lambda} \longrightarrow \lambda_{\text{max}} = \frac{hc}{E_{\text{min}}} = \frac{(4.14 \times 10^{-15}\ \text{eV} \cdot \text{s})(3.00 \times 10^{8}\ \text{m/s})}{2.23 \times 10^{6}\ \text{eV}} = 5.57 \times 10^{-13}\ \text{m}$$

Example 9 The atomic mass of $^{27}_{13}\text{Al}$ is 26.9815 u. What is its nuclear binding energy per nucleon? (Mass of electron = 0.0005486 u)

Solution. The nuclear mass of $^{27}_{13}\text{Al}$ is equal to its atomic mass minus the mass of its electrons. Since an aluminum atom has 13 protons, it must also have 13 electrons. So,

$$\begin{aligned} \text{nuclear mass of } ^{27}_{13}\text{Al} &= (\text{atomic mass of } ^{27}_{13}\text{Al}) - 13m_{\text{e}} \\ &= 26.9815\ \text{u} - 13(0.0005486\ \text{u}) \\ &= 26.9744\ \text{u} \end{aligned}$$

Now, the nucleus contains 13 protons and 27 – 13 = 14 neutrons, so the total mass of the individual nucleons is

$$\begin{aligned} M &= 13m_{\text{p}} + 14m_{\text{n}} \\ &= 13(1.00728\ \text{u}) + 14(1.00867\ \text{u}) \\ &= 27.2160\ \text{u} \end{aligned}$$

and the mass defect of the aluminum nucleus is

$$\Delta m = M - m = 27.2160\ \text{u} - 26.9744\ \text{u} = 0.2416\ \text{u}$$

Converting this mass to energy, we can see that

$$E_{\text{B}} = 0.2416\ \text{u} \times \frac{931\ \text{MeV}}{1\ \text{u}} = 225\ \text{MeV}$$

so the binding energy per nucleon is

$$\frac{225\ \text{MeV}}{27} = 8.3\ \text{MeV/nucleon}$$

NUCLEAR REACTIONS

Matter Cannot be Created or Destroyed
But it *can* undergo other forms—which does not break this rule. The quickest way to determine whether something undergoes alpha, beta, or gamma decay is to find the amount of nucleons that have escaped and then to determine which kind of decay happened.

Natural radioactive decay provides one example of a nuclear reaction. Other examples of nuclear reactions include the bombardment of target nuclei with subatomic particles to artificially induce radioactivity, such as the emission of a particle or the splitting of the nucleus (this is **nuclear fission**), and the **nuclear fusion** of small nuclei at extremely high temperatures. In all cases of nuclear reactions that we'll study, nucleon number and charge must be conserved. In order to balance nuclear reactions, we write ${}^{1}_{1}\text{p}$ or ${}^{1}_{1}\text{H}$ for a proton and ${}^{1}_{0}\text{n}$ for a neutron. Gamma-ray photons can also be produced in nuclear reactions; they have no charge or nucleon number and are represented as ${}^{0}_{0}\gamma$.

Alpha Decay

When a nucleus undergoes alpha decay, it emits an alpha particle, which consists of two protons and two neutrons and is the same as the nucleus of a helium-4 atom. An alpha particle can be represented as

$$\alpha,\ {}^{4}_{2}\alpha,\ \text{or}\ {}^{4}_{2}\text{He}$$

Very large nuclei can shed nucleons quickly by emitting one or more alpha particles; for example, radon-222 (${}^{222}_{86}\text{Rn}$) is radioactive and undergoes alpha decay.

$${}^{222}_{86}\text{Rn} \rightarrow {}^{218}_{84}\text{Po} + {}^{4}_{2}\alpha$$

This reaction illustrates two important features of a nuclear reaction.

(1) Mass number is conserved.
(2) Charge is conserved.

The decaying nuclide is known as the **parent**, and the resulting nuclide is known as the **daughter**. (Here, radon-222 is the parent nuclide and polonium-218 is the daughter.) Alpha decay decreases the mass number by 4 and the atomic number by 2. Therefore, alpha decay looks like the following:

$${}^{A}_{Z}\text{X} \rightarrow {}^{A-4}_{Z-2}\text{X}' + {}^{4}_{2}\alpha$$

Beta Decay

There are three subcategories of **beta** (β) decay, called β^-, β^+, and **electron capture** (**EC**).

β^- Decay When the neutron-to-proton ratio is too large, the nucleus undergoes β^- decay, which is the most common form of beta decay. β^- decay occurs when a neutron transforms into a proton and releases an electron. The expelled electron is called a **beta particle**. The transformation of a neutron into a proton and an electron (and another particle, the **electron-antineutrino**, $\bar{\nu}_e$) is caused by the action of the **weak nuclear force**, another of nature's fundamental forces. A common example of a nuclide that undergoes β^- decay is carbon-14, which is used to date archaeological artifacts.

$$^{14}_{6}\text{C} \rightarrow \, ^{14}_{7}\text{N} + e^- + \bar{\nu}_e$$

The reaction is balanced, since 14 = 14 + 0 and 6 = 7 + (–1).

β^+ Decay When the neutron-to-proton ratio is too small, the nucleus will undergo β^+ decay. In this form of beta decay, a proton is transformed into a neutron and a **positron**, $^{0}_{+1}e$ (the electron's **antiparticle**), plus another particle, the **electron-neutrino**, ν_e, which are then both ejected from the nucleus. An example of a positron emitter is fluorine-17.

$$^{17}_{9}\text{F} \rightarrow \, ^{17}_{8}\text{O} + e^+ + \nu_e$$

Electron Capture Another way in which a nucleus can increase its neutron-to-proton ratio is to capture an orbiting electron and then cause the transformation of a proton into a neutron. Beryllium-7 undergoes this process.

$$^{7}_{4}\text{Be} + e^- \rightarrow \, ^{7}_{3}\text{Li} + \nu_e$$

Gamma Decay

In each of the decay processes defined on the previous pages, the daughter was a different element than the parent. Radon becomes polonium as a result of α decay, carbon becomes nitrogen as a result of β^- decay, fluorine becomes oxygen from β^+ decay, and beryllium becomes lithium from electron capture. By contrast, gamma decay does not alter the identity of the nucleus; it just allows the nucleus to relax and shed energy. Imagine that potassium-42 undergoes β^- decay to form calcium-42.

$$^{42}_{19}\text{K} \rightarrow \, ^{42}_{20}\text{Ca}^* + e^- + \bar{\nu}_e$$

The asterisk indicates that the daughter calcium nucleus is left in a high-energy, excited state. For this excited nucleus to drop to its ground state, it must emit energy in the form of a photon, a **gamma ray**, symbolized by γ.

$$^{42}_{20}\text{Ca}^* \rightarrow \, ^{42}_{20}\text{Ca} + \gamma$$

Let's sum up the three types of radiation:

Type of Radiation		What Happens?	Charge
Alpha		Particle—a helium nucleus, containing 2 protons and 2 neutrons, is released from the nucleus.	Positive 2+
Beta	Beta-minus	Particle—a highly energetic, massless electron is released from the nucleus after a neutron divides into a proton and an electron.	Negative 1–
	Beta-plus	Particle—a highly energetic, massless positron is released from the nucleus after a proton divides into a neutron and a positron.	Positive 1+
Gamma		Wave—energy is emitted in the form of a photon.	No Charge

Example 10 A mercury-198 nucleus is bombarded by a neutron, which causes a nuclear reaction:

$$^{1}_{0}\text{n} + {}^{198}_{80}\text{Hg} \rightarrow {}^{197}_{79}\text{Au} + {}^{?}_{?}\text{X}$$

What's the unknown product particle, X?

Solution. In order to balance the superscripts, we must have $1 + 198 = 197 + A$, so $A = 2$, and the subscripts are balanced if $0 + 80 = 79 + Z$, so $Z = 1$:

$$^{1}_{0}\text{n} + {}^{198}_{80}\text{Hg} \rightarrow {}^{197}_{79}\text{Au} + {}^{2}_{1}\text{X}$$

Therefore, X must be a deuteron, $^{2}_{1}\text{H}$ (or just d).

DISINTEGRATION ENERGY

Nuclear reactions not only produce new nuclei and other subatomic product particles, but they also involve the absorption or emission of energy. Nuclear reactions must conserve total energy, so changes in mass are accompanied by changes in energy according to Einstein's equation

$$\Delta E = (\Delta m)c^2$$

A general nuclear reaction is written

$$\text{A} + \text{B} \rightarrow \text{C} + \text{D} + Q$$

where Q denotes the **disintegration energy**. If Q is positive, the reaction is **exothermic** (or **exoergic**) and the reaction can occur spontaneously; if Q is negative, the reaction is **endothermic** (or **endoergic**) and the reaction cannot occur spontaneously. The energy Q is calculated as follows:

$$Q = [(m_A + m_B) - (m_C + m_D)]c^2 = (\Delta m)c^2$$

For spontaneous reactions—ones that liberate energy—most of the energy is revealed as kinetic energy of the least massive product nuclei.

Example 11 The process that powers the Sun—and upon which all life on Earth is dependent—is the fusion reaction:

$$4\,{}^{1}_{1}\text{H} \rightarrow {}^{4}_{2}\alpha + 2e^{+} + 2\nu_e + \gamma$$

(a) Show that this reaction releases energy.

(b) How much energy is released per proton?

Use the fact that $m_\alpha = 4.0015$ u and ignore the mass of the electron-neutrino, ν_e.

Solution.

(a) We need to find the mass difference between the reactants and products:

$$\begin{aligned}\Delta m &= 4m_p - (m_\alpha + 2m_e)\\ &= 4(1.00728\text{ u}) - [4.0015\text{ u} + 2(0.0005486\text{ u})]\\ &= 0.02652\text{ u}\end{aligned}$$

Since Δm is positive, the reaction is exothermic: energy is released.

(b) Converting the mass difference to energy gives

$$Q = 0.02652\text{ u} \times \frac{931\text{ MeV}}{1\text{ u}} = 24.7\text{ MeV}$$

Since four protons went into the reaction, the energy liberated per proton is

$$\frac{24.7\text{ MeV}}{4\text{ p}} = 6.2\text{ MeV/proton}$$

Example 12 Can the following nuclear reaction occur spontaneously?

$$^{4}_{2}\alpha + ^{14}_{7}\text{N} \rightarrow ^{17}_{8}\text{O} + ^{1}_{1}\text{H}$$

(The mass of the nitrogen nucleus is 13.9992 u, and the mass of the oxygen nucleus is 16.9947 u.)

Solution. We first figure out the mass equivalent of the disintegration energy:

$$\begin{aligned}\Delta m &= (m_\alpha + m_\text{N}) - (m_\text{O} + m_\text{p}) \\ &= (4.0015\text{ u} + 13.9992\text{ u}) - (16.9947\text{ u} + 1.00728\text{ u}) \\ &= -0.00128\text{ u}\end{aligned}$$

Since Δm is negative, this reaction is nonspontaneous; energy must be *supplied* in order for this reaction to proceed. But how much?

$$|Q| = 0.00128\text{ u} \times \frac{931\text{ MeV}}{1\text{ u}} = 1.19\text{ MeV}$$

Want Even More Physics Review?
Check out *High School Physics Unlocked*, our skill-building guide to mastering complex physics concepts.

Chapter 10 Review Questions

Answers and explanations can be found in Chapter 11.

Section I: Multiple Choice

1 Mark for Review

According to the theory put forth by Louis de Broglie, all matter has wave-like properties such as interference, but these properties are seen only at a microscopic scale. Why are these properties not typically observed at a macroscopic scale?

(A) The wavelength of matter is typically too large to observe this interference.

(B) Interference of matter occurs only when these waves interact with other objects comparable to their wavelength and those things are microscopic.

(C) There are no energy level transitions available to allow for this interference to be observed.

(D) At the macroscopic scale, the interference is always destructive so it cannot be observed.

2 Mark for Review

An experiment is conducted of the photoelectric effect. A metal is struck with a beam of light with a certain brightness and wavelength. A photoelectron is ejected from the surface of the metal and is found to have a certain stopping potential. Which of the following changes could be made so that the photoelectron would not be ejected from the surface of the metal?

(A) Decrease the brightness of the light

(B) Decrease the wavelength of the light

(C) Increase the wavelength of the light

(D) Increase the stopping potential

3 Mark for Review

An atom with one electron has an ionization energy of 25.0 eV. An electron in this atom makes a transition from an excited energy level, where $E = -16.0$ eV, to the ground state. What is the wavelength of the emitted photon from this transition?

(A) 138 nm

(B) 112 nm

(C) 77.5 nm

(D) 49.6 nm

4 Mark for Review

The single electron in an atom has an energy of –40 eV when it's in the ground state, and the first excited state for the electron is at –10 eV. What will happen to this electron if the atom is struck by a stream of photons, each of energy 15 eV?

(A) The electron will absorb the energy of one photon and become excited halfway to the first excited state, then quickly return to the ground state, emitting a 15 eV photon in the process.

(B) The electron will absorb the energy of one photon and become excited halfway to the first excited state, then quickly absorb the energy of another photon to reach the first excited state.

(C) The electron will absorb two photons and be excited to the first excited state.

(D) Nothing will happen.

5 Mark for Review

The products of several radioactive decays are being studied. Each particle starts with the same speed and enters into a region with a uniform magnetic field directed perpendicular to the initial velocity of the particles. Which observation could be made?

(A) A neutron and an electron are deflected in the same direction, but with the electron turning with a larger radius.

(B) An alpha particle and an electron are deflected in the same direction, but with the alpha particle turning with a larger radius.

(C) An alpha particle and a neutron are both undeflected.

(D) An alpha particle is deflected and a neutron is undeflected.

6 Mark for Review

A partial energy-level diagram for an atom is shown below. What photon energies could this atom emit if it begins in the $n = 3$ state?

−3 eV	$n = 4$
−5 eV	$n = 3$
−8 eV	$n = 2$
−12 eV	$n = 1$ ground state

(A) 5 eV only

(B) 3 eV or 7 eV only

(C) 2 eV, 3 eV, or 7 eV

(D) 3 eV, 4 eV, or 7 eV

7 Mark for Review

Which of the following transitions between energy levels results in emission of the shortest wavelength photon?

(A) A large energy transition to a higher energy level

(B) A small energy transition to a higher energy level

(C) A large energy transition to a lower energy level

(D) A small energy transition to a lower energy level

8 Mark for Review

The de Broglie Hypothesis, that $\lambda = h/p$, was experimentally confirmed by Davisson, Germer, and Thompson. Which of the following describes this hypothesis?

(A) A photon carries a momentum that depends on the wavelength of that photon.

(B) Momentum will be conserved during quantum mechanical processes.

(C) Photons may behave like particles under certain circumstances.

(D) Electrons will undergo diffraction in certain circumstances.

9 Mark for Review

In nuclear reactions, all of the following are true about mass and mass defect EXCEPT

(A) the mass defect is directly proportional to the binding energy

(B) the mass of the unbound nucleons will be less than their mass when combined

(C) the energy associated with the mass defect is significantly greater than the energy levels for electrons in the atom

(D) the mass defect is significantly less than the rest mass of the nucleons

10 Mark for Review

What's the missing particle in the following nuclear reaction?

$${}^{2}_{1}\text{H} + {}^{63}_{29}\text{Cu} \rightarrow {}^{64}_{30}\text{Zn} + (?)$$

(A) Proton

(B) Neutron

(C) Electron

(D) Positron

11 Mark for Review

A particular isotope of platinum ${}^{175}_{78}\text{Pt}$ has a half-life of just over 2.5 seconds. It can decay either via alpha decay or beta(+) decay. What are the daughter nuclei from each of these processes?

(A) Alpha decay results in ${}^{171}_{76}\text{Os}$ and beta(+) decay results in ${}^{175}_{77}\text{Ir}$.

(B) Alpha decay results in ${}^{171}_{76}\text{Os}$ and beta(+) decay results in ${}^{175}_{79}\text{Au}$.

(C) Alpha decay results in ${}^{179}_{80}\text{Hg}$ and beta(+) decay results in ${}^{175}_{77}\text{Ir}$.

(D) Alpha decay results in ${}^{179}_{80}\text{Hg}$ and beta(+) decay results in ${}^{175}_{79}\text{Au}$.

Section II: Free Response

1 Mark for Review

An experiment is carried out with a series of different colored LED lights (which glow by emitting photoelectrons), which are connected one at a time to a variable voltage supply. When the voltage supply is set to 0 V, none of the LED bulbs light. As the voltage is increased, each LED turns on at a different voltage setting. As the voltage is further increased, the brightness of the LED increases.

Data is collected on the color wavelength (and corresponding frequency) for each LED and the smallest value of the voltage where the light is illuminated.

Table 1

Color	Wavelength (nm)	Frequency (10^{14} Hz)	Lighting Voltage (V)
Red	650	4.62	0.51
Yellow	580	5.17	0.74
Green	532	5.64	0.93
Blue	400	7.50	1.71

A. Explain why an increase of $\Delta V = 0.1$ V in the voltage after the light is illuminated causes the brightness to increase, but the bulb remains dark for any of the LEDs with the same voltage increase $\Delta V = 0.1$ V from 0 V to 0.1 V.

B. Plot the data on the axes in Figure 1. Then calculate the slope and the y-intercept of the plot.

Figure 1

Lighting Voltage vs. Frequency

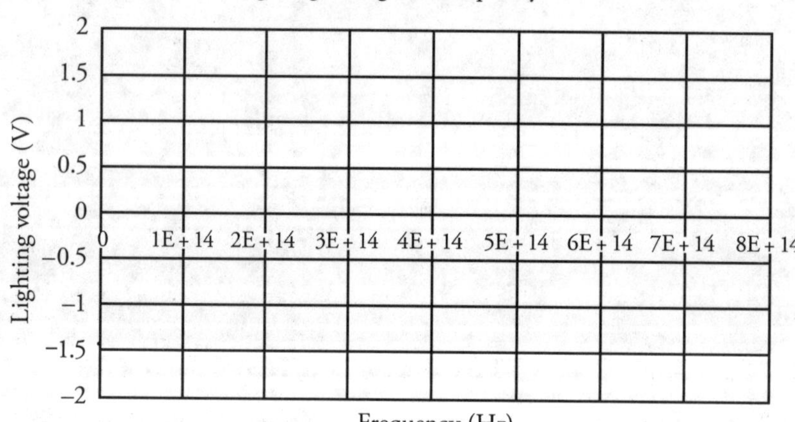

C.

i. Explain how, if possible, this experiment could be repeated to get a plot with the same slope but a different intercept. If this is not possible, explain why.

ii. Explain how, if possible, this experiment could be repeated to get a plot with a different slope but the same intercept. If this is not possible, explain why.

Chapter 10 Summary

- The energy available to liberate electrons near the surface of a metal (the photoelectric effect) is proportional to the frequency of the incident photon. This idea is expressed in $E = hf$, where h is Planck's constant ($h = 6.63 \times 10^{-34}$ J·s).
- The work function (ϕ) indicates the amount of energy needed to liberate the electron. There is a minimum frequency needed to liberate the electrons: $f_0 = \frac{\phi}{h}$. The kinetic energy of the emitted electron is $K_{max} = hf - \phi$.
- Particles in motion have wavelike properties: $\lambda = \frac{h}{p}$, where p is the momentum of the particle. Combining this with $E = hf$ and $c = f\lambda$ yields $E = pc$.
- The standard notation for an element is ${}^{A}_{Z}X$, where A is the mass number, Z is the number of protons in the nucleus, and $A = Z + N$, where N is the number of neutrons in the nucleus.
- A nuclear reaction produces new nuclei, other subatomic particles, and the absorption or emission of energy. The change in mass between the reactants and the products tells how much energy is released (exothermic or $+Q$) or how much energy is needed to produce the reaction (endothermic or $-Q$) in the general equation $\text{A} + \text{B} \rightarrow \text{C} + \text{D} + Q$, where A and B are reactants and C and D are products. This energy released or absorbed is given by $\Delta E = (\Delta m)c^2$.

Chapter 11
Solutions to Chapter Review Questions

CHAPTER 3: THERMODYNAMICS REVIEW QUESTIONS

Section I: Multiple Choice

1. **B** Because the average kinetic energy of a molecule of gas is directly proportional to the temperature of the sample, the fact that the gases are at the same temperature—since they're in the same container at thermal equilibrium—tells you that the molecules have the same average kinetic energy. The ratio of their kinetic energies is therefore equal to 1.

2. **A** P–V diagrams have pressure on the vertical axis and volume on the horizontal axis. You know volume is changing, so the graph will have to show a change in the horizontal axis, so you can eliminate (B). Then, from the Ideal Gas Law, $PV = nRT$, you see that pressure is unchanged if the volume and temperature are both doubled. The vertical value of the P–V diagram remains constant, resulting in a horizontal line on the graph.

3. **A** The work done on the gas during a thermodynamic process is equal to the area of the region in the P–V diagram above the V-axis and below the path the system takes from its initial state to its final state. Since the area below path 1 is the greatest, the work done on the gas during the transformation along path 1 is the greatest.

4. **C** During an isothermal change, ΔU is always zero.

5. **B** There is no way to measure a pressure of 0 Pa, so you can eliminate (A) and (C) as answers. The Ideal Gas Law states $PV = nRT$, so a graph of P versus n will be linear when V and T are both held constant, so you can eliminate (D).

6. **D** In order for the temperature of the gas to remain constant, the added energy from the additional gas needs to be removed. One way to do this is to have heat leave the system. By the Ideal Gas Law, a decrease in pressure would not match with an increase in the amount, n, so (A) is eliminated. Since the volume of the gas is constant in this experiment, no work can be done, so (B) and (C) are eliminated.

7. **A** By convention, work done *on* the gas sample is designated as positive, so in the First Law of Thermodynamics, $\Delta U = Q + W$, you must write $W = +320$ J. Therefore, $Q = \Delta U - W = 560\text{ J} - 320\text{ J} = +240$ J. Positive Q denotes heat *in*.

8. **C** No work is done during the step from state a to state b because the volume doesn't change. Therefore, the work done from a to c is equal to the work done from b to c. Since the pressure remains constant (this step is isobaric), find that

$$W = -P\Delta V = -(3.0 \times 10^5\text{ Pa})[(10 - 25) \times 10^{-3}\text{ m}^3] = 4500\text{ J}$$

9. **C** Choice (A) is wrong because *no heat is exchanged between the gas and its surroundings* is the definition of *adiabatic*, not *isothermal*. Choice (B) cannot be correct since the step described in the question is isothermal; by definition, the temperature does not change. This also eliminates (D) and supports (C). If the sample could be brought back to its initial state *and* have a 100% conversion of heat to work, *that* would violate the Second Law of Thermodynamics, which states that heat cannot be completely converted to work with no other change taking place. In this case, there are changes taking place: the pressure decreases and the volume increases.

10. **C** The Second Law of Thermodynamics indicates that energy will flow from a hot object to a cool object, making the final temperature of the hot object cooler and the cold object warmer.

Section II: Free Response

1. A. First, calculate ΔU_{acb}. Using path *acb*, the question tells you that $Q = +70$ J and $W = -30$ J (W is negative here because it is the *system* that does the work). The first law, $\Delta U = Q + W$, tells you that $\Delta U_{acb} = +40$ J. Because $\Delta U_{a\to b}$ does not depend on the path taken from *a* to *b*, you must have $\Delta U_{ab} = +40$ J, and $\Delta U_{ba} = -\Delta U_{ab} = -40$ J. Thus, $-40 \text{ J} = Q_{ba} + W_{ba}$, where Q_{ba} and W_{ba} are the values along the curved path from *b* to *a*. Since $Q_{ba} = -60$ J, it follows that $W_{ba} = +20$ J. Therefore, the surroundings do 20 J of work on the system.

   B. Again, using the fact that $\Delta U_{a\to b}$ does not depend on the path taken from *a* to *b*, you know that $\Delta U_{adb} = +40$ J, as computed above. Writing $\Delta U_{adb} = QW_{adb} + W_{adb}$, if $W_{adb} = -10$ J, it follows that $Q_{adb} = +50$ J. That is, the system absorbs 50 J of heat.

   C. For the process *db*, there is no change in volume, so $W_{db} = 0$. Therefore, $\Delta U_{db} = Q_{db} + W_{db} = Q_{db}$. Now, since $U_{ab} = +40$ J, the fact that $U_a = 0$ J implies that $U_b = 40$ J, so $\Delta U_{db} = U_b - U_d = 40 \text{ J} - 30 \text{ J} = 10$ J. Thus, $Q_{db} = 10$ J. Now, let's consider the process *ad*. Since $W_{adb} = W_{ad} + W_{db} = W_{ad} + 0 = W_{ad}$, the fact that $W_{adb} = -10$ J [computed in part (b)] tells you that $W_{ad} = -10$ J. Because $\Delta U_{ad} = U_d - U_a = 30$ J, it follows from $\Delta U_{ad} = Q_{ad} + W_{ad}$ that $Q_{ad} = \Delta U_{ad} - W_{ad} = 30 \text{ J} - (-10 \text{ J}) = 40$ J.

   D. The process *adbca* is cyclic, so ΔU is zero. Because this cyclic process is traversed *counterclockwise* in the *P–V* diagram, you know that W is *positive*. Then, since $\Delta U = Q + W$, it follows that Q must be negative.

2. A. i. Use the Ideal Gas Law:

$$T_a = \frac{P_aV_a}{nR} = \frac{(2.4\times10^5 \text{ Pa})(12\times10^{-3} \text{ m}^3)}{(0.4 \text{ mol})(8.31 \text{ J/mol}\cdot\text{K})} = 870 \text{ K}$$

   ii. Since state *b* is on the isotherm with state *a*, the temperature of state *b* must also be 870 K.

   iii. Use the Ideal Gas Law:

$$T_c = \frac{P_cV_c}{nR} = \frac{(0.6\times10^5 \text{ Pa})(12\times10^{-3} \text{ m}^3)}{(0.4 \text{ mol})(8.31 \text{ J/mol}\cdot\text{K})} = 220 \text{ K}$$

B. There is heat flow into the system during step *ab*. As the volume expands, the system does work on the surroundings (uses up some of its internal energy). As a result, the system would drop in temperature if no heat flowed into the gas to make up for this use of energy. During an isothermal process, $\Delta T = 0$ K, so the change in internal energy of the gas, $\Delta U = \frac{3}{2} Nk_b\Delta T$, is zero. From the First Law of Thermodynamics, $\Delta U = W + Q$, any energy that is used by expanding (work done by the system on the surroundings) must be replaced by heat flowing into the system.

C. Using the equation given, find that

$$W_{ab} = -nRT \cdot \ln \frac{V_b}{V_a} = -(0.4 \text{ mol})(8.31 \text{ J/mol} \cdot \text{K})(870 \text{ K}) \cdot \ln \frac{48 \times 10^{-3} \text{ m}^3}{12 \times 10^{-3} \text{ m}^3}$$
$$= -4000 \text{ J}$$

D. The total work done over the cycle is equal to the sum of the values of the work done over each step:

$$\begin{aligned} W_{\text{cycle}} &= W_{ab} + W_{bc} + W_{ca} \\ &= W_{ab} + W_{bc} \\ &= (-4000 \text{ J}) + (2200 \text{ J}) \\ &= -1800 \text{ J} \end{aligned}$$

CHAPTER 4: ELECTRIC FORCES AND FIELDS REVIEW QUESTIONS

Section I: Multiple Choice

1. **D** Electrostatic force obeys an inverse-square law: $F_E \propto 1/r^2$. Therefore, a plot should be made of *F* versus the inverse square of *r* to get a line.

2. **C** The strength of the electric force is given by kq^2/r^2, and the strength of the gravitational force is Gm^2/r^2. Since both of these quantities have r^2 in the denominator, you simply need to compare the numerical values of kq^2 and Gm^2. There's no contest. Since

$$kq^2 = (9 \times 10^9 \text{ N·m}^2/\text{C}^2)(1 \text{ C})^2 = 9 \times 10^9 \text{ N·m}^2$$

and

$$Gm^2 = (6.7 \times 10^{-11} \text{ N·m}^2/\text{kg}^2)(1 \text{ kg})^2 = 6.7 \times 10^{-11} \text{ N·m}^2$$

you can see that $kq^2 >> Gm^2$, so F_E is much stronger than F_G.

3. **C** If the net electric force on the center charge is zero, the electrical repulsion by the $+2q$ charge must balance the electrical repulsion by the $+3q$ charge:

$$\frac{1}{4\pi\varepsilon_0}\frac{(2q)(q)}{x^2} = \frac{1}{4\pi\varepsilon_0}\frac{(3q)(q)}{y^2} \Rightarrow \frac{2}{x^2} = \frac{3}{y^2} \Rightarrow \frac{y^2}{x^2} = \frac{3}{2} \Rightarrow \frac{y}{x} = \sqrt{\frac{3}{2}}$$

4. **D** Since P is equidistant from the two charges, and the magnitudes of the charges are identical, the strength of the electric field at P due to $+Q$ is the same as the strength of the electric field at P due to $-Q$. The electric field vector at P due to $+Q$ points away from $+Q$, and the electric field vector at P due to $-Q$ points toward $-Q$. Since these vectors point in the same direction, the net electric field at P is (E to the right) + (E to the right) = ($2E$ to the right).

5. **C** The acceleration of the small sphere is

$$a = \frac{F_E}{m} = \frac{1}{4\pi\varepsilon_0}\frac{Qq}{mr^2}$$

As r increases (that is, as the small sphere is pushed away), a decreases. However, since a is always positive, the small sphere's speed, v, is always increasing.

6. **B** Since $\mathbf{F}_E$ (on q) = $q\mathbf{E}$, it must be true that $\mathbf{F}_E$ (on $-2q$) = $-2q\mathbf{E} = -2\mathbf{F}_E$.

7. **C** All excess electric charge on a conductor resides on the outer surface.

Section II: Free Response

1. A. i. The force from the charge at A will be repulsive and must act at a 45° angle because the force acts along the bisector of the square. The balancing forces from the charges q will be at right angles to one another, and so each must balance a component of the force from Q alone. Since sin(45°) = cos(45°), the forces from each q must be equal, and therefore the charges must also be equal.

ii. In order for the force on C to be 0, the repulsive force from charge A must be balanced by an attractive force from each q, so the charges at positions B and D must have a negative sign.

A free body diagram of the charge at point C is drawn below.

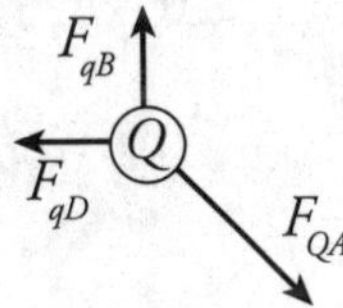

The distance between opposite vertices of a square of side length s is $s\sqrt{2}$. You can look at either the x- or y-components because the equations each direction yields are identical. From the x direction, $F_{QA}\sin(45°) = F_{qD}$. Therefore, $F_{qD} = k\frac{Q^2}{(s\sqrt{2})^2} \times \frac{\sqrt{2}}{2}$.

$$F_{qD} = k\frac{\sqrt{2}Q^2}{4s^2}$$

Now that you know the force from the charge at D, use Coulomb's Law to determine q.

$$F_{qD} = k\frac{Qq}{s^2} = k\frac{\sqrt{2}Q^2}{4s^2}$$

$$Qq = \frac{\sqrt{2}Q^2}{4}$$

$$q = \frac{\sqrt{2}Q}{4}$$

B. Within the square, the field from the charge Q at position A will point into the fourth quadrant (the x-component will be positive and the y-component will be negative). With the charges positive, within the square the field from D will point into the first quadrant (the x-component will be positive and the y-component will be positive), and from B will point into the third quadrant (the x-component will be negative and the y-component will be negative). To cancel the fields, the x-coordinate would have to be closer to point B than to point D, and the y-coordinate would have to be closer to D than to B. Some point which satisfies these two requirements will have a field of 0 N/C.

C. At the center of the square, the net electric field is the vector sum of the contributions from the four charges. The field from the charges at A and C will always balance one another as long as those charges have the same sign, regardless of that sign. First, the charges are equidistant from the center of the square, so r will be equal regardless of the sign or magnitude of the charges. If they are both positive, the field from A will be in the fourth quadrant and the field from C will be in the second quadrant. If both charges are negative, then the field from C will be in the fourth quadrant and the field from A will be in the second quadrant. Therefore, regardless of the magnitude or sign of the charges, the contributions from the charges Q will cancel one another. An identical argument can be made for the charges q showing that their fields will also cancel with one another. Therefore, the net field at the center of the square must be 0 N/C regardless of the signs or magnitudes of Q or q.

2. A. The electric force on Charge 1 is $F_1 = k\dfrac{Q \cdot 2Q}{(a+2a)^2} = k\dfrac{2Q^2}{9a^2}$. The force between the charges is repulsive, so the charge at 1 is pushed upward.

B. $\overrightarrow{E_{net}} = \overrightarrow{E_1} + \overrightarrow{E_2} = k\dfrac{Q}{a^2}$ down $+ k\dfrac{2Q}{(2a)^2}$ up. Making upward positive results in $E_{net} = k\dfrac{Q}{2a^2} - k\dfrac{Q}{a^2} = -k\dfrac{Q}{a^2}$.

The net field is $k\dfrac{Q}{2a}$ in the downward direction.

C. No. The only point on the x-axis where the individual electric field vectors due to each of the two charges point in exactly opposite directions is the origin (0, 0). But at that point, the two vectors are not equal and thus do not cancel.

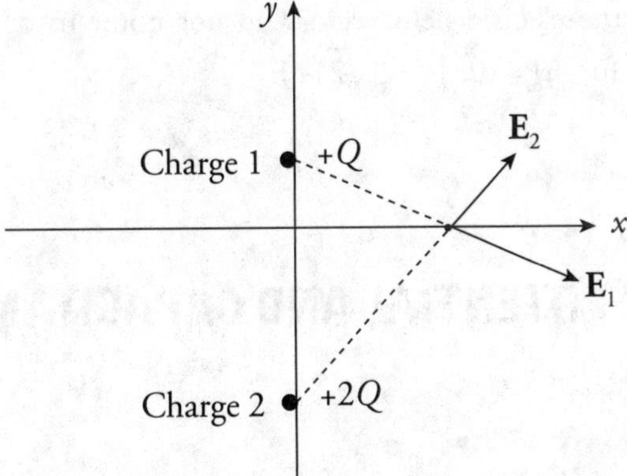

Therefore, at no point on the x-axis could the total electric field be zero.

D. Yes. There will be a Point P on the y-axis between the two charges,

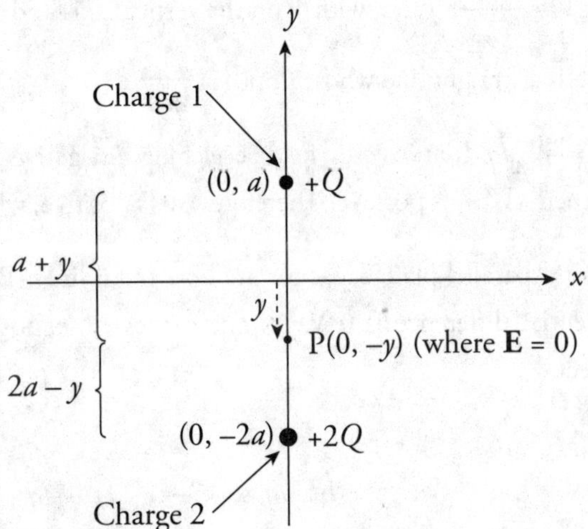

where the electric fields due to the individual charges will cancel each other out.

$$E_1 = E_2$$

$$\frac{1}{4\pi\varepsilon_0}\frac{Q}{(a+y)^2} = \frac{1}{4\pi\varepsilon_0}\frac{2Q}{(2a-y)^2}$$

$$\frac{1}{(a+y)^2} = \frac{2}{(2a-y)^2}$$

$$(2a-y)^2 = 2(a+y)^2$$

$$0 = y^2 + 8ay - 2a^2$$

$$y = \frac{-8a \pm \sqrt{(8a)^2 - 4(-2a^2)}}{2}$$

$$= (-4 \pm 3\sqrt{2})a$$

Disregarding the value $y = (-4-3\sqrt{2})a$ (because it would place the point P below Charge 2 on the y-axis, where the electric field vectors do not point in opposite directions), you find that $\mathbf{E} = 0$ at the point P = $(0, -y) = (0, (4-3\sqrt{2})a)$.

CHAPTER 5: ELECTRIC POTENTIAL AND CAPACITANCE REVIEW QUESTIONS

Section I: Multiple Choice

1. **B** For a point particle, which a uniform sphere will behave like, the relationship between the electric potential and the distance is $V = \frac{1}{4\pi\varepsilon_0}\frac{q}{r}$. So, with V on the vertical axis and $\frac{1}{r}$ on the horizontal axis, a graph will be produced which is a straight line whose slope is $\frac{1}{4\pi\varepsilon_0}q$.

2. **B** Use the definition $\Delta V = -W_E/q$. If an electric field accelerates a negative charge doing positive work on it, then $W_E > 0$. If $q < 0$, then $-W_E/q$ is positive. Therefore, ΔV is positive, which implies that V increases.

3. **D** Equipotential curves for a parallel-plate capacitor are horizontal lines. Both (A) and (B) are at the same potential, so their potential difference is 0 V. As one moves closer to the $+Q$ plate, the potential will increase, so (D) is correct.

4. **D** By definition,

$$V_{A\to B} = \Delta U_E/q, \text{ so } V_B - V_A = \Delta U_E/q$$

5. **B** Because **E** is uniform, the potential varies linearly with distance from either plate ($\Delta V = Ed$). Since Points 2 and 4 are at the same distance from the plates, they lie on the same equipotential. (The equipotentials in this case are planes parallel to the capacitor plates.)

6. **B** The charge Q must be positive because the potential decreases as the distance from the charge increases. A negative charge will be attracted to a positive charge. This means that the electric field of Q would do positive work if q were brought closer, and it would do negative work if q were moved farther away.

Section II: Free Response

1. A. Greatest <u>*A* and *B*</u> <u>*A*</u> <u>*A*, *B*, and *C*</u> <u>All four</u> Least

The work required to move a charge through an existing electric field can be related to the voltage of that field at the initial and final positions of the charge. All voltages are defined to be 0 when the charges are very far away from one another, so V_i is always 0 V. Charge A is placed first and there is no electric field that charge A is moved through, so the charge distribution has no energy. At location B, in the presence of charge A, there is an electric field and you can calculate the voltage at point B as $V_B = \frac{+kQ}{s}$. Moving the charge to point B adds $W = \frac{kQ^2}{s}$ to the charge distribution. At location C, in the presence of charges at A and B, calculate the voltage as the sum of the contributions from A and B as $V_C = \frac{kQ}{\sqrt{2}s} + \frac{kQ}{s}$. Charge C is negative, so putting it at position C removed $W = \frac{kQ^2}{s} + \frac{kQ^2}{\sqrt{2}s}$, making the energy of the distribution $-\frac{kQ^2}{\sqrt{2}s} = -\frac{\sqrt{2}kQ^2}{2s}$. Finally, the voltage at position D from the charges at A, B, and C, respectively, is $V_D = \frac{kQ}{s} + \frac{kQ}{\sqrt{2}s} - \frac{kQ}{s} = \frac{kQ}{\sqrt{2}s}$. Charge D is negative, so putting it at position D removed $W = \frac{kQ^2}{\sqrt{2}s}$ from the distribution. The energy in the distribution of all four charges is $U_E = -\frac{2kQ^2}{\sqrt{2}s} = -\frac{2\sqrt{2}kQ^2}{2s}$.

B. At the exact center of the square, this distance from each charge is d, where from the Pythagorean Theorem, $d^2 = \left(\frac{s}{2}\right)^2 + \left(\frac{s}{2}\right)^2$, so $d = \frac{\sqrt{2}s}{2}$. The voltage at the center is then the sum of the voltages from each charge: $V_{center} = V_A + V_B + V_C + V_D = \frac{+kQ}{d} + \frac{+kQ}{d} + \frac{-kQ}{d} + \frac{-kQ}{d} = 0 \text{ V}$.

C. The equipotential will be described as always the same distance from the pair at *AD* as well as the same distance from the pair at *BD*. Thus, the equipotential line is a horizontal line through the square.

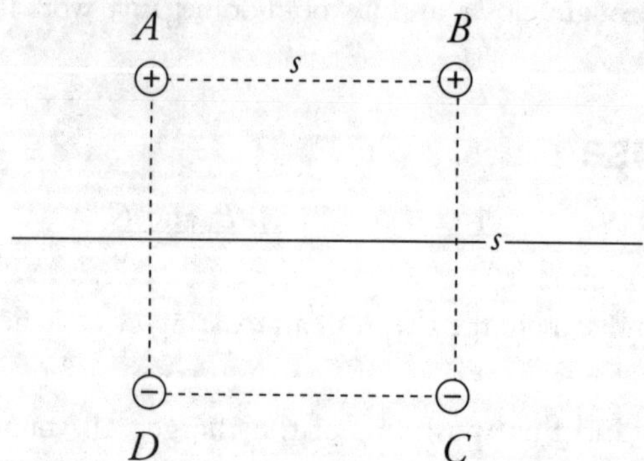

D. The electric field along the center line will be uniform because any diminishment from the influence of the top plate is counteracted by an increased influence from the lower plate. However, at a position not between the plates, this is no longer true, and so for the bottom dots, the lower plate is more influential. This causes the electric field at these locations to tend to point toward the negatively charged plate.

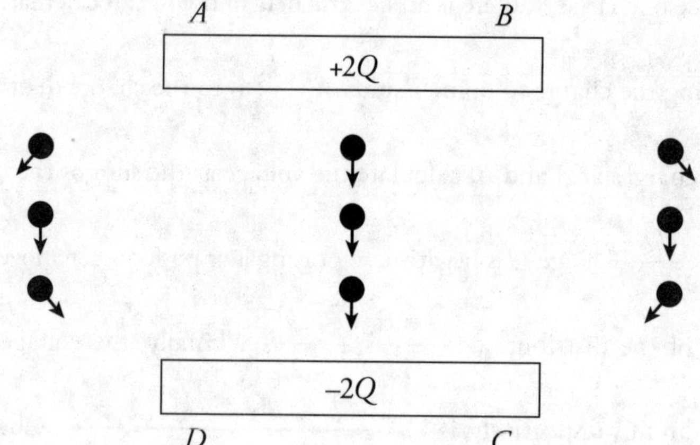

2.

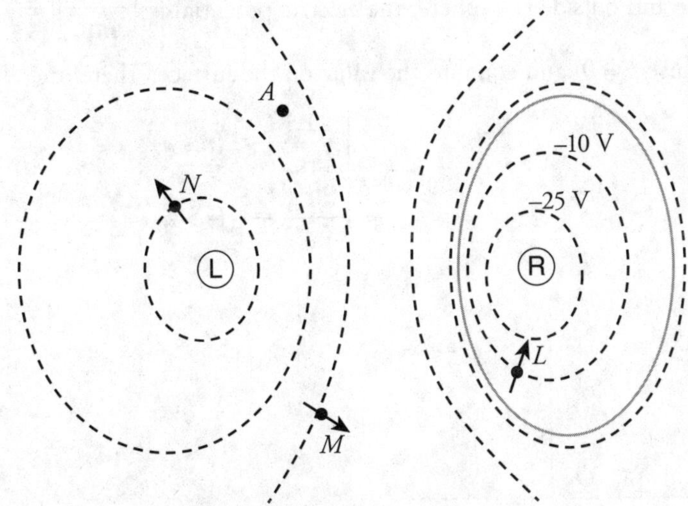

A. Because the potential difference between adjacent lines is $\Delta V = 15$ V and the potential is going from –25 V to –10 V on the two lines closest to R, the 0 V isoline must be in between the line labeled –10 V and the next line in the image. The light gray solid line is the line where the potential is 0 V.

B. 42.5 V

C. i. The potential is increasing as the isolines get farther away from R, so R must be negative and L must be positive.

ii. R carries the greater magnitude of charge, as the isolines are closer together in the neighborhood of R, which indicates a stronger magnitude of electric field.

D. i. See image above.

ii. $L > M > N$. At L, the isolines are closest together, and at N, they are farthest apart.

E. L carries positive charge, and potential is larger (relative to positions near negative charges) at positions near positive charges. Making L negatively charged would therefore make the potential at point N lower than it is when L is positive.

3. A. Outside the sphere, the sphere behaves as if all the charge were concentrated at the center. Inside the sphere, the electrostatic field is zero:

$$E(r) = \begin{cases} 0 & (r < a) \\ \dfrac{1}{4\pi\varepsilon_0}\dfrac{Q}{r^2} & (r > a) \end{cases}$$

B. On the surface and outside the sphere, the electric potential is $\left(\frac{1}{4\pi\varepsilon_0}\right)\left(\frac{Q}{r}\right)$. Within the sphere, V is constant (because $E = 0$) and equal to the value on the surface. Therefore,

$$V(r) = \begin{cases} \frac{1}{4\pi\varepsilon_0}\frac{Q}{a} & (r \le a) \\ \frac{1}{4\pi\varepsilon_0}\frac{Q}{r} & (r > a) \end{cases}$$

C. See diagrams:

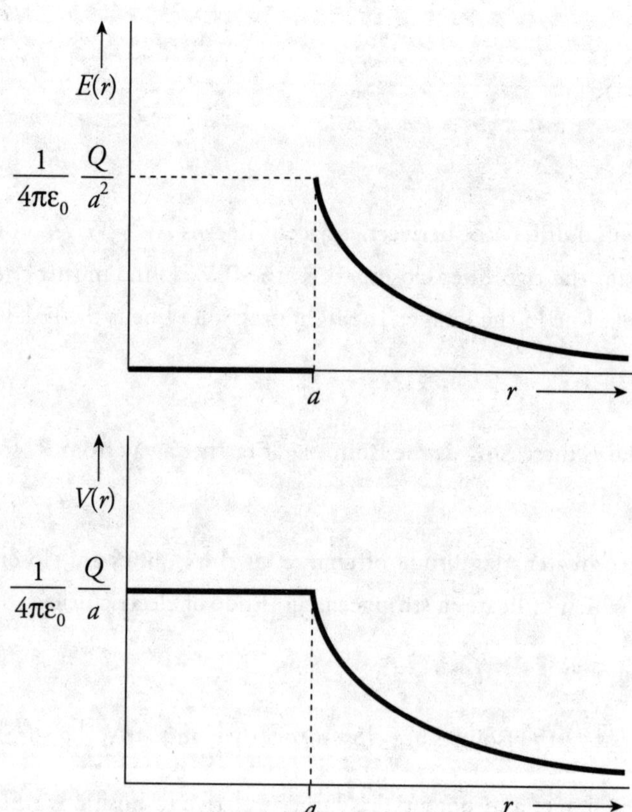

CHAPTER 6: ELECTRIC CIRCUITS REVIEW QUESTIONS

Section I: Multiple Choice

1. **A** Let ρ_S denote the resistivity of silver and let A_S denote the cross-sectional area of the silver wire. Then

$$R_B = \frac{\rho_B L}{A_B} = \frac{(5\rho_S)L}{4^2 A_S} = \frac{5}{16}\frac{\rho_S L}{A_S} = \frac{5}{16}R_S$$

2. **C** The equation $I = V/R$ implies that increasing V by a factor of 2 will cause I to increase by a factor of 2.

3. **D** Initially the uncharged capacitor will have no voltage across it and will therefore have essentially zero resistance. This means the charge flows freely through it and all the current will flow through the path with the capacitor rather than the path with the resistor. After a long time, the capacitor will become fully charged and will not allow any more charge to add to or leave the plates, so it essentially has infinite resistance. Therefore, none of the current will flow through that path and all of the current flows through the path with the resistor.

4. **C** The resistivity ρ is found from the resistance $R = \rho L/A$. To find the resistance, Ohm's Law $V = IR$ is needed. In addition to V and L, you need to find both I and A. The ammeter will yield I, and the diameter of the wire will allow you to determine A.

5. **C** Because the voltage has been applied for a long time, the capacitor will be fully charged, so it behaves like an open switch. This puts the 4 Ω resistor and the top 3 Ω resistor in series with one another, making an equivalent resistor of 7 Ω. That combination is in parallel with the other 3 Ω resistor, so $\frac{1}{R_{eq}} = \frac{1}{7\Omega} + \frac{1}{3\Omega} \rightarrow \frac{1}{R_{eq}} = \frac{0.48}{\Omega} \rightarrow R_{eq} = 2.1\Omega$.

6. **C** If each of the identical bulbs has resistance R, then the current through each bulb is ε/R. This is unchanged if the middle branch is taken out of the parallel circuit. (What *will* change is the total amount of current supported by the battery.)

7. **B** The three parallel resistors are equivalent to a single 2 Ω resistor, because $\frac{1}{8\,\Omega} + \frac{1}{4\,\Omega} + \frac{1}{8\,\Omega} = \frac{1}{2\,\Omega}$. This 2 Ω resistance is in series with the given 2 Ω resistor, so their equivalent resistance is 2 Ω + 2 Ω = 4 Ω. Therefore, three times as much current will flow through this equivalent 4 Ω resistance in the top branch as through the parallel 12 Ω resistor in the bottom branch, which implies that the current through the bottom branch is 3 A, and the current through the top branch is 9 A. The voltage drop across the 12 Ω resistor is therefore $V = IR$ = (3 A)(12 Ω) = 36 V.

8. **A** Ohm's Law states that $V = IR$, so a graph of V versus I will have a slope equal to the resistance. If the battery has an internal resistance, then the circuit will have a larger resistance than anticipated, which will lead to a steeper slope. The graph, however, will still show a directly proportional relationship between V and I.

9. **D** The equation $P = I^2R$ gives

$$P = (0.5\text{ A})^2(100\ \Omega) = 25\text{ W} = 25\text{ J/s}$$

Therefore, in 20 s, the energy dissipated as heat is

$$E = Pt = (25\text{ J/s})(20\text{ s}) = 500\text{ J}$$

10. **D**

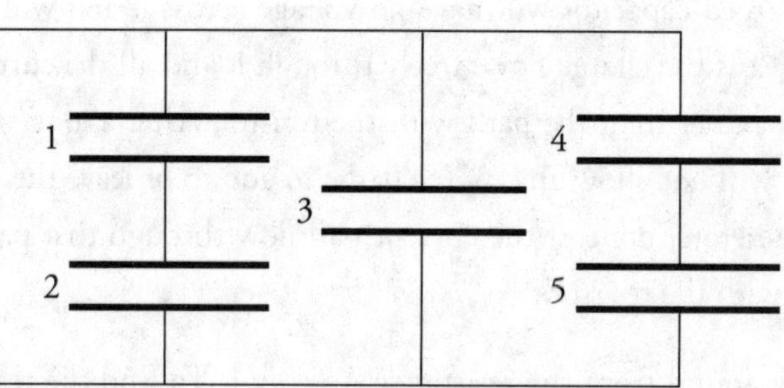

Capacitors 1 and 2 are in series, so their equivalent capacitance is $C_{1\text{-}2} = C/2$. (This is obtained from the equation $1/C_{1\text{-}2} = 1/C_1 + 1/C_2 = 1/C + 1/C = 2/C$.) Capacitors 4 and 5 are also in series, so their equivalent capacitance is $C_{4\text{-}5} = C/2$. The capacitances $C_{1\text{-}2}$, C_3, and $C_{4\text{-}5}$ are in parallel, so the overall equivalent capacitance is $(C/2) + C + (C/2) = 2C$.

Section II: Free Response

1. Begin by labeling each resistor and making a chart of the resistance, voltage, current, and power for each resistor.

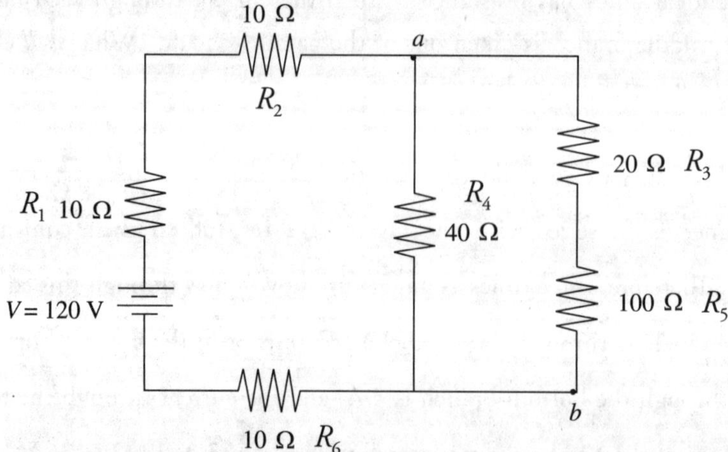

	Resistance (Ω)	Voltage (V)	Current (A)	Power (W)
R_1	10			
R_2	10			
R_3	20			
R_4	40			
R_5	100			
R_6	10			
Entire Circuit		120		

The two parallel branches, the one containing the 40 Ω resistor and the other a total of 120 Ω, are equivalent to a single 30 Ω resistance. This 30 Ω resistance is in series with the three 10 Ω resistors, giving an overall circuit resistance of 10 Ω + 10 Ω + 30 Ω + 10 Ω = 60 Ω.

	Resistance (Ω)	Voltage (V)	Current (A)	Power (W)
R_1	10			
R_2	10			
R_3	20			
R_4	40			
R_5	100			
R_6	10			
Entire Circuit	**60**	120	120/60 = 2	$120^2/60 = 240$

R_1, R_2, and R_6 are all in series with the battery, so they all have the same current as the current supported by the battery.

	Resistance (Ω)	Voltage (V)	Current (A)	Power (W)
R_1	10	2 * 10 = 20	**2**	$2^2 * 10 = 40$
R_2	10	2 * 10 = 20	**2**	$2^2 * 10 = 40$
R_3	20			
R_4	40			
R_5	100			
R_6	10	2 * 10 = 20	**2**	$2^2 * 10 = 40$
Entire Circuit	60	120	2	240

There is a loop $V - V_1 - V_2 - V_4 - V_6 = 0$, so $120 - 20 - 20 - V_4 - 20 = 0$ and $V_4 = 60$ V.

	Resistance (Ω)	Voltage (V)	Current (A)	Power (W)
R_1	10	20	2	40
R_2	10	20	2	40
R_3	20			
R_4	40	**60**	60/40 = 1.5	$60^2/40 = 90$
R_5	100			
R_6	10	20	2	40
Entire Circuit	60	120	2	240

Finally, you know that 2 A of current flows into the junction point labeled *a* and 1.5 A flows into R_4, which leaves 2 A – 1.5 A = 0.5 A to travel down the other branch.

	Resistance (Ω)	Voltage (V)	Current (A)	Power (W)
R_1	10	20	2	40
R_2	10	20	2	40
R_3	20	0.5 * 20 = 10	**0.5**	$0.5^2 * 20 = 5$
R_4	40	60	1.5	90
R_5	100	0.5 * 100 = 50	**0.5**	$0.5^2 * 100 = 25$
R_6	10	20	2	40
Entire Circuit	60	120	2	240

Check the power sum: 40 + 40 + 5 + 90 + 25 + 40 = 240 is correct. Use the completed chart to answer the questions.

A. The battery delivers 240 W of power.

B. There is 1.5 A of current through the 40 Ω resistor.

C. i. The potential difference between points *a* and *b* is the voltage across R_4, so it is 60 V.

ii. The higher potential is near the long pole of the battery in the diagram, so *a* is 60 V higher in potential than *b*.

D. Using $R = \rho L/A$ gives $100 = 0.45(0.04)/\pi r^2$ so $r = \sqrt{\dfrac{0.45(0.04)}{100\pi}} = 0.0076\text{ m}$.

E. In order for there to be no current flow supported by the battery after the capacitor is fully charged, the capacitor would have to be in series with the battery. Then, once the capacitor is full, it would have the same voltage drop as the battery supplies and the current in the circuit would go to 0 A. This happens if you replace any of the 10 Ω resistors with a capacitor.

2. Begin by making a chart of the resistance, voltage, current, and power for each resistor.

	Resistance (Ω)	Voltage (V)	Current (A)	Power (W)
R_1	10			
R_2	60			
R_3	20			
R_4	40			
R_5	60	0.5 * 60 = 30	0.5	0.5^2 * 60 = 15
Entire Circuit				

To calculate the equivalent resistance of the entire circuit, there are three parallel branches each with a resistance of 60 Ω. Those are equivalent to a single 20 Ω resistor. That is in series with a 10 Ω resistor.

	Resistance (Ω)	Voltage (V)	Current (A)	Power (W)
R_1	10			
R_2	60			
R_3	20			
R_4	40			
R_5	60	30	0.5	15
Entire Circuit	**30**			

As the standard approach did not result in a row with two entries in a column, you must examine the circuit to see whether the Loop Rule or Junction Rule will be more useful as the next step. You can see that the three parallel branches all have the same resistance, and therefore they must all have the same current of 0.5 A flowing through them. This is a result of the Loop Rule stating that parallel branches must have equal voltages.

	Resistance (Ω)	Voltage (V)	Current (A)	Power (W)
R_1	10			
R_2	60	0.5 * 60 = 30	**0.5**	0.5^2 * 60 = 15
R_3	20	0.5 * 20 = 10	**0.5**	0.5^2 * 20 = 5
R_4	40	0.5 * 40 = 20	**0.5**	0.5^2 * 40 = 10
R_5	60	30	0.5	15
Entire Circuit	30			

As you just used the Loop Rule, you can now use the Junction Rule to see that $I_1 = I_2 + I_3 + I_5 = 1.5$ A. Also, since R_1 is in series with the battery, you can complete the chart.

	Resistance (Ω)	Voltage (V)	Current (A)	Power (W)
R_1	10	1.5 * 10 = 15	**1.5**	1.5^2 * 10 = 22.5
R_2	60	30	0.5	15
R_3	20	10	0.5	5
R_4	40	20	0.5	10
R_5	60	30	0.5	15
Entire Circuit	30	1.5 * 30 = 45	**1.5**	1.5^2 * 30 = 67.5

A., B., C. The answers can now simply be read off the chart.

D. i. When the capacitor is uncharged, the rightmost 60 Ω resistor (along with the 20 Ω and 40 Ω series combination) will be shorted out and will have a current of 0 A flowing through them.

ii. The current will be greater than 0.5 A. The full capacitor acts like a resistor with infinite resistance, which essentially removes a parallel branch from the circuit. Removing a parallel branch lowers the current throughout the circuit, so it lowers the voltage drop across the 10 Ω resistor and increases the voltage drop across the remaining parallel branches. Thus, it increases the current through the remaining 60 Ω resistor.

3. A. First, find the equivalent resistance for the whole circuit. First, the total resistance of the two parallel resistors can be found by using $\frac{1}{R_T} = \frac{1}{R_2} + \frac{1}{R_3} = \frac{R_2 + R_3}{R_2R_3}$, which means $R_T = \frac{R_2R_3}{R_2 + R_3}$. Then this would be in series with the first resistor, so the equivalent resistance of the whole circuit, R_{eq}, would be $R_{eq} = R_1 + R_T = R_1 + \frac{R_2R_3}{R_2 + R_3} = \frac{R_1R_2 + R_1R_3 + R_2R_3}{R_2 + R_3}$.

Next, you know from Ohm's Law that $V = IR$, so you can say $I = V/R = \frac{V}{\frac{R_1R_2 + R_1R_3 + R_2R_3}{R_2 + R_3}} = \frac{V(R_2 + R_3)}{R_1R_2 + R_1R_3 + R_2R_3}$. This will be the total current of the circuit. Because there are no branches before R_1, this will also be the current flowing through that resistor. So the voltage drop of that resistor, V_1, can be found by again using Ohm's Law for that location rather than the whole resistor. This gives $V_1 = I_1R_1 = \frac{V(R_2 + R_3)}{R_1R_2 + R_1R_3 + R_2R_3}R_1$.

B. The diagram should look like this:

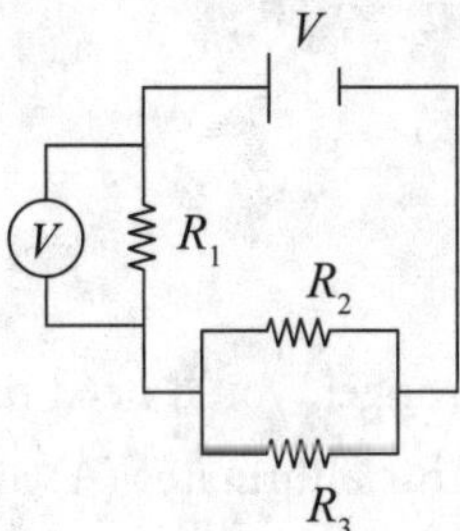

In order to measure the voltage loss of a particular resistor, the voltmeter must be arranged in parallel with that resistor. This is because parallel elements always have equal voltage drops. Thus, whatever voltage drop the device measures will be the same as the voltage drop in the resistor being measured.

C. The diagram should look like this:

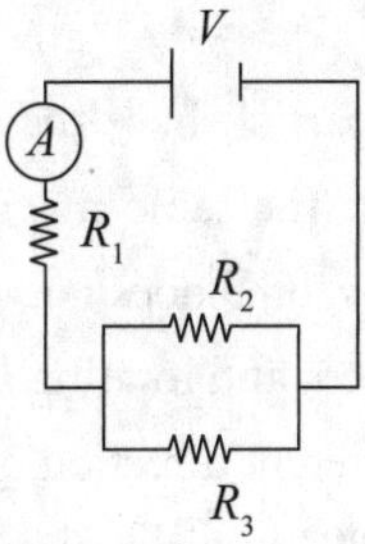

In order to measure the current passing through a particular resistor, the ammeter must be arranged in series with that resistor. This is because elements in series always have equal currents. Thus, whatever current the device measures will be the same as the current flowing through the resistor being measured.

D. For the voltmeter, infinite resistance would be ideal. As in all measurements, you don't want to disturb the system in any way as you take the measurement. Otherwise, your readings would not be accurate for the original system. Thus, in order to leave the circuit as it was, you need to maintain the flow of current that existed before you started your measurement. Having an infinite resistance would ensure no current flows into the voltmeter, leaving it all on its original path.

For the ammeter, 0 resistance would be ideal. Again, you want to have minimal disturbance on the system. If the ammeter did have resistance, then it would be a source of voltage loss that did not previously exist. In essence, it would be an additional resistor that would need to be considered, forcing you to recalculate everything from the ground up.

CHAPTER 7: MAGNETISM AND ELECTROMAGNETISM REVIEW QUESTIONS

Section I: Multiple Choice

1. **B** Because the current in wire 1 is stronger, you'll need to be further from that wire so that the magnitudes of the fields end up equal. This eliminates (A) and (C). Current-carrying wires produce fields which are circles around the wire. If the currents in each wire are going opposite directions, then in between the wires, the magnetic fields would go the same direction. Therefore, the currents must be in the same direction so that the magnetic fields in between the wires are in the opposite directions and (D) is also eliminated.

2. **D** The magnetic field strength for a current-carrying wire is given by $\mathbf{B} = \frac{\mu_0}{2\pi}\frac{I}{r}$. A graph of **B** versus I will give a line with a slope of $\frac{\mu_0}{2\pi}\frac{1}{r}$.

3. **B** All of the conditions for the second particle have the same magnitude of force as the original. In order to have the new experiment result in the same magnitude of the force, $F_B = |q||\mathbf{v}||\mathbf{B}|\sin(\theta)$ must be unchanged. Changing the signs of q, **v**, or **B** does not change the strength of the force. Rotating **v** and **B** by the same amount leaves θ unchanged, and rotating **B** by 180° results in the same value for $\sin(\theta)$. Thus, the invalid conditions must have a different direction of the magnetic force. Using the Right-Hand Rule, rotating your thumb and fingers by 90° into the plane of the paper causes the magnetic force to rotate from the original direction out of the plane (recalling q is negative) to finally point to the left.

4. **C** Relative to a current-carrying wire, the measured value of the magnetic field is directly proportional to the current and inversely proportional to the distance the measurement is taken. Therefore, a large current at a small distance will result in the largest measured magnetic field value.

5. **D** The strength of the magnetic field at a distance r from a long, straight wire carrying a current I is given by the equation $B = (\mu_0/2\pi)(I/r)$. Therefore, the strength of the magnetic field at Point P due to either wire is $B = (\mu_0/2\pi)(I/\frac{1}{2}d)$. By the Right-Hand Rule, the direction of the magnetic field at P due to the top wire is into the plane of the page and the direction of the magnetic field at P due to the bottom wire is out of the plane of the page. Since the two magnetic field vectors at P have the same magnitude but opposite directions, the net magnetic field at Point P is zero.

6. **C** Use the Right-Hand Rule for wires. If you point your thumb to the right and wrap your fingers along the wire, you will note that the magnetic field goes into the page below the wire and comes out of the page above the wire. This allows you to eliminate (A) and (B). Because $B = \frac{\mu_o I}{2\pi r}$, the closer you are to the wire, the stronger the magnetic field. Choice (C) is closer, so it is the correct answer.

7. **C** Magnetic fields point from north to south. Therefore, the magnetic field between the two magnets is toward the right of the page. Use the Right-Hand Rule. Because the B field is to the right and the charges through the wire flow to the bottom of the page, the force must be out of the page.

8. **C** Since **v** is upward and **B** is into the page, the direction of **v** × **B** is to the left. Therefore, free electrons in the wire will be pushed to the right, leaving an excess of positive charge at the left. Therefore, the potential at point a will be higher than that at point b, by $\varepsilon = BLv$.

9. **A** The magnitude of the emf induced between the ends of the rod is $\varepsilon = B\ell v = (0.5\text{ T})(0.2\text{ m})(3\text{ m/s}) = 0.3\text{ V}$. Since the resistance is 10 Ω, the current induced will be $I = \varepsilon / R = (0.3\text{ V})/(10\ \Omega) = 0.03\text{ A}$. To determine the direction of the current, note that since positive charges in the rod are moving to the left and the magnetic field points into the plane of the page, the Right-Hand Rule states that the magnetic force, $q\boldsymbol{v} \times \boldsymbol{B}$, points downward. Since the resulting force on the positive charges in the rod is downward, so is the direction of the induced current.

10. **C** First, you can eliminate (A) and (B). By definition, magnetic field lines emerge from the north pole and enter at the south pole. The magnetic field from the bar magnet will always point toward a viewer who is looking down at the loop from above. As the north pole gets closer to the loop, the field at the loop grows in strength, but as the south pole recedes from the loop, the magnetic field strength shrinks. Because of this, the flux changes from growing to shrinking as the magnet passes through the loop and the induced current must also change directions. Therefore, (A) and (B) are wrong. To determine whether (C) or (D) is correct, look at the first half of the motion. As the north pole gets closer to the loop, the magnetic flux increases. To oppose an increasing flux, the direction of the magnetic field generated by induced current must be downward. Looking from above, a clockwise induced current generates a downward magnetic field. Therefore, (C) is correct.

11. **C** An induced emf requires that the magnetic flux through a loop of wire change over time. Changing the area of the loop of wire will always cause the flux to change. Choice (B) can cause an emf, but will not necessarily, depending on the axis of rotation. Choice (D) is also incorrect because while a time varying field will cause an emf, it is not necessary to have such a field to produce an emf.

Section II: Free Response

1. A. The acceleration of an ion of charge q is equal to F_E/m. The electric force is equal to qE, where $E = V/d$. Therefore, $a = qV/(dm)$.

   B. Using $a = qV/(dm)$ and the equation $v^2 = v_0^2 + 2ad = 2ad$, you get

$$v^2 = 2\frac{qV}{dm}d \quad \Rightarrow \quad v = \sqrt{\frac{2qV}{m}}$$

   As an alternate solution, notice that the change in the electrical potential energy of the ion from the source S to the entrance to the magnetic-field region is equal to qV; this is equal to the gain in the particle's kinetic energy.

Therefore,

$$qV = \frac{1}{2}mv^2 \quad \Rightarrow \quad v = \sqrt{\frac{2qV}{m}}$$

C. i. and ii. Use the Right-Hand Rule. Since **v** points to the right and **B** is into the plane of the page, the direction of **v** × **B** is upward. Therefore, the magnetic force on a positively charged particle (cation) will be upward, and the magnetic force on a negatively charged particle (anion) will be downward. The magnetic force provides the centripetal force that causes the ion to travel in a circular path. Therefore, a cation would follow Path 1 and an anion would follow Path 2.

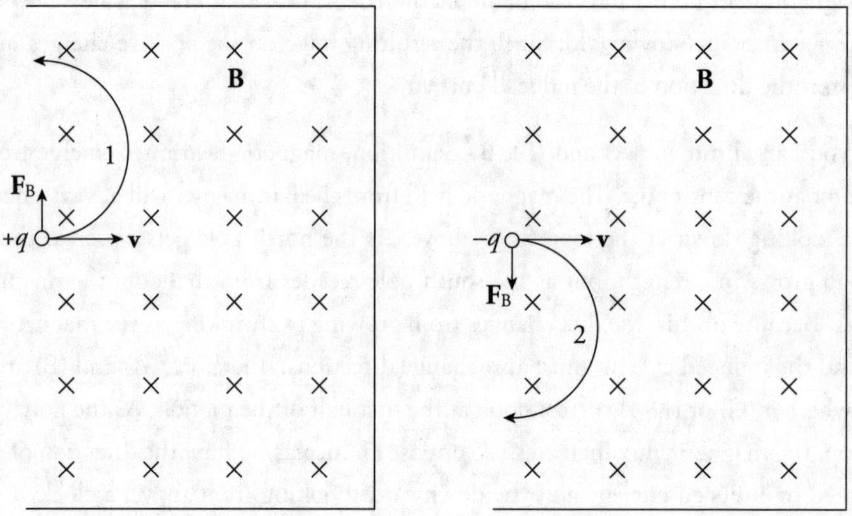

D. Since the magnetic force on the ion provides the centripetal force,

$$qvB = \frac{mv^2}{r} \quad \Rightarrow \quad qvB = \frac{mv^2}{\frac{1}{2}y} \quad \Rightarrow \quad m = \frac{qBy}{2v}$$

Now, by the result of part B,

$$m = \frac{qBy}{2\sqrt{\frac{2qV}{m}}} \quad \Rightarrow \quad m^2 = \frac{q^2B^2y^2}{\frac{8qV}{m}} \quad \Rightarrow \quad m^2 = \frac{mq^2B^2y^2}{8qV} \quad \Rightarrow \quad m = \frac{qB^2y^2}{8V}$$

E. Since the magnetic force cannot change the speed of a charged particle, the time required for the ion to hit the photographic plate is equal to the distance traveled (the length of the semicircle) divided by the speed computed in part B:

$$t = \frac{s}{v} = \frac{\pi \cdot \frac{1}{2}y}{\sqrt{\frac{2qV}{m}}} = \frac{1}{2}\pi y\sqrt{\frac{m}{2qV}}$$

F. Since the magnetic force $\mathbf{F}_B$ is always perpendicular to a charged particle's velocity vector $\mathbf{v}$, it can do no work on the particle. Thus, the answer is zero.

2. A. The force is directed in the $+y$ direction. The particle is charged positively and is moving in the $+x$ direction, so the thumb points to the right. With your fingers facing inward toward the paper, your palm is toward the top of the paper.

   B. i. The new speed is equal to the old speed because the magnetic force cannot do work, and therefore is unable to change the kinetic energy, and thus the speed, of the particle.

   ii. The new velocity is different from the old velocity as the direction has changed. The particle is traveling along a circular arc, and at any instant the velocity direction will be tangential to that arc. The new velocity will have the same magnitude as the initial velocity, but its x-component will be smaller as the y-component becomes larger. If the particle passes through 90° of the circular arc, the y-component will again begin to shrink as the x-component grows.

   C. Both A and B could be exit points from the magnetic field region.

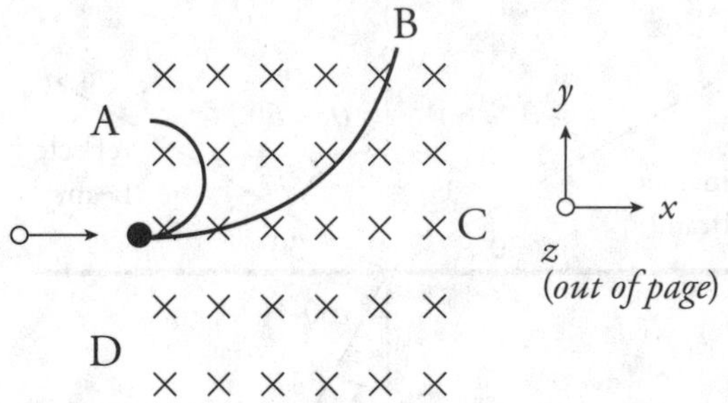

3. A. $\varepsilon = B\ell v = (2\text{ T})(0.4\text{ m})(1\text{ m/s}) = 0.8\text{ V}$

   B. $V = IR$ becomes $I = \frac{V}{R} = \frac{0.8\text{ V}}{20\,\Omega} = 0.04\text{ A}$. The direction is given by Lenz's Law. Current flows in order to oppose the change in the magnetic flux. Because there is suddenly a new flux "in," current flows to produce an outward flux. This would be in the counterclockwise direction.

   C. As the edge of the loop first encounters the magnetic field region, the induced current is established at 0.04 A, counterclockwise (found in part B). This continues as long as the flux is changing, which occurs until the left edge of the loop enters the magnetic field region. During the time the entire loop is within the region of the magnetic field, the flux is constant and the emf and current are both 0. Once the loop begins to leave the region of the magnetic field, the flux begins to change again and an induced current is established at 0.04 A. As the loop leaves the region of the magnetic field, the flux is shrinking instead of growing, so the current must be clockwise as the loop exits the magnetic field region.

D. The only thing that has changed is that the length of the wire at the leading edge has decreased. Thus, to maintain the same induced current, the speed at which the wire is pulled would have to increase. Moreover, as the length is halved, the speed would need to double in order for the induced current to remain constant.

CHAPTER 8: GEOMETRIC OPTICS REVIEW QUESTIONS

Section I: Multiple Choice

1. **B** As light travels into an optically dense medium, it will refract in toward the normal and away from the surface. So, the light in the glass will always be at a greater angle from the surface than the light in the air, so (B) is correct. Note that total internal reflection, (C), will not occur in this situation because it happens only with the initial medium being more optically dense.

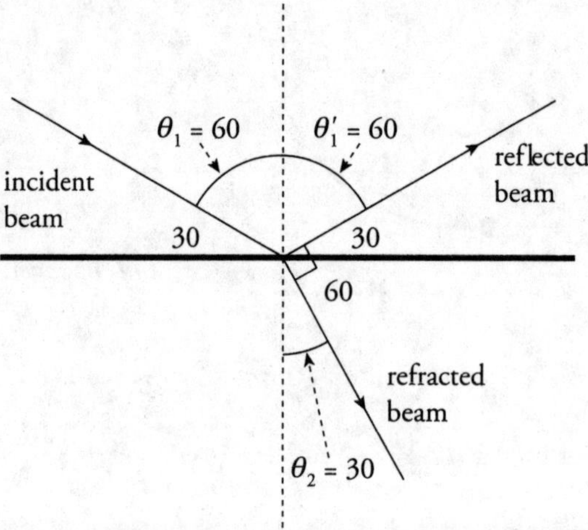

2. **B** Replacing the lens with one of higher index will mean that the light will bend more as it passes through the lens. This has the effect of shortening the focal length. A virtual image can only be made by having the object closer than the focal length of the lens, so shortening the focal length will not cause it to become virtual and (D) is eliminated. Shortening the focal length with the same d_o also means that d_i will have to be smaller by the lens equation $\frac{1}{f} = \frac{1}{d_o} + \frac{1}{d_i}$, so (B) is correct.

3. **D** Choice (A) is false because a change in speed at an interface does not cause intensity to change. Choice (B) is false because had there been total internal reflection, the light would have remained in Medium 1. Choice (C) is false because the problem states that Medium 2 has a lower index of refraction, and from Snell's Law, $n_1 \sin(\theta_1) = n_2 \sin(\theta_2)$, if $n_2 < n_1$, then $\sin(\theta_2) > \sin(\theta_1)$ and the angles are not equal. Choice (D) is true because when light arrives at an interface, it must be absorbed, reflected, or transmitted. Therefore, any intensity not detected as transmitted into Medium 2 was either reflected back into Medium 1 or absorbed.

4. **A** The critical angle for total internal reflection is computed as follows:

$$n_1 \sin \theta_c = n_2 \sin 90^\circ$$

$$\sin \theta_c = \frac{n_2}{n_1} = \frac{1.45}{2.90} = \frac{1}{2} \quad \Rightarrow \quad \theta_c = 30^\circ$$

Total internal reflection can happen only if the incident beam originates in the medium with the higher index of refraction, and therefore the beam must originate in the solid. The refracted ray must strike the interface of the other medium at an angle of incidence greater than the critical angle.

5. **C** Every ray, regardless of whether it is a principal ray or not, which originates at the tip of an object and strikes a mirror, will converge at the tip of the image.

6. **A** All of the rays which originate at the tip of the image and travel along paths that remain above the optic axis will be blocked. However, all of the rays that originate at the tip of the image and travel along paths that are below the optic axis at the plane of the block will be unaltered. As a result, fewer rays will converge at the image location, resulting in a less bright image.

7. **D** Diverging lenses always create virtual images.

Section II: Free Response

1. A. i. Student 1 is correct that "According to Snell's Law, the light will bend at the first interface" and also "unbend at the second interface and travel parallel to its original path."

ii. Student 2 is correct that "Snell's Law is $n_{in} \sin(\theta_{in}) = n_{out} \sin(\theta_{out})$" and that "$\theta_{out}$ is different from θ_{in} at the first interface."

iii. Student 1 is incorrect that "The spot on the screen will still be at point P."

iv. Student 2 is incorrect that "The beam after the glass cannot be parallel to the beam before the glass."

B.

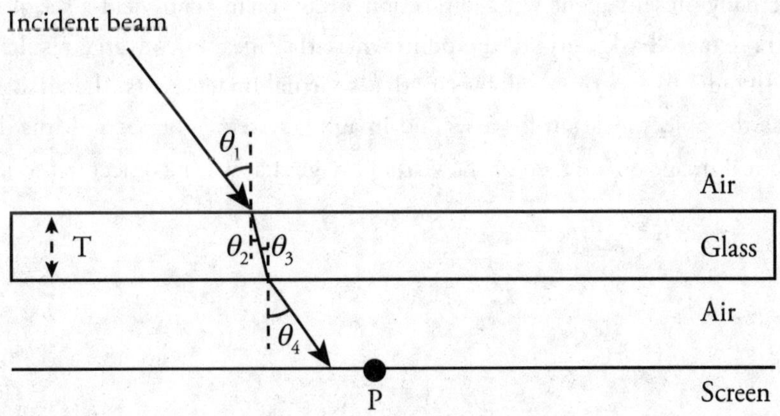

Here, $\theta_1 = \theta_4$ and $\theta_2 = \theta_3$.

C. If the area between the glass and screen were filled with water, it would still be the case that $\theta_2 = \theta_3$ because there is no change to the glass. However, upon leaving the glass and entering the water, the amount that the light would bend away from the normal would be less than in the case of air because the difference in the indices of refraction is smaller with glass-to-water than glass-to-air. As a result, the beams in air and water would not be parallel and the spot that would appear on the screen would be even farther away from point P than in the original experiment.

2. A.

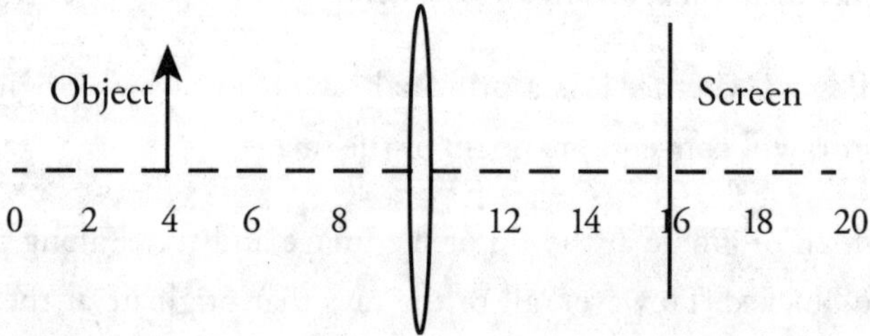

In order for the object and image to be the same size, the magnitude of the magnification must be 1, so $|M| = \left|\left(\frac{s_i}{s_o}\right)\right| = 1$ and therefore $|s_o| = |s_i|$. The only way to satisfy the equation $\frac{1}{f} = \frac{1}{s_i} + \frac{1}{s_o}$ when the image distance is equal to the object distance is for both to be $2f$. The object should be placed $2f$ before the lens and the screen should be placed $2f$ after the lens. The image will be below the optical axis in this case, so the part of the screen where the image forms needs to also be placed below the optic axis. Note: switching the positions of the object and screen (so that the object is on the right and the screen is on the left) is also correct.

B. A converging lens is similar to a convex mirror, but with the mirror, the light always stays on one side of the mirror. Both the object and the screen should be placed at the 4 cm mark. The image will be below the optical axis in this case.

C. A virtual image is a position in space where the light rays appear to have converged, although no light is actually present at the "image position." Ray tracing can yield a set of non-converging rays. Tracing those rays "backward" to the point where the lines do converge results in the virtual image position. In order to experimentally verify that a virtual image exists, the virtual image location may be used as the object location for a second imaging system. The second imaging system may then produce a real image on a screen of the virtual image that was its object.

CHAPTER 9: WAVES, SOUND, AND PHYSICAL OPTICS REVIEW QUESTIONS

Section I: Multiple Choice

1. **C** From the equation $\lambda = v/f$, find that

$$\lambda = \frac{v}{f} = \frac{10 \text{ m/s}}{5 \text{ Hz}} = 2 \text{ m}$$

2. **C** The speed of a transverse traveling wave on a stretched rope is given by the equation $v = \sqrt{\frac{F_T}{m/L}}$. Therefore,

$$v = \sqrt{\frac{F_T}{m/L}} = \sqrt{\frac{80 \text{ N}}{(1 \text{ kg})/(5 \text{ m})}} = \sqrt{400 \text{ m}^2/\text{s}^2} = 20 \text{ m/s}$$

3. **C** The time interval from a point moving from its maximum displacement above $y = 0$ (equilibrium) to its maximum displacement below equilibrium is equal to one-half the period of the wave. In this case,

$$T = \frac{1}{f} = \frac{\lambda}{v} = \frac{8 \text{ m}}{2 \text{ m/s}} = 4 \text{ s}$$

so the desired time is $\frac{1}{2}$ (4 s) = 2 s.

4. **D** The distance between successive nodes is always equal to $\frac{1}{2}\lambda$. If a standing wave on a string fixed at both ends has a total of 4 nodes, the string must have a length L equal to $3(\frac{1}{2}\lambda)$. If $L = 6$ m, then λ must equal 4 m.

5. **B** The previous question showed that $\lambda = 4$ m. Since $v = 40$ m/s, the frequency of this standing wave must be

$$f = \frac{v}{\lambda} = \frac{40 \text{ m/s}}{4 \text{ m}} = 10 \text{ Hz}$$

6. **A** In general, sound travels faster through solids than through gases. Therefore, when the wave enters the air from the metal rod, its speed will decrease. The frequency, however, will not change. Since $\lambda = v/f$ must always be satisfied, a decrease in v implies a decrease in λ.

7. **A** The distance from S_2 to P is 5 m (it's the hypotenuse of a 3-4-5 triangle), and the distance from S_1 to P is 4 m. The difference between the path lengths to Point P is 1 m, which is half the wavelength. Therefore, the sound waves are always exactly out of phase when they reach Point P from the two speakers, causing destructive interference there. By contrast, since Point Q is equidistant from the two speakers, the sound waves will always arrive in phase at Q, interfering constructively. Since there is destructive interference at P and constructive interference at Q, the amplitude at P will be less than at Q.

8. **B** An air column (such as an organ pipe) with one closed end resonates at frequencies given by the equation $f_n = nv/(4L)$ for odd integers n. The fundamental frequency corresponds, by definition, to $n = 1$. Therefore,

$$f_1 = \frac{v}{4L} = \frac{340 \text{ m/s}}{4(0.17 \text{ m})} = 500 \text{ Hz}$$

9. **B** The speed of the chirp is

$$v = \lambda f = (8.75 \times 10^{-3} \text{ m})(40 \times 10^{3} \text{ Hz}) = 350 \text{ m/s}$$

If the distance from the bat to the tree is d, then the wave travels a total distance of $d + d = 2d$ (round-trip distance). If T is the time for this round-trip, then

$$2d = vT \quad \Rightarrow \quad d = \frac{vT}{2} = \frac{(350 \text{ m/s})(0.4 \text{ s})}{2} = 70 \text{ m}$$

10. **C** In the equation for a wave, $x = A\cos(\omega t)$, the amplitude A will always have the same units as the wave itself. This is because the value of $\cos(\omega t)$ is a number between –1 and 1 and has no units. An electric field, whether stationary or oscillating, is always measured in N/C (or equivalently V/m), so (B) and (D) are incorrect. In $\cos(\omega t)$, ω is the angular frequency, found using $\omega = 2\pi f$.

11. **C** Since the fringe is bright, the waves must interfere constructively. This implies that the difference in path lengths must be a whole number times the wavelength, eliminating (A) and (B). The central maximum is equidistant from the two slits, so $\Delta\ell = 0$ there. At the first bright fringe above the central maximum, you have $\Delta\ell = \lambda$.

12. **B** In the double-slit experiment, $y = \frac{m\lambda D}{d}$. Using a shorter wavelength, as in the second experiment, results in a decrease in the fringe spacing, so (C) and (D) are false. Using slits which are closer together causes the fringe spacing to increase, so (A) is false.

Section II: Free Response

1. A. The speed of a transverse traveling wave on a stretched rope is given by the equation $v = \sqrt{\frac{F_T}{m/L}}$. Therefore,

$$v^2 = \frac{F_T}{m/L} \quad \Rightarrow \quad F_T = \mu v^2 = (0.4 \text{ kg/m})(12 \text{ m/s})^2 = 58 \text{ N}$$

B. Use the fundamental equation $\lambda = v/f$:

$$f = \frac{v}{\lambda} = \frac{12 \text{ m/s}}{2 \text{ m}} = 6 \text{ Hz}$$

C. i. Because higher harmonic numbers correspond to shorter wavelengths, the harmonic number of the 3.2 m standing wave must be higher than that of the 4 m standing wave. You're told that these harmonic numbers are consecutive integers, so if n is the harmonic number of the 4 m standing wave, then $n + 1$ is the harmonic number of the 3.2 m wave. Therefore,

$$\frac{2L}{n} = 4 \text{ m} \quad \text{and} \quad \frac{2L}{n+1} = 3.2 \text{ m}$$

The first equation says that $2L = 4n$, and the second one says that $2L = 3.2(n + 1)$. Therefore, $4n$ must equal $3.2(n + 1)$; solving this equation gives $n = 4$. Substituting this into either one of the displayed equations then gives $L = 8$ m.

ii. Because the mass per unit length of the rope is 0.4 kg/m, the mass of the rope must be

$$m = \mu L = (0.4 \text{ kg/m})(8 \text{ m}) = 3.2 \text{ kg}$$

D. You determined this in the solution to part C (i). The 4 m standing wave has harmonic number $n = 4$.

E.

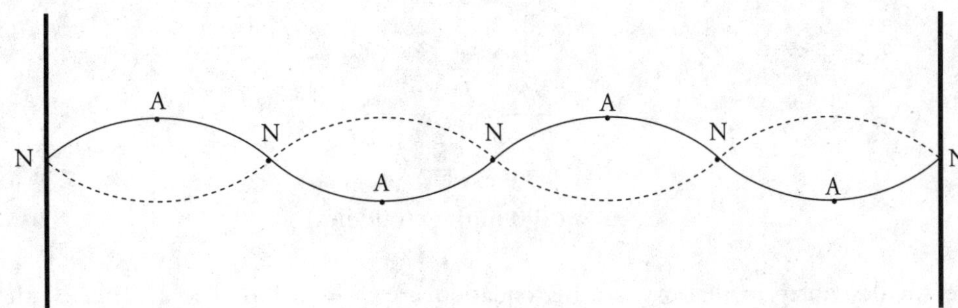

2. A. Use the equation for the beat frequency:

$$f_{\text{beat}} = |f_1 - f_2| = |400 \text{ Hz} - 440 \text{ Hz}| = 40 \text{ Hz}$$

B. As sound travels from air into water, the speed of sound will increase, as sound travels faster in liquids than in gases. By Wave Rule #2, when a wave passes into another medium, its speed changes but its frequency does not. By the equation $v = f\lambda$, it can be observed that if speed increases and frequency does not change, then the wavelength must increase.

C. As the sound source (tuning fork) is traveling toward the stationary detector (the student on the second floor), the observed frequency will be higher than the source frequency. However, the question is asking what happens to the observed frequency as the tuning fork travels upward. Due to acceleration from gravity, the speed of the tuning fork as it moves up will decrease. As the relative speed between the source and detector is decreasing, the impact of the Doppler effect will decrease. This will lead to a decrease in the observed frequency as the tuning fork travels upward (although the observed frequency will still be higher than the source frequency).

3. A. Interference effects can be observed only if the light is coherent. Using two independent light sources at the slits in Barrier 2 would generate incoherent light waves. The setup shown guarantees that the light reaching the two slits will be coherent.

B. Maxima are located at positions given by the equation $y = m\lambda D/d$, where m is an integer ($y_0 = 0$ is the central maximum). The first-order maximum for red light occurs at

$$y_{1,\text{ red}} = \frac{1 \cdot \lambda_{\text{red}} D}{d} = \frac{1 \cdot (750 \times 10^{-9}\text{ m})(4.0\text{ m})}{0.50 \times 10^{-3}\text{ m}} = 0.006\text{ m} = 6.0\text{ mm}$$

and the first-order maximum for violet light occurs at

$$y_{1,\text{ violet}} = \frac{1 \cdot \lambda_{\text{violet}} D}{d} = \frac{1 \cdot (400 \times 10^{-9}\text{ m})(4.0\text{ m})}{0.50 \times 10^{-3}\text{ m}} = 0.0032\text{ m} = 3.2\text{ mm}$$

Therefore, the vertical separation of these maxima on the screen is $\Delta y = 6.0\text{ mm} - 3.2\text{ mm} = 2.8\text{ mm}$.

C. Maxima are located at positions given by the equation $y = m\lambda D/d$, where m is an integer. Therefore, you want to solve the equation

$$\begin{aligned} y_{m,\text{ violet}} &= x_{n,\text{ orange-yellow}} \\ \frac{m\lambda_{\text{violet}} D}{d} &= \frac{n\lambda_{\text{orange-yellow}} D}{d} \\ m\lambda_{\text{violet}} &= n\lambda_{\text{orange-yellow}} \\ m(400\text{ nm}) &= n(600\text{ nm}) \end{aligned}$$

The smallest integers that satisfy this equation are $m = 3$ and $n = 2$. That is, the third-order maximum for violet light coincides with the second-order maximum for orange-yellow light. The position on the screen (relative to the central maximum at $x = 0$) of these maxima is

$$y_{3,\text{ violet}} = \frac{m\lambda_{\text{violet}} D}{d} = \frac{3(400 \times 10^{-9}\text{ m})(4.0\text{ m})}{0.50 \times 10^{-3}\text{ m}} = 0.0096\text{ m} = 9.6\text{ mm}$$

D. Within the glass, the wavelength is reduced by a factor of $n = 1.5$ from the wavelength in air. Therefore, the difference in path lengths, $d \sin \theta$, must be equal to $m(\lambda/n)$ in order for constructive interference to occur. This implies that the maxima are located at positions given by $y_m = m\lambda D/(nd)$. The distance between adjacent bright fringes is therefore

$$y_{m+1} - y_m = \frac{(m+1)\lambda D}{nd} - \frac{m\lambda D}{nd} = \frac{\lambda D}{nd} = \frac{(500 \times 10^{-9}\text{ m})(4.0\text{ m})}{1.5(0.50 \times 10^{-3}\text{ m})} = 0.0027\text{ m} = 2.7\text{ mm}$$

CHAPTER 10: MODERN PHYSICS REVIEW QUESTIONS

Section I: Multiple Choice

1. **B** Waves will interfere with apertures that are approximately the same size as the wave. The wavelengths of matter from the de Broglie Theory are on the same order as the size of molecule spacing in solids, and therefore these effects are not visible at macroscopic scales.

2. **C** Photoelectrons will be emitted in the photoelectric effect when the energy of a photon of the incident light is greater than the work function of the metal. Thus, decreasing the energy of the photon below the work function of the metal would prevent photoelectrons from being ejected by the metal. The energy of a photon is $E = hf$, so decreasing the frequency of the photon decreases its energy. Since the frequency of light is inversely proportional to the wavelength, decreasing the frequency corresponds to increasing the wavelength, so (C) is correct. The brightness of the light does not correlate with photon energy, which eliminates (A). The stopping potential is something observed from photoelectrons that are ejected, not something that is independently controlled, so (D) can be eliminated.

3. **A** The ground state of an atom is (–1) times its ionization energy, so the ground state would have to be at an energy level of –25 eV. To transition from –16 eV to –25 eV, a photon with an energy of 9 eV would have to be produced in order for energy to be conserved. To find the wavelength of the photon, $E = hf = h(c/\lambda) = hc/\lambda$, so $\lambda = hc/E$. From the equation sheet, $hc = 1240 \text{ nm} \cdot \text{eV}$ and $\lambda = \dfrac{1240 \text{ nm} \cdot \text{eV}}{9 \text{ eV}} = 138 \text{ nm}$.

4. **D** The gap between the ground-state and the first excited state is

$$-10 \text{ eV} - (-40 \text{ eV}) = 30 \text{ eV}$$

Therefore, the electron must absorb the energy of a 30 eV photon (at least) in order to move even to the first excited state. Since the incident photons have only 15 eV of energy, the electron will be unaffected.

5. **D** When a charged particle goes into a magnetic field, it will be deflected, but an uncharged particle will not be deflected. Particles with opposite signs will deflect in opposite directions. An alpha particle is charged and a neutron is not, so (D) is the correct answer.

6. **D** If the atom begins in the $n = 3$ state, it could lose energy by making any of the following transitions: $3 \rightarrow 2$, $3 \rightarrow 1$, or $3 \rightarrow 2 \rightarrow 1$. The $3 \rightarrow 2$ transition would result in the emission of –5 eV – (–8 eV) = 3 eV; the $3 \rightarrow 1$ transition would emit a –5 eV – (–12 eV) = 7 eV photon; and the $2 \rightarrow 1$ transition would result in the emission of –8 eV – (–12 eV) = 4 eV. Therefore, if the atom is initially in the $n = 3$ state, it could emit photons of energy 3 eV, 4 eV, or 7 eV.

7. **C** To emit a photon, an energy level transition must go to a lower energy level (a transition to a higher level requires absorbing a photon), so (A) and (B) are incorrect. A long wavelength corresponds to a small energy photon, so (D) is also incorrect, and (C) is the correct answer.

8. **D** At the time that Louis de Broglie made his hypothesis, it was known that light, which often behaves like a wave, also has particle-like behavior. The de Broglie Hypothesis is that while electrons (and other matter) most often demonstrate particle-like behaviors, they would also demonstrate wave-like behaviors. Diffraction is a phenomenon associated with wave interference, so de Broglie's Hypothesis can be summarized as electrons will diffract in certain circumstances.

9. **B** The binding energy is given by $E_B = (\Delta m)c^2$, so (A) is true. The binding energy, which is the energy associated with the mass defect, is typically measured in MeV, while typical electronic transitions are measured in eV, so (C) is also true. Mass defect is also typically 4 orders of magnitude smaller than the mass of the involved nucleons, so (D) is true. Choice (B) is false because the binding energy is always positive, so that the mass of the bound nucleons plus the binding energy equals the mass of the constituent nucleons.

10. **B** In order to balance the mass number (the superscripts), you must have $2 + 63 = 64 + A$, so $A = 1$. In order to balance the charge (the subscripts), you need $1 + 29 = 30 + Z$, so $Z = 0$. A particle with a mass number of 1 and no charge is a neutron, ${}^{1}_{0}\text{n}$.

11. **A** The alpha decay process produces an alpha particle ${}^{4}_{2}\text{He}$ and a daughter nucleus. The daughter nucleus must have a mass number 4 below the original isotope and a charge number 2 below the original isotope, resulting in ${}^{171}_{76}\text{Os}$. The beta(+) decay process produces a positive beta particle ${}^{0}_{1}e$ and a daughter nucleus. The daughter nucleus must have a mass number equal to the original isotope and a charge number 1 below the original isotope, resulting in ${}^{175}_{77}\text{Ir}$.

Section II: Free Response

1. A. The work function is the amount of energy that must be supplied before a photoelectron is released. For any amount of energy supplied by the power supply below the work function of the material that the LED light is made of, there will be no production of photoelectrons. Therefore, the work function of the material must be greater than 0.1 eV since there were no photoelectrons emitted at the voltage of 1 V.

B.

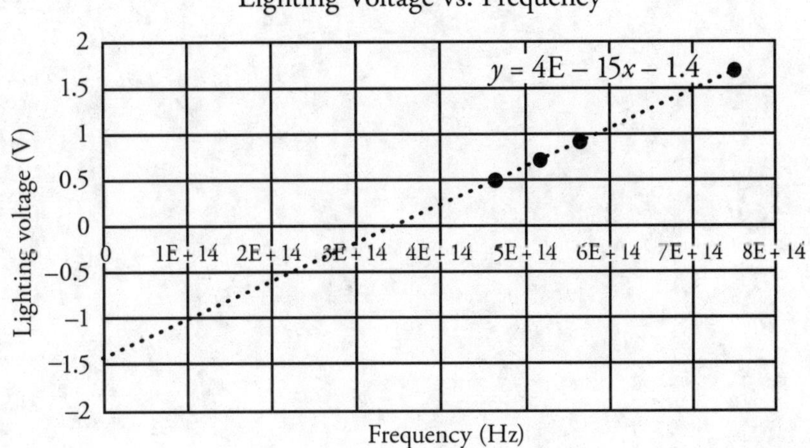

The slope of the graph is 4×10^{-15}V · s and the y-intercept is –1.4 V. The slope has the value of Planck's constant, and the intercept is the work function of the material the LED lights are made of.

C. i. To change the intercept, the work function would have to be different, so a supply of LEDs made from a different material would have to be obtained.

ii. It would not be possible to change the slope, as the slope determined from this experiment is Planck's constant.

Part VI
Practice Test 2

- Practice Test 2
- Practice Test 2: Answers and Explanations
- Practice Test 2: Free-Response Questions Grading Rubrics
- How to Score Practice Test 2

Practice Test 2

AP® Physics 2 Exam

SECTION I: Multiple-Choice Questions

At a Glance

Total Time
80 minutes
Number of Questions
40
Percent of Total Grade
50%

DISCLAIMER: The official AP Physics 2 Exam will be administered in a hybrid digital format. Instructions for the digital exam may differ from this practice test.

Instructions

Section I of this examination contains 40 multiple-choice questions. Fill in only the ovals for numbers 1 through 40 on your answer sheet.

CALCULATORS MAY BE USED IN BOTH SECTIONS OF THE EXAMINATION.

Indicate all of your answers to the multiple-choice questions on the answer sheet. Give only one answer to each question.

Use your time effectively, working as quickly as you can without losing accuracy. Do not spend too much time on any one question. Go on to other questions and come back to the ones you have not answered if you have time. It is not expected that everyone will know the answers to all the multiple-choice questions.

GO ON TO THE NEXT PAGE.

ADVANCED PLACEMENT PHYSICS 2 TABLE OF INFORMATION

CONSTANTS AND CONVERSION FACTORS	
Avogadro's number,	$N_0 = 6.02 \times 10^{23}$ mol^{-1}
Universal gas constant,	$R = 8.31$ J/(mol • K)
Boltzmann's constant,	$k_B = 1.38 \times 10^{-23}$ J/K
1 atmosphere of pressure,	1 atm = 1.0×10^5 N/m^2 = 1.0×10^5 Pa
Coulomb's law constant,	$k = \dfrac{1}{4\pi\varepsilon_0} = 9.0 \times 10^9 \dfrac{\text{N} \cdot \text{m}^2}{\text{C}^2}$
Proton mass,	$m_p = 1.67 \times 10^{-27}$ kg
Neutron mass,	$m_n = 1.67 \times 10^{-27}$ kg
Electron mass,	$m_e = 9.11 \times 10^{-31}$ kg
Elementary charge,	$e = 1.60 \times 10^{-19}$ C
Vacuum permittivity,	$\varepsilon_0 = 8.85 \times 10^{-12}$ C^2/(N • m^2)
Vacuum permeability,	$\mu_0 = 4\pi \times 10^{-7}$ (T • m)/A
1 electron volt,	1 eV = 1.60×10^{-19} J
Planck's constant,	$h = 6.63 \times 10^{-34}$ J • s = 4.14×10^{-15} eV • s
	$hc = 1.99 \times 10^{-25}$ J • m = 1.24×10^3 eV • nm
Speed of light,	$c = 3.00 \times 10^8$ m/s
Wien's constant,	$b = 2.90 \times 10^{-3}$ m • K
Stefan-Boltzmann constant	$\sigma = 5.67 \times 10^{-8}$ W/(m^2 • K^4)
1 unified atomic mass unit,	1 u = 1.66×10^{-27} kg = 931 MeV/c^2

Universal gravitational constant, $G = 6.67 \times 10^{-11}$ m^3/(kg • s^2) = 6.67×10^{-11} N • m^2/kg^2
Magnitude of the acceleration due to gravity at Earth's surface, $g = 9.8$ m/s^2
Magnitude of the gravitational field strength at the Earth's surface, $g = 9.8$ N/kg

UNIT SYMBOLS	
ampere,	A
coulomb,	C
degree Celsius,	°C
electron volt,	eV
farad,	F
hertz,	Hz
joule,	J
kelvin,	K
kilogram,	kg
meter,	m
mole,	mol
newton,	N
ohm,	Ω
pascal,	Pa
second,	s
tesla,	T
volt,	V
watt,	W

PREFIXES

Factor	Prefix	Symbol
10^{12}	tera	T
10^9	giga	G
10^6	mega	M
10^3	kilo	k
10^{-2}	centi	c
10^{-3}	milli	m
10^{-6}	micro	μ
10^{-9}	nano	n
10^{-12}	pico	p

VALUES OF TRIGONOMETRIC FUNCTIONS FOR COMMON ANGLES

θ	0°	30°	37°	45°	53°	60°	90°
sin θ	0	1/2	3/5	$\sqrt{2}/2$	4/5	$\sqrt{3}/2$	1
cos θ	1	$\sqrt{3}/2$	4/5	$\sqrt{2}/2$	3/5	1/2	0
tan θ	0	$\sqrt{3}/3$	3/4	1	4/3	$\sqrt{3}$	∞

The following conventions are used in this exam.

- The frame of reference of any problem is assumed to be inertial unless otherwise stated.
- Air resistance is assumed to be negligible unless otherwise stated.
- In all situations, positive work is defined as work done **ON** a system.
- Springs and strings are assumed to be ideal unless otherwise stated.
- The electric potential is zero at an infinite distance from an isolated point charge.
- The direction of current is the direction in which positive charges would drift.
- All batteries and meters are assumed to be ideal unless otherwise stated.

GO ON TO THE NEXT PAGE.

ADVANCED PLACEMENT PHYSICS 2 EQUATIONS

ELECTRICITY

$$\left|\vec{F}_E\right| = \frac{1}{4\pi\varepsilon_0}\frac{|q_1 q_2|}{r^2} = k\frac{|q_1 q_2|}{r^2}$$

$$\vec{E} = \frac{\vec{F}_E}{q}$$

$$\left|\vec{E}\right| = \frac{1}{4\pi\varepsilon_0}\frac{|q|}{r^2} = k\frac{|q|}{r^2}$$

$$U_E = \frac{1}{4\pi\varepsilon_0}\frac{q_1 q_2}{r} = k\frac{q_1 q_2}{r}$$

$$\Delta V = \frac{\Delta U_E}{q}$$

$$V = \frac{1}{4\pi\varepsilon_0}\sum_i \frac{q_i}{r_i}$$

$$\left|\vec{E}\right| = \left|\frac{\Delta V}{\Delta r}\right|$$

$$C = \frac{Q}{\Delta V}$$

$$C = \kappa\varepsilon_0\frac{A}{d}$$

$$E_C = \frac{Q}{\kappa\varepsilon_0 A}$$

$$U_C = \frac{1}{2}Q\Delta V$$

$$I = \frac{\Delta Q}{\Delta t}$$

$$R = \frac{\rho\ell}{A}$$

$$P = I\Delta V$$

$$I = \frac{\Delta V}{R}$$

$$R_{eq,s} = \sum_i R_i$$

$$\frac{1}{R_{eq,p}} = \sum_i \frac{1}{R_i}$$

$$\frac{1}{C_{eq,s}} = \sum_i \frac{1}{C_i}$$

$$C_{eq,p} = \sum_i C_i$$

$$\tau = R_{eq}C_{eq}$$

A = area
C = capacitance
d = distance
E = electric field
F = force
I = current
ℓ = length
P = power
q = charge
Q = charge
r = distance, radius, or position
R = resistance
t = time
U = potential energy
V = electric potential
κ = dielectric constant
ρ = resistivity
τ = time constant

MAGNETISM

$$F_B = qvB\sin\theta$$

$$B = \frac{\mu_0 I}{2\pi r}$$

$$F_B = I\ell B\sin\theta$$

$$\Phi_B = \vec{B}\cdot\vec{A}$$

$$\Phi_B = \left|\vec{B}\right|\cos\theta\left|\vec{A}\right|$$

$$|\varepsilon| = \left|\frac{\Delta\Phi_B}{\Delta t}\right|$$

$$\varepsilon = B\ell v$$

A = area
B = magnetic field
F = force
I = current
ℓ = length
q = charge
r = distance, radius, or position
t = time
v = velocity or speed
ε = emf
θ = angle
Φ = flux

THERMAL PHYSICS

$$P = \frac{F_\perp}{A}$$

$$K_{avg} = \frac{3}{2}k_B T = \frac{1}{2}mv_{rms}^2$$

$$\frac{Q}{\Delta t} = \frac{kA\Delta T}{L}$$

$$PV = nRT = Nk_B T$$

$$U = \frac{3}{2}nRT = \frac{3}{2}Nk_B T$$

$$W = -P\Delta V$$

$$\Delta U = Q + W$$

$$Q = mc\Delta T$$

A = area
c = specific heat
F = force
k = thermal conductivity
K = kinetic energy
L = length
m = mass
n = number of moles
N = number of atoms
P = pressure
Q = energy transferred to a system by heating
t = time
T = temperature
U = internal energy
v = velocity or speed
V = volume
W = work done on a system

GO ON TO THE NEXT PAGE.

ADVANCED PLACEMENT PHYSICS 2 EQUATIONS

WAVES, SOUND, AND OPTICS

$$\lambda = \frac{v}{f}$$

$$n = \frac{c}{v}$$

$$n_1 \sin\theta_1 = n_2 \sin\theta_2$$

$$\frac{1}{s_i} + \frac{1}{s_o} = \frac{1}{f}$$

$$|M| = \left|\frac{h_i}{h_o}\right| = \left|\frac{s_i}{s_o}\right|$$

$$\Delta D = m\lambda$$

$$\Delta D = a\sin\theta$$

$$a\left(\frac{y_{min}}{L}\right) \approx m\lambda$$

$$\Delta D = d\sin\theta$$

$$d\left(\frac{y_{max}}{L}\right) \approx m\lambda$$

$$v_{string} = \sqrt{\frac{F_T}{m/\ell}}$$

$$T = \frac{1}{f}$$

$$x(t) = A\cos(\omega t) = A\cos(2\pi ft)$$

$$y(x) = A\cos\left(2\pi \frac{x}{\lambda}\right)$$

$$|f_{beat}| = |f_1 - f_2|$$

a = width
A = amplitude
d = separation
D = path length
f = frequency or focal length
F = force
h = height
ℓ = length
L = distance
m = order or mass
M = magnification
n = index of refraction
s = position
t = time
T = period
v = speed
x = position
y = position
λ = wavelength
θ = angle
ω = angular frequency

MODERN PHYSICS

$$E = hf$$

$$\lambda = \frac{h}{p}$$

$$\lambda = \frac{c}{f}$$

$$\lambda_{max} = \frac{b}{T}$$

$$P = A\sigma T^4$$

$$K_{max} = hf - \phi$$

$$\Delta\lambda = \frac{h}{m_e c}(1 - \cos\theta)$$

$$E = mc^2$$

$$N = N_0 e^{-\lambda t}$$

$$\lambda = \frac{\ln 2}{t_{1/2}}$$

A = area
b = constant
E = energy
f = frequency
K = kinetic energy
m = mass
N = number of particles
p = momentum
P = power
t = time
T = absolute temperature
θ = angle
λ = wavelength or decay constant
σ = constant
ϕ = work function

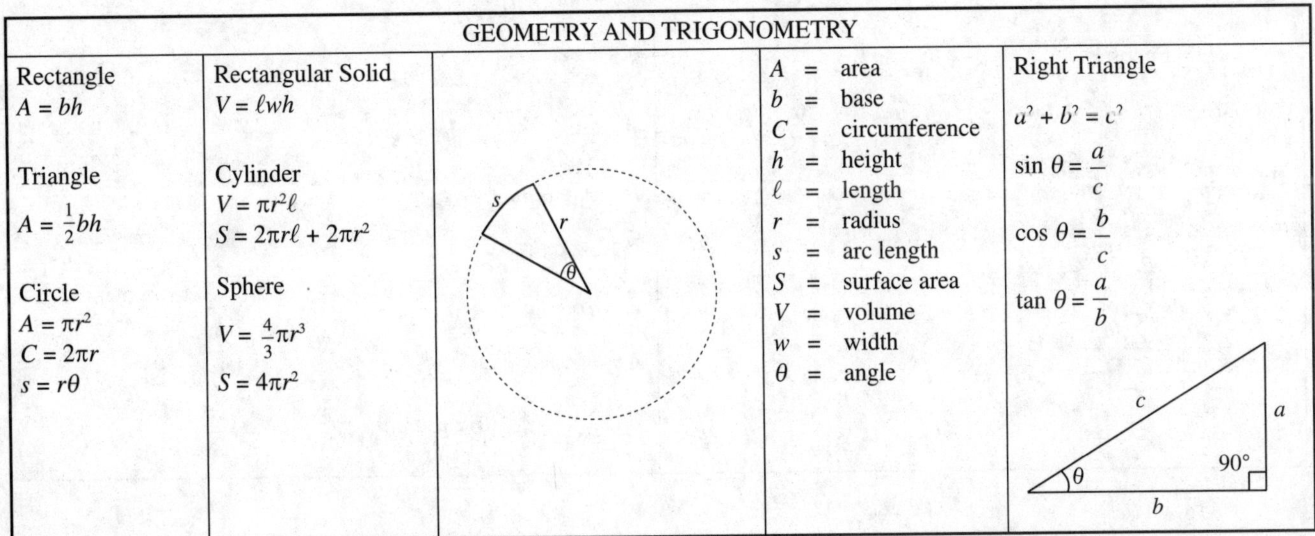

GEOMETRY AND TRIGONOMETRY

Rectangle
$A = bh$

Triangle
$A = \frac{1}{2}bh$

Circle
$A = \pi r^2$
$C = 2\pi r$
$s = r\theta$

Rectangular Solid
$V = \ell wh$

Cylinder
$V = \pi r^2 \ell$
$S = 2\pi r\ell + 2\pi r^2$

Sphere
$V = \frac{4}{3}\pi r^3$
$S = 4\pi r^2$

A = area
b = base
C = circumference
h = height
ℓ = length
r = radius
s = arc length
S = surface area
V = volume
w = width
θ = angle

Right Triangle
$a^2 + b^2 = c^2$
$\sin\theta = \frac{a}{c}$
$\cos\theta = \frac{b}{c}$
$\tan\theta = \frac{a}{b}$

GO ON TO THE NEXT PAGE.

ADVANCED PLACEMENT PHYSICS 2 EQUATIONS

MECHANICS AND FLUIDS

$$v_x = v_{x0} + a_x t$$

$$x = x_0 + v_{x0}t + \frac{1}{2}a_x t^2$$

$$v_x^2 = v_{x0}{}^2 + 2a_x(x - x_0)$$

$$\vec{x}_{cm} = \frac{\sum m_i \vec{x}_i}{\sum m_i}$$

$$\vec{a}_{sys} = \frac{\sum \vec{F}}{m_{sys}} = \frac{\vec{F}_{net}}{m_{sys}}$$

$$\left|\vec{F}_g\right| = G\frac{m_1 m_2}{r^2}$$

$$\left|\vec{F}_f\right| \le \left|\mu \vec{F}_n\right|$$

$$\vec{F}_S = -k\Delta\vec{x}$$

$$a_c = \frac{v^2}{r}$$

$$K = \frac{1}{2}mv^2$$

$$W = F_{\parallel}d = Fd\cos\theta$$

$$\Delta K = \sum W_i = \sum F_{\parallel,i}d_i$$

$$\Delta U_S = \frac{1}{2}k(\Delta x)^2$$

$$U_G = -\frac{Gm_1m_2}{r}$$

$$\Delta U_g = mg\Delta y$$

$$P_{avg} = \frac{W}{\Delta t} = \frac{\Delta E}{\Delta t}$$

$$P_{inst} = F_{\parallel}v = Fv\cos\theta$$

$$\vec{p} = m\vec{v}$$

$$\vec{F}_{net} = \frac{\Delta\vec{p}}{\Delta t} = m\frac{\Delta\vec{v}}{\Delta t} = m\vec{a}$$

$$\vec{J} = \vec{F}_{avg}\Delta t = \Delta\vec{p}$$

$$\vec{v}_{cm} = \frac{\sum \vec{p}_i}{\sum m_i} = \frac{\sum m_i \vec{v}_i}{\sum m_i}$$

a = acceleration
d = distance
E = energy
F = force
J = impulse
k = spring constant
K = kinetic energy
m = mass
p = momentum
P = power
r = radius, distance, or position
t = time
U = potential energy
v = velocity or speed
W = work
x = position
y = height
θ = angle
μ = coefficient of friction

$$\omega = \omega_0 + \alpha t$$

$$\theta = \theta_0 + \omega_0 t + \frac{1}{2}\alpha t^2$$

$$\omega^2 = \omega_0^2 + 2\alpha(\theta - \theta_0)$$

$$v = r\omega$$

$$a_T = r\alpha$$

$$\tau = r_{\perp}F = rF\sin\theta$$

$$I = \sum m_i r_i^2$$

$$I' = I_{cm} + Md^2$$

$$\alpha_{sys} = \frac{\sum \tau}{I_{sys}} = \frac{\tau_{net}}{I_{sys}}$$

$$K = \frac{1}{2}I\omega^2$$

$$W = \tau\Delta\theta$$

$$L = I\omega$$

$$L = rmv\sin\theta$$

$$\Delta L = \tau\Delta t$$

$$\Delta x_{cm} = r\Delta\theta$$

$$T = \frac{1}{f}$$

$$T_s = 2\pi\sqrt{\frac{m}{k}}$$

$$T_P = 2\pi\sqrt{\frac{\ell}{g}}$$

$$x = A\cos(2\pi ft)$$

$$x = A\sin(2\pi ft)$$

$$\rho = \frac{m}{V}$$

$$P = \frac{F_{\perp}}{A}$$

$$P = P_0 + \rho gh$$

$$P_{gauge} = \rho gh$$

$$F_b = \rho Vg$$

$$A_1v_1 = A_2v_2$$

$$P_1 + \rho gy_1 + \frac{1}{2}\rho v_1^2 = P_2 + \rho gy_2 + \frac{1}{2}\rho v_2^2$$

a = acceleration
A = amplitude or area
d = distance
f = frequency
F = force
h = height
I = rotational inertia
k = spring constant
K = kinetic energy
ℓ = length
L = angular momentum
m = mass
M = mass
P = pressure
r = radius, distance, or position
t = time
T = period
v = velocity or speed
V = volume
W = work
x = position
y = vertical position
α = angular acceleration
θ = angle
ρ = density
τ = torque
ω = angular speed

GO ON TO THE NEXT PAGE.

AP PHYSICS 2

SECTION I

Time—80 minutes

40 Questions

Note: To simplify calculations, you may use $g = 10 \text{ m/s}^2$ in all problems.

Directions: Each of the questions or incomplete statements below is followed by four suggested answers or completions. Select the one that is best in each case and mark it on your sheet.

1 Mark for Review

A person moves a pan that is at 200°C using both hands while wearing oven mitts. Assuming the oven mitts stay at 20°C throughout the motion. Which of the following changes would cause the greatest reduction in the amount of heat that flows through the mitts?

(A) Moving the pan in half the amount of time

(B) Holding onto the pan with one hand instead of two

(C) Reducing the temperature of the pan to 100°C

(D) Doubling the thickness of the oven mitt

2 Mark for Review

An object with a mass of 20 g and a charge of +4 mC is traveling through space at a speed of 4.0×10^4 m/s. A stationary point charge with a charge of +2 C is initially a distance of 3 m away. The object passes close to the point charge, without hitting it, is deflected, and continues traveling. After it has moved very far away, what is the object's speed?

(A) 0 m/s

(B) 2.8×10^4 m/s

(C) 4.0×10^4 m/s

(D) 4.9×10^4 m/s

3 Mark for Review

A sound wave in a pipe with one end closed has a fundamental frequency of 400 Hz. How many nodes would a standing sound wave at the frequency of 2800 Hz have in the pipe?

(A) 3

(B) 4

(C) 5

(D) 6

4 Mark for Review

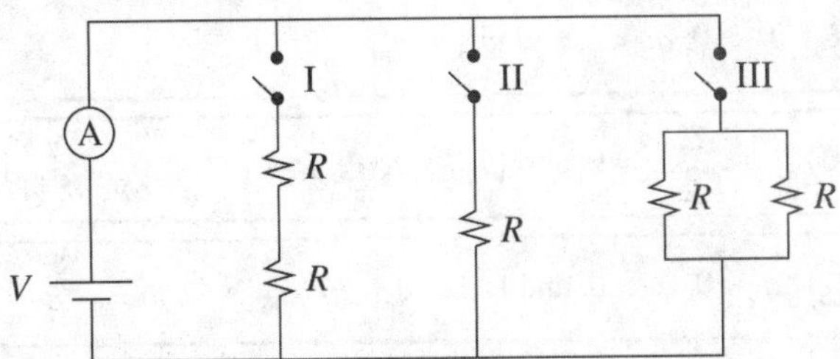

In the above circuit, all resistors have the same resistance, R. Which option would result in the greatest current through the ammeter?

(A) Closing only switch I

(B) Closing only switch II

(C) Closing only switch III

(D) Closing all three switches

GO ON TO THE NEXT PAGE.

5 Mark for Review

If a convex lens with a focal length of 5 cm is used to create an image of an object placed 3 cm in front of the lens, where will the image form?

(A) 1.9 cm in front of the lens

(B) 1.9 cm behind the lens

(C) 7.5 cm in front of the lens

(D) 7.5 cm behind the lens

6 Mark for Review

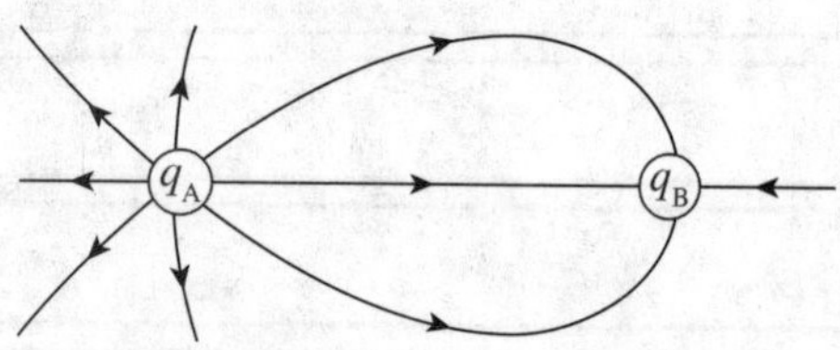

The diagram above shows the electric field lines created by two charges, q_A and q_B. Which of the following accurately describes the properties of the two charges?

(A) $q_A > 0$, $q_B < 0$, and $|q_A| < |q_B|$

(B) $q_A > 0$, $q_B < 0$, and $|q_A| > |q_B|$

(C) $q_A < 0$, $q_B > 0$, and $|q_A| < |q_B|$

(D) $q_A < 0$, $q_B > 0$, and $|q_A| > |q_B|$

7 Mark for Review

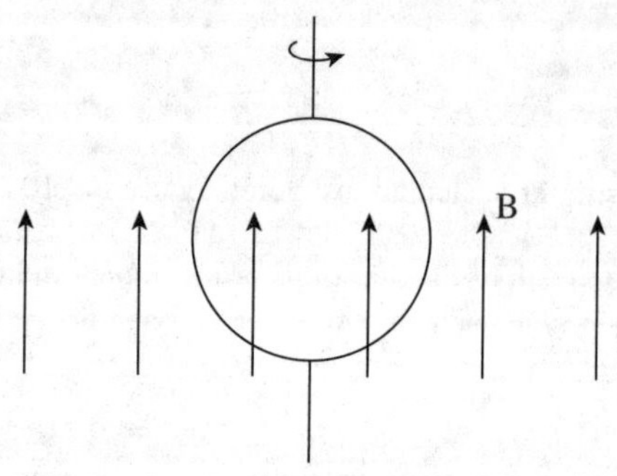

A loop of wire is oriented so that the plane of the loop is parallel to the page in a uniform upward magnetic field. It is then rotated about a vertical axis at a constant angular speed, as shown above. When is the induced emf in the wire zero?

(A) Only when the plane of the loop is parallel to the page

(B) Only when the plane of the loop is perpendicular to the page

(C) The induced emf is always non-zero.

(D) The induced emf is always zero.

8 Mark for Review

The five lowest energy states of an electron orbiting an atom are as follows: $E_1 = -15$ eV, $E_2 = -11$ eV, $E_3 = -8.5$ eV, $E_4 = -6.5$ eV $E_5 = -5$ eV. If a single electron starts in the ground state, which of the following interactions are NOT possible?

(A) A 6.5 eV photon is absorbed, then a 2 eV photon is emitted.

(B) A 40 eV photon Compton scatters off of the electron, then a 2.5 eV photon is emitted.

(C) A 17 eV photon is absorbed and the electron is ejected from the atom.

(D) An 10 eV photon is absorbed, then a 3.5 eV photon is emitted.

GO ON TO THE NEXT PAGE.

9 Mark for Review

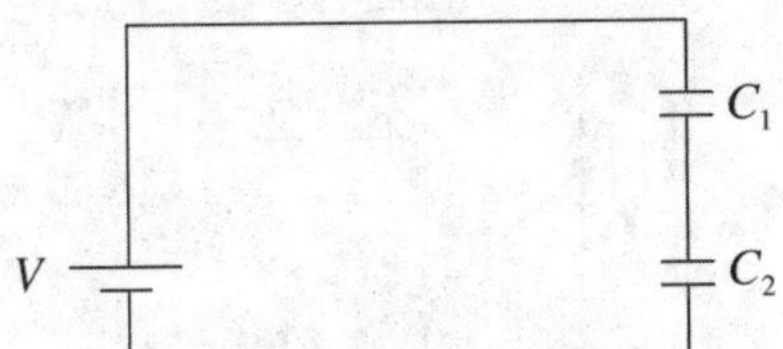

The above circuit is in its steady state with $V = 12$ V, $C_1 = 20\ \mu F$, and the charge on C_2 is 180 μC. What is the capacitance of the capacitor C_2?

(A) 5 μF

(B) 20 μF

(C) 30 μF

(D) 60 μF

10 Mark for Review

A ball emitting a sound with a constant frequency, f, is held by an observer. The observer drops the ball, which bounces off the ground and comes back to the same height. Which of the following most accurately describes the frequency of the sound heard by the observer?

(A) The frequency starts at f, decreases, then increases.

(B) The frequency starts at f, increases, then decreases.

(C) The frequency starts at a frequency lower than f, then suddenly changes to a frequency higher than f.

(D) The frequency starts at f, decreases, suddenly changes to a frequency higher than f, then decreases.

11 Mark for Review

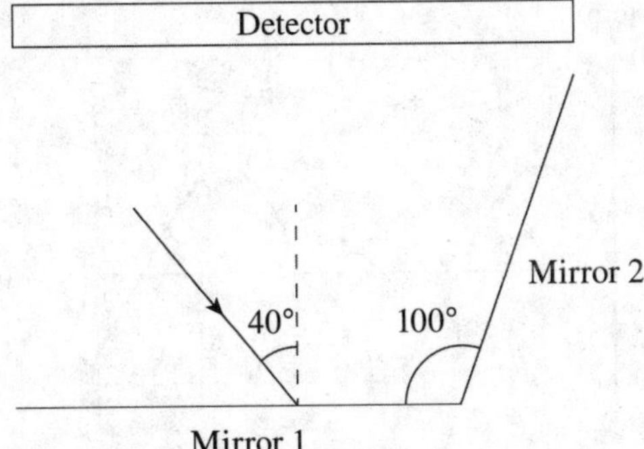

A light ray is incident on a system of two plane mirrors, as shown in the diagram above. It reflects off of mirror 1, then off of mirror 2, and finally hits a detector positioned above the system. If the detector is parallel to mirror 1, at what angle from the normal will the light ray hit the detector?

(A) 10°

(B) 20°

(C) 50°

(D) 60°

GO ON TO THE NEXT PAGE.

12 Mark for Review

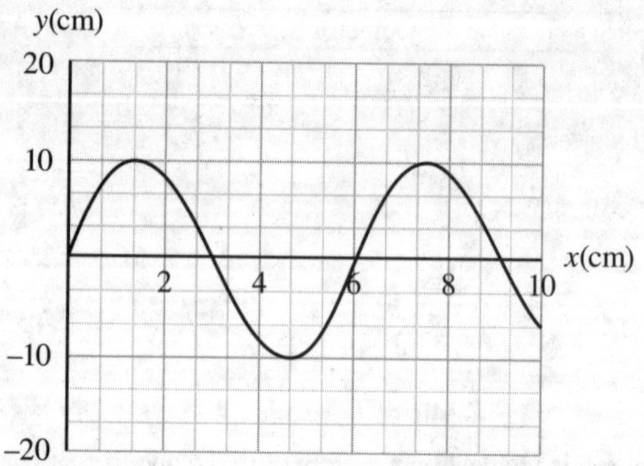

Above is the graph of a wave on a string at a certain point in time. Which of the following equations matches with the graph?

(A) $y = 6\sin\left(\pi\frac{x}{10}\right)$

(B) $y = 6\sin\left(\pi\frac{x}{5}\right)$

(C) $y = 10\sin\left(\pi\frac{x}{6}\right)$

(D) $y = 10\sin\left(\pi\frac{x}{3}\right)$

Questions 13 through 15 refer to the following.

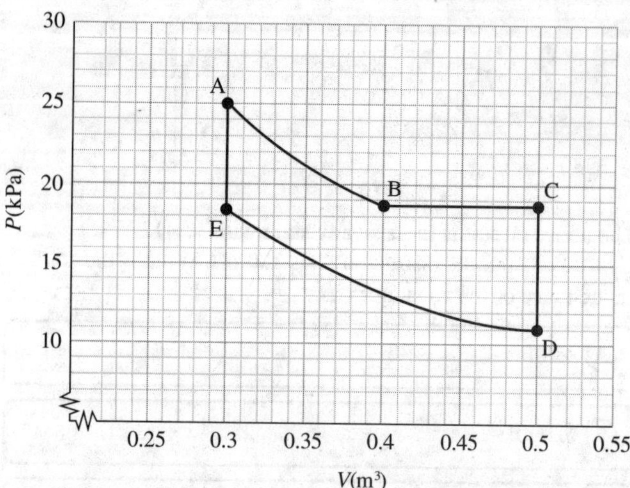

The above PV diagram shows an ideal gas in a closed system going through a thermodynamic cycle. The paths A→B and D→E are both isotherms.

13 Mark for Review

If the temperature of the gas at point A is 500 K, approximately what is the temperature at point D?

(A) 220 K

(B) 370 K

(C) 500 K

(D) 830 K

14 Mark for Review

After the gas undergoes the transition A→B→C, approximately what is the magnitude of work done on the gas?

(A) 150 J

(B) 300 J

(C) 3800 J

(D) 4100 J

GO ON TO THE NEXT PAGE.

15 Mark for Review

Which of the following most accurately describes what would be different if this cycle were attempted with a real gas instead of an ideal gas?

(A) More heat would be output due to the increase in total entropy.

(B) Energy would need to be added through additional heat or work due to the increase in total entropy.

(C) The system would do more work each cycle.

(D) There would be no significant difference.

16 Mark for Review

Which combination of waves would create the largest beat frequency?

(A) Two light waves traveling through space, one with wavelength 550 nm, the other with wavelength 700 nm

(B) Two sound waves, one with a frequency of 600 Hz, the other a frequency of 660 Hz

(C) Two waves on a string, one with a period of 10 ms, the other with a period of 80 ms

(D) Two water waves with the same frequency, one with an amplitude of 1 cm, the other with an amplitude of 40 cm

17 Mark for Review

Pair production is a process where a photon spontaneously converts into an electron-positron pair. Given that the positron is the antiparticle of the electron (same mass, opposite charge) which of the following frequencies of light could result in pair production?

(A) 5.2×10^{12} Hz

(B) 2.0×10^{20} Hz

(C) 7.1×10^{22} Hz

(D) Both B and C

GO ON TO THE NEXT PAGE.

18 Mark for Review

A 30 V battery with internal resistance of 5 Ω is attached in series to a lightbulb with a resistance of 40 Ω by using wires with a total length of 120 cm and a cross-sectional area of 1 cm^2. If the material of the wires has a resistivity of $8 \times 10^{-6}\ \Omega \cdot m$, how much power will be dissipated by the lightbulb?

Ⓐ 12 W

Ⓑ 16 W

Ⓒ 20 W

Ⓓ 23 W

19 Mark for Review

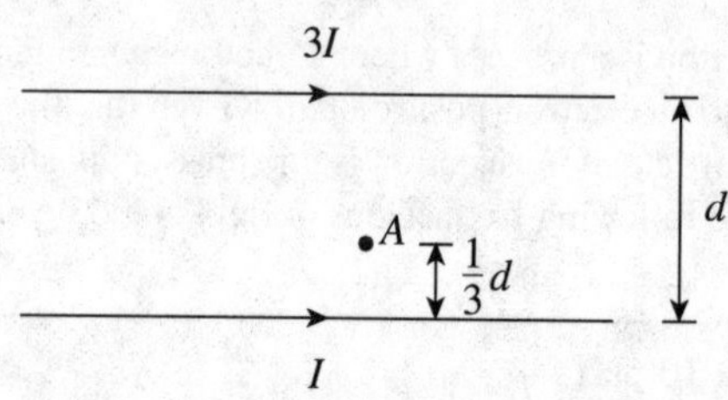

Two long parallel wires, each in the plane of the page, are separated by a distance *d*, as shown. The top wire carries a current of 3*I* directed to the right, and the bottom wire carries a current of *I* directed to the right. Point A is also in the plane of the page, located a distance *d*/3 above the bottom wire. Which of the following correctly represents the magnitude of the magnetic field due to both wires at Point A?

Ⓐ $\frac{1}{3}\frac{\mu_0 I}{\pi d}$

Ⓑ $\frac{3}{4}\frac{\mu_0 I}{\pi d}$

Ⓒ $3\frac{\mu_0 I}{\pi d}$

Ⓓ $\frac{15}{4}\frac{\mu_0 I}{\pi d}$

20 Mark for Review

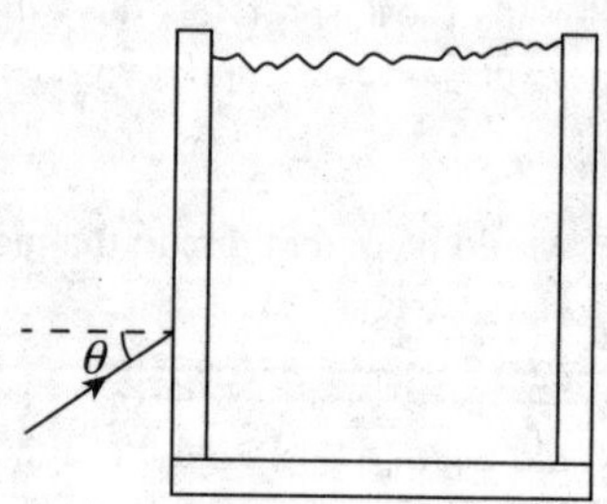

A light ray is incident on a glass tank filled with water at an angle, $\theta = 70°$, as shown above. Given that the index of the glass is $n_{glass} = 1.5$ and the index of the water is $n_{water} = 4/3$, at what angle from the normal will the light ray enter the water?

Ⓐ 34°

Ⓑ 39°

Ⓒ 45°

Ⓓ The light ray will not enter the water.

GO ON TO THE NEXT PAGE.

21 Mark for Review

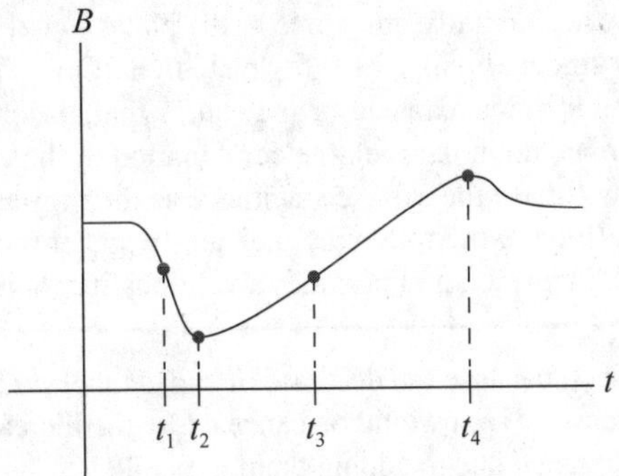

A loop of conducting material is placed in a magnetic field such that the plane of the loop is perpendicular to the magnetic field. The strength of the magnetic field is then varied according to the graph above. At which of the times marked on the graph will the magnitude of the induced emf be the largest?

(A) t_1

(B) t_2

(C) t_3

(D) t_4

22 Mark for Review

A radioactive isotope has a half-life of 2 minutes. After 5 minutes, 100 g of a sample of this isotope has decayed. Approximately how much mass was in the original sample?

(A) 20 g

(B) 120 g

(C) 410 g

(C) 560 g

23 Mark for Review

Which of the following nuclear decays is NOT possible?

(A) ${}^{238}_{92}\text{U} \rightarrow {}^{234}_{90}\text{Th} + {}^{4}_{2}\text{He}$

(B) ${}^{11}_{6}\text{C} \rightarrow {}^{11}_{5}\text{B} + \beta^+ + \nu_e$

(C) ${}^{14}_{7}\text{N} \rightarrow {}^{14}_{6}\text{C} + \beta^- + \bar{\nu}_e$

(D) ${}^{137}_{55}\text{Cs} \rightarrow {}^{137}_{56}\text{Ba} + \beta^- + \bar{\nu}_e + \gamma$

24 Mark for Review

Coherent light with a wavelength of 200 nm passes through a pinhole with a diameter of 1 μm and creates a diffraction pattern on a screen positioned 5 m away from the hole. What is the distance from the center point of the diffraction pattern to the 3rd dark fringe?

(A) 1 m

(B) 3 m

(C) 3.5 m

(D) 5 m

25 Mark for Review

0.5 kg of an oil with a specific heat of 2100 J/(kg·K) is added to an insulated container with 2 kg of water at a temperature of 280 K. The two liquids come to equilibrium at a temperature at 290 K. The specific heat of water is 4184 J/(kg·K). What was the initial temperature of the oil?

(A) 295 K

(B) 300 K

(C) 330 K

(D) 370 K

GO ON TO THE NEXT PAGE.

26 Mark for Review

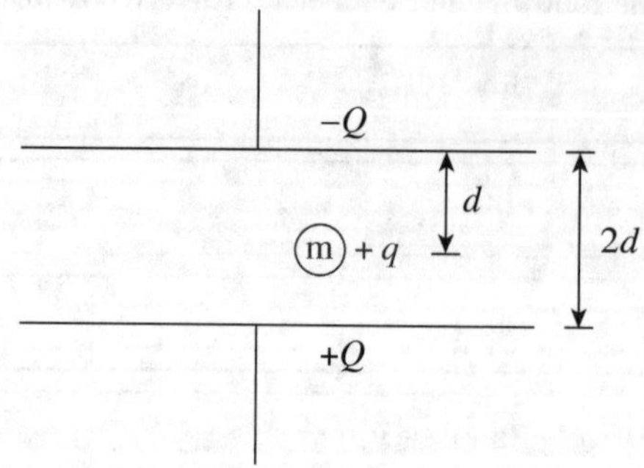

A charge with a mass *m* and positive charge +*q* is placed in the center of a charged capacitor, as shown above. The capacitor has a charge *Q*, the distance between the plates is 2*d*, the area of each plate is *A*, and there is no dielectric between the plates. How fast will the charge be moving when it impacts one of the plates of the capacitor?

(A) $\frac{1}{2}\sqrt{\frac{dqQ}{\varepsilon_0 Am}}$

(B) $\sqrt{\frac{1}{2}\frac{dqQ}{\varepsilon_0 Am}}$

(C) $\sqrt{2\frac{dqQ}{\varepsilon_0 Am}}$

(D) $2\sqrt{\frac{dqQ}{\varepsilon_0 Am}}$

27 Mark for Review

A student is constructing a capacitor using two conductive plates. They accidentally break one of the plates in half. To fix this, the student breaks the other plate in half to make two separate identical capacitors and claims that, as long as they are connected in parallel, the combination of the two capacitors will have the same capacitance as the original capacitor. Which of the following indicates whether the claim is correct or incorrect, and provides a valid justification?

(A) Correct, because the decrease of capacitance due to the decrease of area would be canceled by the increase in capacitance due to adding them in parallel.

(B) Correct, because the increase of capacitance due to the decrease of area would be canceled by the decrease in capacitance due to adding them in parallel.

(C) Incorrect, because adding them in parallel would not affect the electric field between the plates.

(D) Incorrect, because adding them in parallel would not change the distance between the plates.

28 Mark for Review

A long straight wire has a current of 0.5 A. An object with a negative charge of –3 C is a distance of 20 cm from the wire and moving directly away from the wire with a speed of 40 m/s. What is the magnitude and direction of the magnetic force on this object?

(A) 3×10^{-4} N in the same direction as the current

(B) 6×10^{-5} N in the same direction as the current

(C) 3×10^{-4} N in the opposite direction of the current

(D) 6×10^{-5} N in the opposite direction of the current

GO ON TO THE NEXT PAGE.

29 Mark for Review

1.8 mol of a monoatomic ideal gas is in a closed container with a piston. When 80 J of work is done on the gas by compressing the piston, the temperature of the gas changes from 250 K to 252 K. How much heat was released by the gas during this compression?

Ⓐ 23 J

Ⓑ 35 J

Ⓒ 45 J

Ⓓ Cannot be determined without more information

30 Mark for Review

A charged object with a mass of 2.0 kg orbits a stationary positive point charge of 2.5 mC. The radius of the orbit is 3.0 m and the object moves with a speed of 120 m/s. What is the magnitude of the charge of the orbiting object?

Ⓐ 1.0 mC

Ⓑ 1.9 mC

Ⓒ 3.8 mC

Ⓓ 12 mC

Questions 31 through 32 refer to the following.

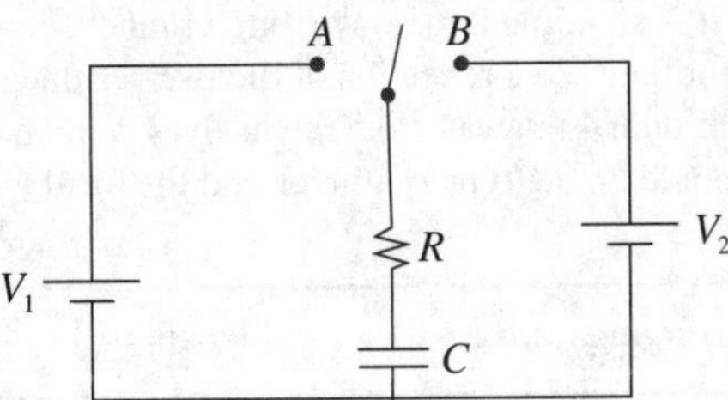

The circuit shown consists of two ideal batteries with voltages $V_1 = 20$ V and $V_2 = 15$ V, a resistor with resistance $R = 10\ \Omega$, a capacitor with capacitance $C = 40\ \mu$F, and a switch. The switch can be either set to position A or position B to complete the circuit. The capacitor is initially uncharged and the switch is initially open.

31 Mark for Review

The switch is set to position A and a long time passes so that the circuit reaches its steady state. What is the charge on the capacitor?

Ⓐ 0.4 mC

Ⓑ 0.7 mC

Ⓒ 0.8 mC

Ⓓ 1.4 mC

32 Mark for Review

After spending a long time in position A, the switch is moved to position B. Which of the following is the best description of the current flowing through the resistor after the switch is moved to position B.

Ⓐ The current starts at 0 A and increases over a long period of time to 0.5 A.

Ⓑ The current starts at 0 A and increases over a long period of time to 3.5 A.

Ⓒ The current starts at 0.5 A and decreases over a long period of time to 0 A.

Ⓓ The current starts at 3.5 A and decreases over a long period of time to 0 A.

GO ON TO THE NEXT PAGE.

33 Mark for Review

An observer is standing 2 m away from a funhouse mirror. In the mirror they see a reflection of themselves that appears to be farther than 4 m away from themselves. Which of the following could be the type of mirror and the focal length of the mirror?

(A) A diverging mirror with a focal length of 1 m

(B) A converging mirror with a focal length of 1 m

(C) A diverging mirror with a focal length of 4 m

(D) A converging mirror with a focal length of 4 m

34 Mark for Review

A material with a work function of 5 eV is exposed to light with a wavelength of 155 nm. What is the maximum speed of electrons emitted due to the photoelectric effect?

(A) 7.3×10^5 m/s

(B) 1.0×10^6 m/s

(C) 1.7×10^6 m/s

(D) 3.0×10^6 m/s

35 Mark for Review

A monatomic gas in a closed container undergoes isobaric compression from a volume of 2 m^3 to a volume of 1.5 m^3. During this compression, 6000 J of heat is released from the gas and the internal energy of the gas decreases by 1200 J. What is the pressure of the gas?

(A) 4800 Pa

(B) 9600 Pa

(C) 7200 Pa

(D) 14400 Pa

36 Mark for Review

(+) $2q$

A •

(−) $-q$ (+) $2q$

Two positive charges with charge $2q$ and one negative charge with charge $-q$ are arranged as shown in the diagram above. Point A is equidistant from all three charges. Which of the following is the most accurate representation of the direction of the net electric field at Point A?

(A) ← (arrow pointing left and slightly down)

(B) ↖ (arrow pointing left and slightly up)

(C) ↘ (arrow pointing right and slightly down)

(D) ↗ (arrow pointing right and slightly up)

37 Mark for Review

An object with a positive charge of 2 mC is place 2 m from another object with a negative charge of –5 mC. What is the electric potential energy of the two charged objects?

(A) –45 kJ

(B) –22.5 kJ

(C) 22.5 kJ

(D) 45 kJ

GO ON TO THE NEXT PAGE.

38 Mark for Review

Two spherical objects are at the same temperature and emit blackbody radiation. Object 1 has half the radius of object 2. Which of the following correctly describes the relationship of the rate of energy emitted of the two objects' blackbody radiation?

(A) Object 2 emits energy at half the rate of object 1.

(B) Object 2 emits energy at twice the rate of object 1.

(C) Object 2 emits energy at four times the rate of object 1.

(D) Object 2 emits energy at 16 times the rate of object 1.

39 Mark for Review

A converging mirror has a focal length of *f*. How far from the mirror should an object be placed in order to produce an upright image that is the same size as the object?

(A) *f*/2

(B) *f*

(C) 2*f*

(D) It is impossible to create such an image with a converging mirror.

40 Mark for Review

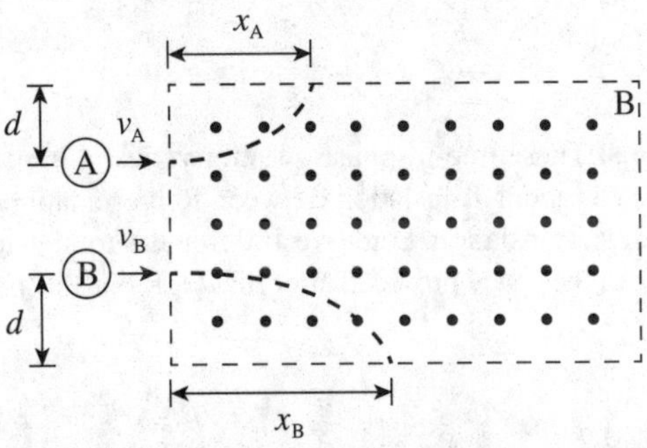

The above diagram shows two objects, object A and object B, as they enter a region with uniform magnetic field pointing out of the page. Both objects are initially moving directly to the right. Object A enters the field a distance of *d* away from the top of the region and object B enters the field a distance of *d* away from the bottom of the region. The paths of the two objects are shown above. Object A exits the field a horizontal distance x_A from the left side and object B exits the field a horizontal distance x_B from the left side. Which of the following is a valid reason that $x_A < x_B$? Ignore any interactions between the two objects.

(A) The speed of object A was slower than that of object B.

(A) The mass of object A was greater than that of object B.

(C) The magnitude of charge on object A was less than that on object B.

(D) The magnetic flux through object A was greater than that through object B.

GO ON TO THE NEXT PAGE.

AP PHYSICS 2
SECTION II
Time—100 minutes
4 Questions

Directions: These free-response questions cover 4 important areas of Physics 2. Question 1 is about Mathematical Routines. Question 2 is about Translation Between Representations. Question 3 is about Experimental Design. Question 4 is about Qualitative/Quantitative Translation. You have 100 minutes total to answer all 4 questions. On test day, you will be asked to show your work for each part in the space provided after that part. For this practice test, you may use scrap paper.

1 Mark for Review

Figure 1

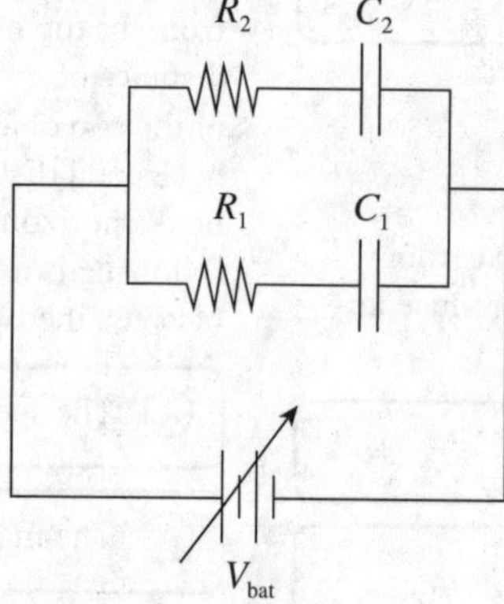

The circuit shown in Figure 1 consists of two resistors with different resistances R_1 and R_2, two capacitors with different capacitances C_1 and C_2, and a battery that is able to supply a variable voltage with negligible internal resistance. Initially the battery is set to 0 V and the capacitors are uncharged.

A. The battery is set to an initial voltage of V_i and a long time is allowed to pass. The battery's voltage is then suddenly increased to a voltage of V_f.

i. Derive an expression for the current, I_1, through the resistor with resistance R_1 immediately after the voltage is increased. Express your answer in terms of V_i, V_f, R_1, R_2, C_1, C_2, and physical constants, as appropriate. Begin your derivation by writing a fundamental physics principle or an equation for the reference booklet.

ii. Will your answer to part A (i) also be I_2, the current through the resistor with resistance R_2? Briefly justify your answer.

GO ON TO THE NEXT PAGE.

Figure 2

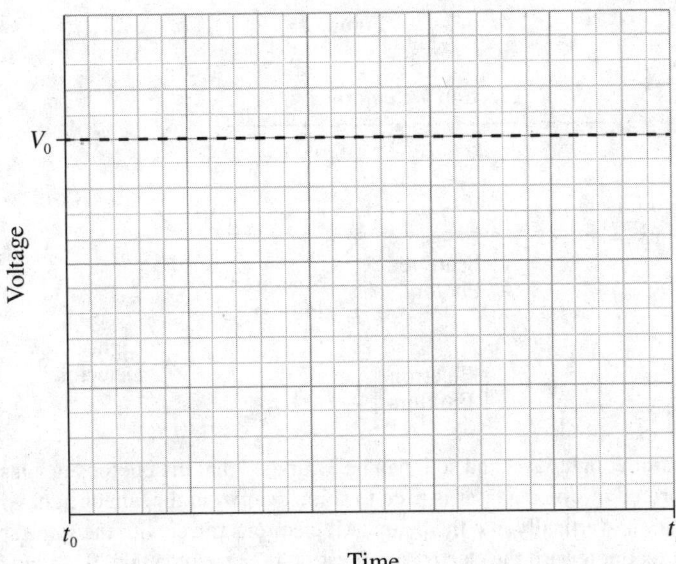

B. The battery is reset to 0 V and the capacitor is allowed to discharge. At time t_0, the battery is then suddenly set to a new voltage V_0. After a long period of time, the time is now t_1 and the circuit has nearly reached its steady-state. In Figure 2, draw two lines to qualitatively plot both the voltage across the resistor with resistance R_1, V_R, and the voltage across the capacitor with capacitance C_1, V_C, between time t_0 and time t_1. Clearly label each line. You do not need to add anything to either axis.

C. The battery is still set at the voltage V_0, and the circuit has reached its steady-state. The battery is also able to flip polarity, so that the positive and negative terminals are reversed. This is done so that the voltage provided by the battery suddenly changes from V_0 to $-V_0$. Describe what would happen to the charge on the capacitor with capacitance C_1 after this change up until the circuit reaches its new steady-state.

GO ON TO THE NEXT PAGE.

2 Mark for Review

Figure 1

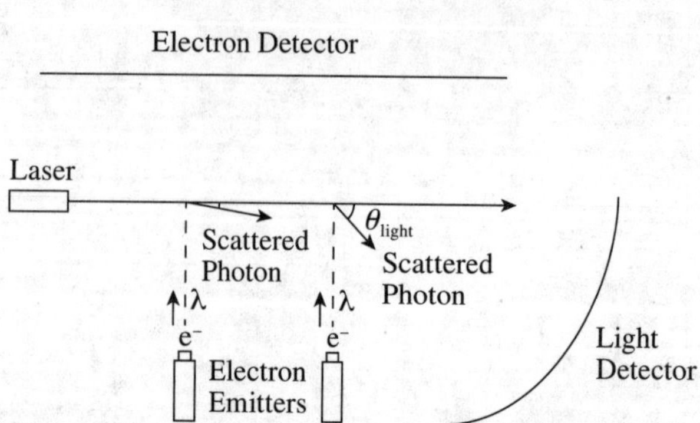

A laser is a collection of nearly parallel light rays, and so it can be estimated that the photons in a laser are all traveling in the same direction. An experiment is performed in which a laser is used to shine a horizontal beam of light with wavelength λ, and electron emitters emit single electrons traveling vertically into the beam. All electrons travel with the same speed v. Detectors are set up to detect any photons that Compton scatter toward the electron emitters at a scattering angle, θ_{light}, and any electrons passing through the light beam. Figure 1 shows two of the electrons as they enter the light beam and scatter photons.

A. On the diagram of the experiment in Figure 2, sketch the paths of the electrons after they pass through the light.

Figure 2

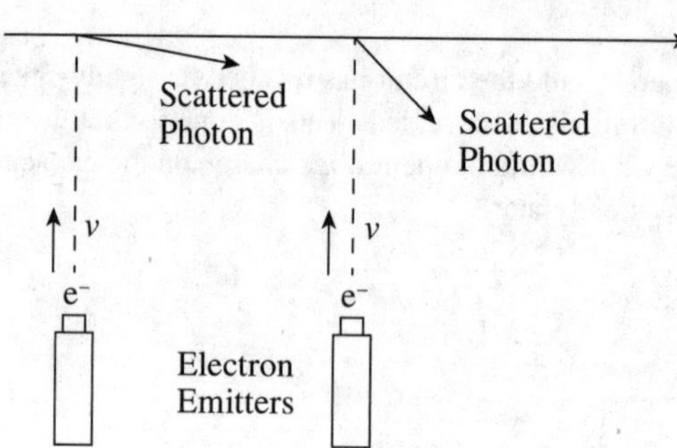

B. Derive an equation for the change in kinetic energy of an electron after it passes through the light beam, ΔK_e. Express your answer in terms of λ, θ_{light}, v, and fundamental constants, as appropriate. Begin your derivation by writing a fundamental physics principle or an equation from the reference booklet.

C. Indicate whether the equation derived in part B is consistent with the paths drawn in part A. Briefly justify your response.

GO ON TO THE NEXT PAGE.

D. This experiment is repeated many times, measuring the scattering angles of the light, θ_{light}, and the wavelength of the scattered light, λ_f, and $\cos\theta_{light}$ is calculated.

On the axes below in Figure 3, sketch λ_f, as a function of $\cos\theta_{light}$.

Figure 3

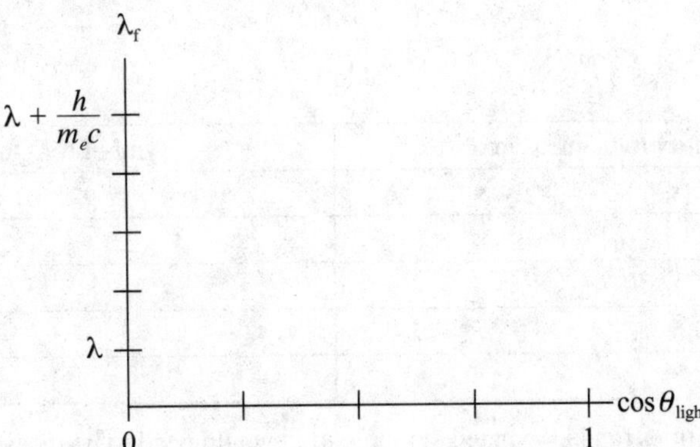

GO ON TO THE NEXT PAGE.

3 Mark for Review

A student is provided with an object made of an unknown metal, an insulated container with a gauge that shows the temperature of any fluid inside the container filled with water, an oven, a thermometer, and a scale. He is told that the specific heat of water is c_w and that the mass of the container is m_c.

A. The student is asked to design an experimental procedure to determine the specific heat of the unknown metal, c_m, using the equipment provided.

Table 1

Quantity to be measured	Symbol for Quantity

i. In Table 1, list the quantities (and associated symbols) that would need to be measured during the procedure. You do not need to fill in every row. If you need additional rows, you may add them to the space just below the table.

ii. Referring to the symbols listed in part A (i), describe the overall procedure, providing enough detail so that students could replicate the experiment, including any steps necessary to reduce experimental uncertainty.

B. Describe how the data collected in part A could be graphed and how that graph would be analyzed to determine the specific heat of the unknown metal cm.

In a different experiment, the insulated container is filled with a monatomic ideal gas. Another gauge is added to the container that is able to measure the root-mean-squared velocity, v_{rms}, of the molecules of the gas. The student is asked to determine the molar mass, M, (total mass divided by the number of moles) of the gas. The student heats the metal object to a high temperature and places it in the gas. Over a period of time, the student records the temperature of the gas and the root-mean-squared velocity. The student's data are represented in Table 2.

Table 2

T (K)	v_{rms} (m/s)			
300	1362			
315	1410			
340	1455			
360	1501			
410	1600			
435	1644			

C. i. Indicate two quantities that could be graphed to yield a straight line that could be used to determine the molar mass, M, of the ideal gas. Use the blank columns in the table to list any calculated quantities you will graph other than the data provided.

Vertical Axis: __________ Horizontal Axis: __________

GO ON TO THE NEXT PAGE.

ii. Plot the data points for the quantities indicated in part C (i) on the graph in Figure 1. Clearly scale and label all axes, including units, as appropriate.

Figure 1

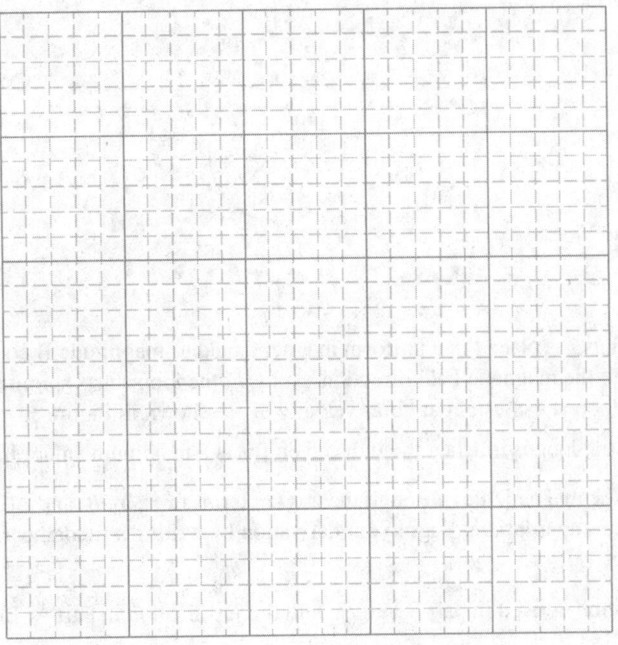

iii. Draw a best-fit line to the data graphed in part C (ii).

D. Using the best-fit line, calculate the molar mass of the ideal gas.

GO ON TO THE NEXT PAGE.

4 Mark for Review

Figure 1

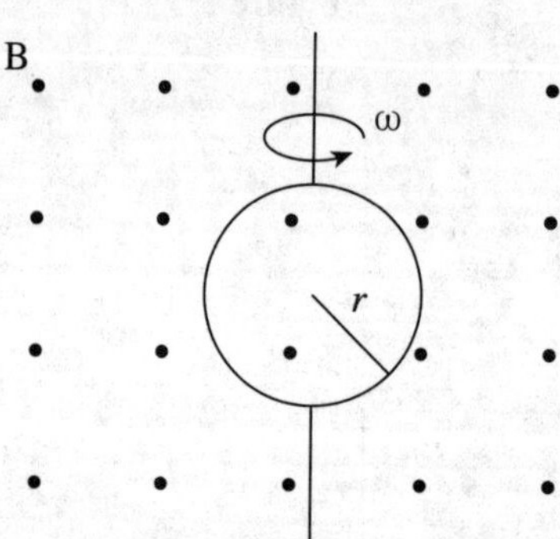

A conductive loop of wire with radius r is placed in a uniform magnetic field of magnitude B coming out of the page so that the plane of the loop is perpendicular to the magnetic field direction. At $t = t_0$, the loop starts spinning about an axis through the center of the loop and in the plane of the loop at a constant angular velocity ω_0, as shown in Figure 1.

A. Indicate what orientation the loop is in when the induced emf is at a maximum value. Justify your response.

B. Derive an expression for the magnetic flux through the loop in terms of r, ω_0, B, and Δt, where Δt is the time passed since t_0, and physical constants, as appropriate. Begin your derivation by writing a fundamental physics principle or an equation from the reference booklet.

C. Indicate whether the equation derived in part B agrees with the justification in part A. Briefly justify your reasoning.

STOP

END OF EXAM

Practice Test 2: Answers and Explanations

PRACTICE TEST 2: ANSWER KEY

1.	C	21.	A
2.	D	22.	B
3.	B	23.	C
4.	D	24.	B
5.	C	25.	D
6.	B	26.	C
7.	D	27.	A
8.	A	28.	D
9.	D	29.	B
10.	D	30.	C
11.	B	31.	C
12.	D	32.	C
13.	B	33.	D
14.	D	34.	B
15.	B	35.	B
16.	A	36.	A
17.	C	37	A
18.	A	38.	C
19.	B	39.	D
20.	C	40.	A

PRACTICE TEST 2: ANSWERS AND EXPLANATIONS

Section I: Multiple-Choice Questions

1. **C** From the conduction equation $\frac{Q}{\Delta t} = \frac{kA\Delta T}{L}$, moving the pan in half the time would reduce the heat transferred by half, holding the pan with one hand instead of two would reduce the area by half and so reduce the heat transferred by half, and doubling the thickness of the mitt would double the length the heat flowed through and so reduce the heat transferred by half. Reducing the temperature from 200°C to 100°C would change the difference in temperature from ΔT = 200°C – 20°C = 180°C to ΔT = 100°C – 20°C = 80°C. This would reduce the heat transfer by a factor of 80/180 = 0.44, so option (C) causes the greatest reduction of heat transfer.

2. **D** To apply Conservation of Energy, start by finding the electric potential created by the stationary charge at the object's initial position:

$$\phi = k\frac{Q}{r} = (9\times10^{9}\ \text{N·m}^2/\text{C}^2)\frac{2\ \text{C}}{(3\ \text{m})^2} = 2\times10^{9}\ \text{V}$$

At the final position, after the object has moved very far away, the electric potential will be 0 V. Therefore, the change in potential is $V = 0 - 2\times10^9$ V = -2×10^9 V and the energy equation becomes

$$\Delta K = -\Delta U$$

$$\frac{1}{2}mv_f^{\,2} - \frac{1}{2}mv_i^{\,2} = -qV$$

$$v_f = \sqrt{v_i^{\,2} - \frac{2qV}{m}} = \sqrt{(4\times10^4\ \text{m/s})^2 - \frac{2(.004\ \text{C})(-2\times10^9\ \text{V})}{.02\ \text{kg}}} \approx 4.9\times10^4\ \text{m/s}$$

3. **B** The fundamental frequency of the wave is the lowest frequency and therefore the longest wavelength that is resonant. For a pipe with one end closed, this occurs when ¼ of the wavelength matches with the length of the pipe, L, as shown below.

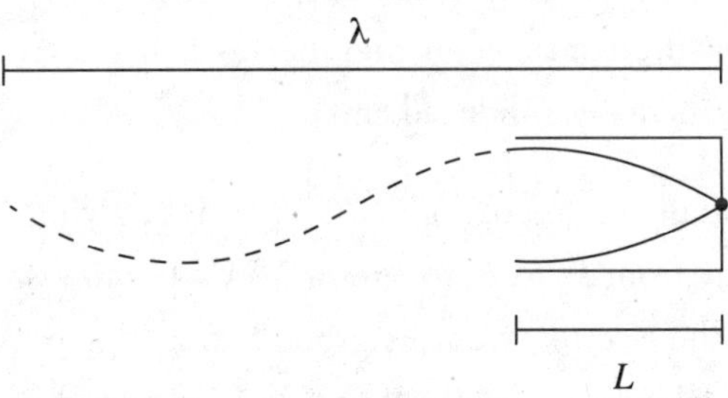

So, at the fundamental frequency of 400 Hz, $\lambda = 4L$. Increasing the frequency to 2800 Hz is an increase by a factor of 7, and so the wavelength will decrease by a factor of 7, $\lambda = 4L/7$, which corresponds to the diagram below:

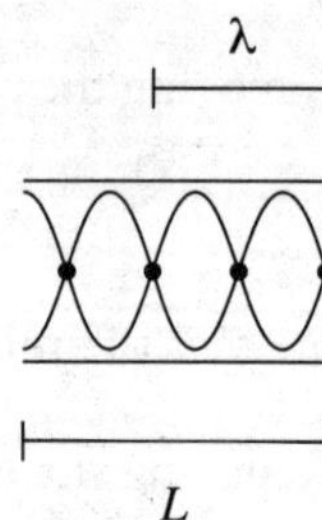

This resonance has 4 nodes.

4. **D** The path with switch I has an equivalent resistance of $2R$, the path with switch II has an equivalent resistance of R, and the path with switch II has an equivalent resistance of $R/2$. By closing all three switches, all these paths would be in parallel and the equivalent resistance of the circuit would be $\frac{1}{R_{eq}} = \frac{1}{2R} + \frac{1}{R} + \frac{1}{R/2} = \frac{1+2+4}{2R} = \frac{7}{2R} \rightarrow R_{eq} = \frac{2R}{7}$. As this is less than the resistance of any one path, it would result in the greatest current.

5. **C** From the lens equation,

$$\frac{1}{s_i} + \frac{1}{s_o} = \frac{1}{f} \rightarrow \frac{1}{s_i} = \frac{1}{f} - \frac{1}{s_o} = \frac{1}{5\text{ cm}} - \frac{1}{3\text{ cm}} = -\frac{2}{15\text{ cm}} \rightarrow s_i = -7.5\text{ cm}$$

For a lens, a negative image distance means the image is on the same side as the object.

6. **B** The electric field points away from positive charge and toward negative charge, so q_A is positive and q_B is negative. The density of the field lines, how close together they are, indicates relative strength, so there's a stronger field near q_A and it must have a greater magnitude of charge.

7. **D** The induced emf is created by a changing magnetic flux. However, in this case, the loop is never oriented so that the magnetic field passes through the loop, and therefore the magnetic flux is always zero. Since there is no change in the magnetic flux, there is no induced emf.

8. **A** A 40 eV photon could potentially undergo Compton scattering, giving some amount of energy to the electron. The electron could then decay from E_3 to E_2 to emit a 2.5 eV photon, so (B) is possible. A 17 eV photon would be enough energy to ionize the electron from its ground state of –15 eV, which would free it from the atom, so (C) is possible. If a 10 eV photon is absorbed, the electron would be in the state $E_5 = -5$ eV. It could then decay to $E_3 = -8.5$ eV, emitting a 3.5 eV photon, so (D) is possible. However, if a 6.5 eV photon is absorbed, the electron would be in the state $E_3 = -8.5$ eV, and there is no state 2 eV below that, so (A) is not possible.

9. **D** C_2 has a charge of 180 μC, and, since capacitors in series have the same charge, C_1 would also have a charge of 180 μC. The voltage across C_1 is therefore $\Delta V = \frac{Q}{C_1} = \frac{180\ \mu\text{C}}{20\ \mu\text{F}} = 9\text{ V}$. The remaining 3 V from the battery would drop across C_2, and so $C_2 = \frac{Q}{\Delta V} = \frac{180\ \mu\text{C}}{3\text{ V}} = 60\ \mu\text{F}$.

10. **D** The Doppler shift causes a lower detected frequency when things are moving away from each other and a higher detected frequency when things are moving toward each other. Since, when dropped, the ball is initially at rest, the detected frequency should start at the emitted frequency of *f*; this eliminates (C). When the ball falls, it accelerates away from the observer, which means the frequency should be decreasing; this eliminates (B). When the ball hits the ground, it will bounce and suddenly change from moving away from the observer to moving toward the observer, which causes a sudden change in the detected frequency; this eliminates (A). As the ball rises, it will slow down, which causes the detected frequency to decrease as it approaches the observer, so (D) is correct.

11. **B** Look at the diagram below:

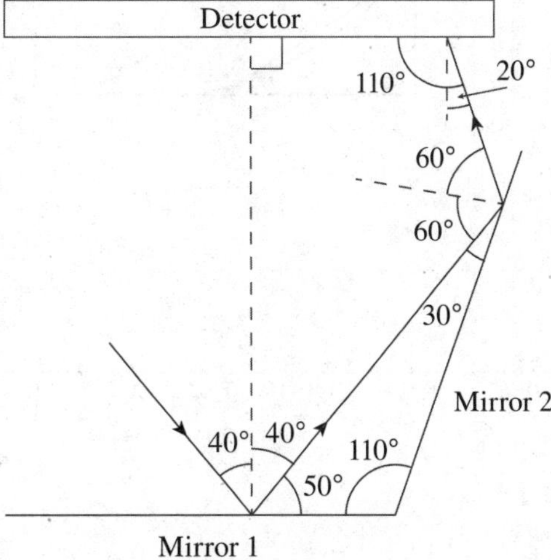

By the Law of Reflection, the incident angle is the same as the reflected angle. This means the angle from the surface of mirror 1 to the reflected ray is 50°, and since the three angles in the triangle created by that reflected ray need to add to 180°, the angle from the surface of mirror 2 to the ray reflected off of mirror 1 is 30°. Therefore, the angle from the normal to the ray incident on mirror 2 is 60°. The angles of the quadrilateral drawn using the normal from mirror 1 must add to 360°, which means the angle from the surface of the detector to the incident ray is 110°. Finally, the angle from the incident ray to the normal would be 110° – 90° = 20°.

12. **D** The equation for a wave is $y(x) = A\cos\left(2\pi\frac{x}{\lambda}\right)$. This wave has been phase shifted so that it starts at zero, which corresponds to a sine function. The wavelength of the wave is 6 cm and its amplitude is 10 cm. Therefore, the equation becomes $y(x) = 10\sin\left(2\pi\frac{x}{6}\right) = 10\sin\left(\pi\frac{x}{3}\right)$.

13. **B** From the Ideal Gas Law, $PV = nRT$, you can create a ratio between points A and D:

$$\frac{P_D V_D}{P_A V_A} = \frac{nRT_D}{nRT_A} \rightarrow T_D = \frac{P_D V_D}{P_A V_A} T_A = \frac{(11000\text{ Pa})(0.5\text{ m}^3)}{(25000\text{ Pa})(0.3\text{ m}^3)}(500\text{ K}) \approx 370\text{ K}$$

14. **D** The absolute value of the work is given by the area under the curve. You can estimate this by using triangles and rectangles, as seen below

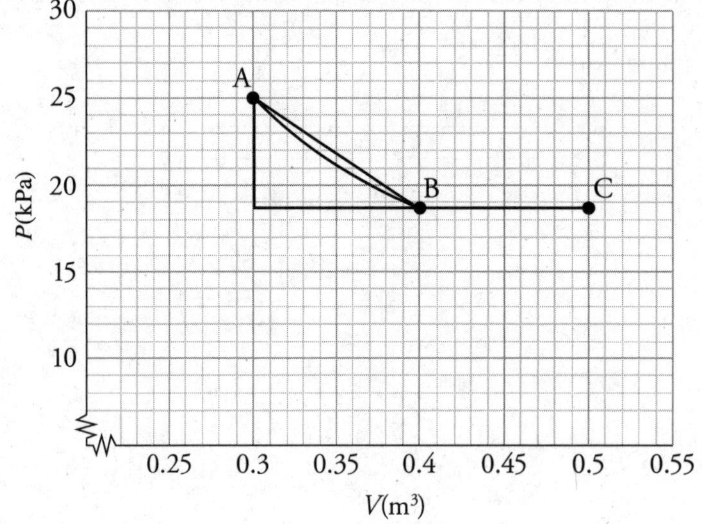

The area of the triangle is $\frac{1}{2}bh \approx \frac{1}{2}(0.1\text{ m}^3)(6000\text{ Pa}) = 300\text{ J}$ and the area of the rectangle is $bh \approx (0.2\text{ m}^3)(19000\text{ Pa}) = 3800\text{ J}$; therefore, the absolute value of the total work done is 300 J + 3800 J = 4100 J.

15. **B** If a real gas is used, these processes would no longer be reversible and the total entropy (system + surroundings) must increase. This increase of entropy corresponds to some of the energy of the system being unavailable to do work; thus, additional energy would need to be added in order to complete the cycle.

16. **A** The beat frequency is the difference between the frequencies of the two waves. For the sound waves, this would be $|660\text{ Hz} - 600\text{ Hz}| = 60\text{ Hz}$. For the waves on a string, $f = \frac{1}{T}$, so the difference would be $|1/(0.01\text{ s}) - 1/(0.08\text{ s})| = 87.5\text{ Hz}$. For the two water waves, since they have the same frequency, there would be no beat frequency. Finally, for the light waves, using the wave equation $\lambda = \frac{v}{f}$ or $f = \frac{c}{\lambda}$, the difference would be $\left|\frac{3\times10^{8}\text{ m/s}}{550\times10^{-9}\text{ m}} - \frac{3\times10^{8}\text{ m/s}}{700\times10^{-9}\text{ m}}\right| \approx 1.17\times10^{14}\text{ Hz}$.

17. **C** In order to undergo pair production, the energy of the photon must be at least enough to produce both an electron and a positron. Since they have the same mass, the energy required would be $E = mc^2 = (m_e + m_e)c^2 = 2(9.11\times10^{-31})(3\times10^{8})^2 \approx 1.6\times10^{-13}\text{ J}$. The energy of a photon is given by $E = hf$; therefore, the required frequency is at least $f = \frac{E}{h} = \frac{1.6\times10^{-13}}{6.63\times10^{-34}} \approx 2.4\times10^{20}$. The only answer with a frequency higher than this is (C).

18. **A** First, the resistance of the wire can be found using

$$R = \frac{\rho\ell}{A} = \frac{(8\times10^{-6}\ \Omega\cdot\text{m})(0.12\text{ m})}{1\times10^{-4}\text{ m}^2} = 9.6\ \Omega$$

Then, since all of these things are in series, you can find the equivalent resistance using

$$R_{\text{eq,s}} = \sum_i R_i = 5\ \Omega + 40\ \Omega + 9.6\ \Omega = 54.6\ \Omega$$

The current in the circuit would be $I = \frac{\Delta V}{R} = \frac{30}{54.6} \approx 0.55\text{ A}$, so the power in the lightbulb would be

$$P = I\Delta V = I^2R = (0.55\text{ A})^2(40\ \Omega) \approx 12\text{ W}$$

19. **B** By the Right-Hand Rule, at Point A, the magnetic field created by the top wire will be going into the page, and the magnetic field created by the bottom wire will be coming out of the page. If the distance from the bottom wire is $d/3$, the distance from the top wire will be $2d/3$. The two fields will therefore combine like so:

$$\left|\frac{\mu_0 I}{2\pi(\frac{1}{3}d)} - \frac{\mu_0(3I)}{2\pi(\frac{2}{3}d)}\right| = \frac{3}{4}\frac{\mu_0 I}{\pi d}$$

20. **C** Since the index of both the water and the glass is larger than the index of air, the light will not undergo total internal reflection, so it will enter into the glass and subsequently enter the water. From Snell's Law,

$$n_{\text{air}} \sin\theta_{\text{air}} = n_{\text{glass}} \sin\theta_{\text{glass}} = n_{\text{water}} \sin\theta_{\text{water}}$$

and so

$$\theta_{\text{water}} = \sin^{-1}\left(\frac{n_{\text{air}}}{n_{\text{water}}}\sin\theta_{\text{air}}\right) = \sin^{-1}\left(\frac{1}{4/3}\sin(70°)\right) \approx 45°$$

21. **A** According to Faraday's Law, $|\varepsilon| = \left|\frac{\Delta\Phi_B}{\Delta t}\right|$, the magnitude of the induced emf is equal to the change in the magnetic flux. Since the magnetic flux is directly related to the magnetic field through $\Phi_B = |B|\cos\theta|A|$, the time when the magnetic field is changing the fastest would result in the largest emf. This corresponds to the steepest slope on the graph, which happens at time t_1.

22. **B** Since 100 g has decayed, the original sample must be larger than that, so (A) is eliminated. 5 minutes is longer than the half-life of 2 minutes, so more than half of the sample would have decayed. This means 100 g is larger than half the sample, which eliminates (C) and (D). Alternatively, you can solve the decay equation:

$$\begin{aligned} N &= N_0 e^{-\lambda t} \\ N_0 - 100\text{ g} &= N_0 e^{-(\ln 2/2\text{ min})(5\text{ min})} \approx 0.177 N_0 \\ N_0 - 0.177 N_0 &= 100\text{ g} \\ N_0 &\approx 120\text{ g} \end{aligned}$$

23. **C** Nuclear decay must conserve charge and mass, but the decay ${}^{14}_{7}\text{N} \rightarrow {}^{14}_{6}\text{C} + \beta^- + \bar{\nu}_e$ starts with a charge of +7e from the nitrogen and decays to a carbon with a charge of +6e and a negative beta particle with a charge of –e, resulting in a total of +5e (the antineutrino has no charge.) Since this is not balanced with the +7e, this decay is not possible.

24. **B** The position of the mth minimum in the diffraction pattern from single slit interference is given by

$$a\left(\frac{y_{\min}}{L}\right) = m\lambda \rightarrow y_{\min} = \frac{L}{a}m\lambda = \frac{5\text{ m}}{1\times10^{-6}\text{ m}}3(200\times10^{-9}\text{ m}) = 3\text{ m}$$

25. **D** Since the container is insulated, the amount of heat lost from the added liquid must be the same as the amount of heat added to the water. The amount of heat is given by $Q = mc\Delta T$ and so

$$m_{\text{oil}}c_{\text{oil}}\Delta T_{\text{oil}} = m_{\text{water}}c_{\text{water}}\Delta T_{\text{water}} \rightarrow \Delta T_{\text{oil}} = \frac{m_{\text{water}}c_{\text{water}}\Delta T_{\text{water}}}{m_{\text{oil}}c_{\text{oil}}} = \frac{(4\text{ kg})(4184\ ^{\text{J}}\!/_{\text{kg}\cdot\text{K}})(10\text{ K})}{(0.5\text{ kg})(2100\ ^{\text{J}}\!/_{\text{kg}\cdot\text{K}})} \approx 80\text{ K}$$

Therefore, the oil started 80 K above the equilibrium temperature, so the initial temperature is 290 K + 80 K = 370 K.

26. **C** The speed can be found using Conservation of Energy, $\Delta K = -\Delta U_{\text{E}}$. The change in potential energy for a charge is found from $\Delta U_{\text{E}} = q\Delta V$. The change in electric potential in an electric field is found from $|E| = \left|\frac{\Delta V}{\Delta r}\right| \rightarrow |\Delta V| = |\Delta r E|$, and since the positive charge would move toward the negative plate, this change in potential would be negative. The electric field from a capacitor is given by $E_{\text{C}} = \frac{Q}{\kappa\varepsilon_0 A}$. Combining all of these results in

$$\Delta K = -\Delta U_{\text{E}}$$

$$\frac{1}{2}mv^2 = -q\Delta V$$

$$v = \sqrt{-2\frac{q\Delta V}{m}} = \sqrt{-2\frac{q(-dE)}{m}} = \sqrt{2\frac{qd(\frac{Q}{\varepsilon_0 A})}{m}} = \sqrt{2\frac{dqQ}{\varepsilon_0 Am}}$$

27. **A** Decreasing the area by half would half the capacitance according to $C = \kappa\varepsilon_0\frac{A}{d}$, and adding the identical capacitors in parallel would double the capacitance according to $C_{\text{eq,p}} = \sum_i C_i = C + C = 2C$. These two changes would therefore cancel out.

28. **D** Consider the diagram below

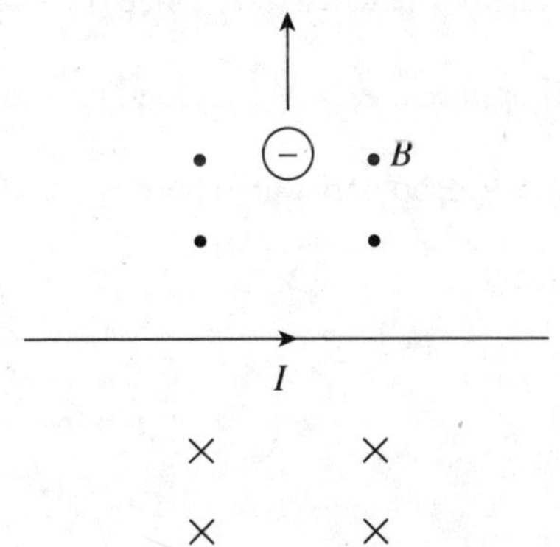

Applying the Right-Hand Rule to the charge in this diagram would result in a force to the left, which is opposite to the direction of the current. This eliminates (A) and (B). Plugging in the magnetic field created by a wire to the magnetic force equation results in

$$F_B = qvB\sin\theta = qv\left(\frac{\mu_0 I}{2\pi r}\right)\sin\theta = (3\text{ C})(40\ \text{m/s})\left(\frac{(4\pi\times10^{-7}\ \text{T·m/A})(0.5\text{ A})}{2\pi(0.2\text{ m})}\right)(1) = 6\times10^{-5}\text{ N}$$

29. **B** The energy of a monoatomic ideal gas is given by $U = \frac{3}{2}nRT$, so applying the First Law of Thermodynamics gives

$$\Delta U = Q + W$$
$$\frac{3}{2}nRT_f - \frac{3}{2}nRT_i = Q + W$$

$$Q = \frac{3}{2}nR(T_f - T_i) - W = \frac{3}{2}(1.8\text{ mol})(8.314\ \text{J/K·mol})(252\text{ K} - 250\text{ K}) - 80\text{ J} \approx -35\text{ J}$$

Therefore, 35 J of heat left the system.

30. **C** For the object to orbit the point charge, the electrostatic force must be equal to the required centripetal force

$$k\frac{|q_1q_2|}{r^2} = m\frac{v^2}{r} \rightarrow |q_2| = m\frac{v^2 r}{k|q_1|} = (2\text{ kg})\frac{(120\ \text{m/s})^2(3\text{ m})}{(9\times10^9\ \text{N·m}^2/\text{C}^2)(2.5\times10^{-3}\text{ C})} \approx 3.8\times10^{-3}\text{ C}$$

31. **C** After a long time, the capacitor will become fully charged and the voltage across the capacitor will match that of the battery, 20 V. The charge on the capacitor is therefore given by $Q = C\Delta V = (40\times10^{-6}\text{ F})(20\text{ V}) = 8\times10^{-4}\text{ C} = 0.8\text{ mC}$.

32. **C** While the switch is in position A, the capacitor will charge to have the same voltage as the battery and will oppose the battery. This means the voltage across the capacitor will be 20 V with the higher potential on the top side of the capacitor. Once the switch is moved to position B, the 20 V of the charged capacitor will oppose the 15 V provided by the other battery, and so the resistor will have a total of 20 V – 15V = 5 V drop across it. Therefore, initially the current through the resistor would be $I = \frac{\Delta V}{R} = \frac{5\text{ V}}{10\ \Omega} = 0.5\text{ A}$.

33. **D** From the lens equation $\frac{1}{f}=\frac{1}{s_i}+\frac{1}{s_o}\rightarrow\frac{1}{s_i}=\frac{1}{f}-\frac{1}{s_o}$. For converging mirrors, f is positive and for diverging mirrors f is negative. Trying each option yields the following:

$$\frac{1}{s_i}=\frac{1}{-4\text{ m}}-\frac{1}{2\text{ m}}=-\frac{3}{4\text{ m}}\rightarrow s_i=-\frac{4}{3}\text{ m}$$
$$\frac{1}{s_i}=\frac{1}{-1\text{ m}}-\frac{1}{2\text{ m}}=-\frac{3}{2\text{ m}}\rightarrow s_i=-\frac{2}{3}\text{ m}$$
$$\frac{1}{s_i}=\frac{1}{1\text{ m}}-\frac{1}{2\text{ m}}=\frac{3}{2\text{ m}}\rightarrow s_i=\frac{2}{3}\text{ m}$$
$$\frac{1}{s_i}=\frac{1}{4\text{ m}}-\frac{1}{2\text{ m}}=-\frac{1}{4\text{ m}}\rightarrow s_i=-4\text{ m}$$

Choice (D) implies the image will be 6 m away from the observer on the other side of the mirror. All other options would result in too small of a distance from the observer.

34. **B** The energy of a photon with a wavelength of 155 nm is

$$E=hf=\frac{hc}{\lambda}=\frac{(4.14\times10^{-15}\text{ eV}\cdot\text{s})(3\times10^{8}\text{ m/s})}{155\times10^{-9}\text{ m}}\approx 8\text{ eV}$$

The maximum kinetic energy would then be $K_{max}=hf-\phi=8\text{ eV}-5\text{ eV}=3\text{ eV}$, and the speed of the electron can be found using

$$K=\frac{1}{2}mv^2\rightarrow v=\sqrt{\frac{2K}{m}}=\sqrt{\frac{2(3\text{ eV})(1.6\times10^{-19}\text{ J/eV})}{9.11\times10^{-31}\text{ kg}}}\approx1.0\times10^{6}\text{ m/s}$$

35. **B** For an isobaric process, the work done on the gas is $W=-P\Delta V$. Applying this with the First Law of Thermodynamics results in

$$\Delta U=Q+W=Q-P\Delta V\rightarrow P=\frac{Q-\Delta U}{\Delta V}=\frac{-6000\text{ J}-(-1200\text{ J})}{1.5\text{ m}^3-2\text{ m}^3}=9600\text{ Pa}$$

36. **A** Consider the following diagrams

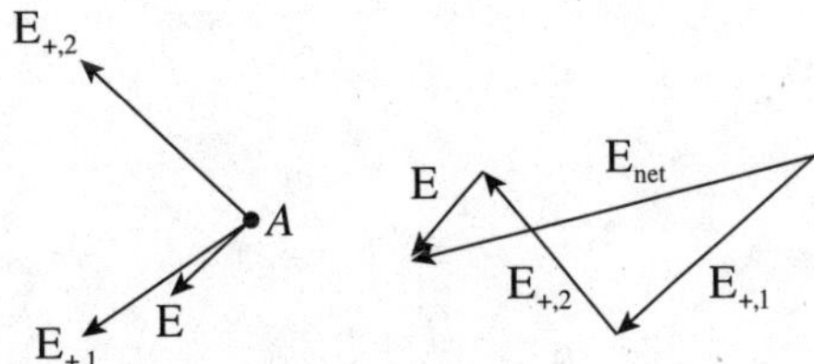

The left diagram shows the electric fields from each individual charge. Note that since the negative charge has half the charge, the magnitude of the electric field is also half. The right diagram shows the vector sum of the three fields.

37. **A** The electric potential energy of a system of two charges is given by $U_E = k\frac{q_1 q_2}{r} = (9\times10^9 \text{ Nm}^2/\text{C}^2)\frac{(2\times10^{-3}\text{ C})(-5\times10^{-3}\text{ C})}{2\text{ m}} = -45000\text{ J}$.

38. **C** The rate of energy emission, or power, of a blackbody is given by $P = A\sigma T^4$, so power is proportional to surface area. For a sphere, the surface area is $A = 4\pi r^2$. Therefore, power is proportional to the radius squared. Since object 2 has twice the radius of object 1, it will emit energy at four times the rate.

39. **D** In order for the image to be the same size, $M = \frac{|s_i|}{|s_o|} = 1 \rightarrow |s_i| = |s_o|$, and in order for the image to be upright, s_i must be negative. From the lens equation, $\frac{1}{f} = \frac{1}{s_i} + \frac{1}{s_o} = -\frac{1}{s_o} + \frac{1}{s_o} = 0 \rightarrow f = \infty$, but an infinite focal length corresponds to a plane mirror, which would not be converging.

40. **A** When traveling perpendicular through a magnetic field, the magnetic force acts as a centripetal force. The radius of the motion can be found using

$$F_B = F_c \rightarrow qvB\sin\theta = m\frac{v^2}{r} \rightarrow r = \frac{mv}{qB\sin\theta}$$

Therefore, having a slower speed would result in a smaller radius of motion, leading object A to exit the field at a smaller distance compared to object B.

Section II: Free Response Questions

1. A. i. Use Ohm's Law: $I = \frac{\Delta V}{R}$. Since the circuit was in steady-state, immediately after the change the voltage on the capacitor is still V_i and it will be opposing the voltage provided by the battery. Therefore, applying the loop rule to the loop with the battery, resistor with resistance R_1, and capacitor with capacitance with C_1 gives you $V_f = V_R + V_i \rightarrow V_R = V_f - V_i$, where V_R is the voltage across the resistor. Therefore, $I_1 = \frac{V_f - V_i}{R_1}$.

ii. No, I_2 and I_1 will be different. While the resistors will have the same voltage across them, $V_f - V_i$, the resistances are different, so by Ohm's Law the currents will be different.

B. The two voltages must add to V_0 by the loop rule and initially all the voltage will drop across the resistor, since an uncharged capacitor is equivalent to a wire. They will asymptotically approach the steady-state of $V_R = 0$ and $V_C = V_0$.

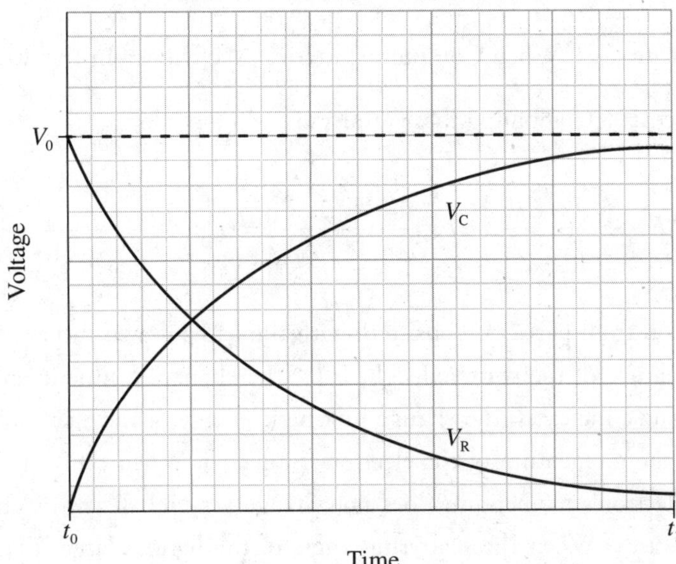

C. Initially, the capacitor would quickly discharge. It would continue to discharge until the charge on each plate is zero, and then start to charge with the reverse charge on each plate. The charging would slow down as it approaches its steady state where the magnitude of charge on each plate is the same as its initial state, but the sign of the charge on each plate is reversed.

2. A.

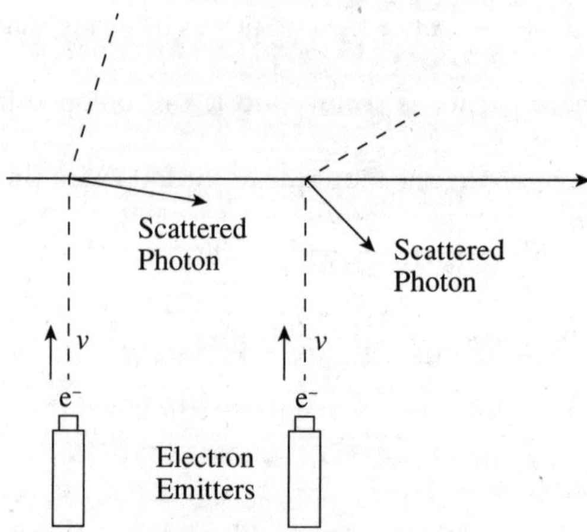

B. By Conservation of Energy, the energy lost by the photon must be equal to the energy gained by the electron. Applying the wave equation to a photon yields $\lambda = \frac{v}{f} \rightarrow f = \frac{c}{\lambda}$, so the energy of a photon is $E = hf = \frac{hc}{\lambda}$. When Compton scattering, the change in the wavelength of a photon is $\Delta\lambda = \frac{h}{m_e c}(1 - \cos\theta)$. Combining these results in

$$\Delta K_e = -\Delta E = -\left(\frac{hc}{\lambda_f} - \frac{hc}{\lambda_i}\right) = hc\left(\frac{1}{\lambda} - \frac{1}{\lambda + \Delta\lambda}\right) = hc\left(\frac{1}{\lambda} - \frac{1}{\lambda + \frac{h}{m_e c}(1 - \cos\theta_{\text{light}})}\right)$$

C. Yes, the equation from part B is consistent with the paths drawn in part A. The equation shows that when the angle of the scattered light is 0°, the change in kinetic energy of the electron is 0, and as the angle increases, the change in kinetic energy of the electron increases. This is seen by how much the momentum of the electron is affected. When the scattering angle of the light is small, the electron's momentum does not change very much and it continues heading close to straight upwards. When the scattering angle of the light is large, the electron's momentum changes significantly and the path of the electron is more noticeably altered.

D. The change in wavelength of a Compton scattered photon is $\Delta\lambda = \frac{h}{m_e c}(1 - \cos\theta)$. Therefore

$$\lambda_f = \lambda + \frac{h}{m_e c}(1 - \cos\theta_{\text{light}})$$
$$= -\frac{h}{m_e c}\cos\theta_{\text{light}} + \lambda + \frac{h}{m_e c}$$

Therefore, the graph should be a linear graph with a negative slope and a y-intercept at $\lambda + \frac{h}{m_e c}$. Additionally, when $\cos\theta_{\text{light}} = 1$, $\lambda_f = \lambda$.

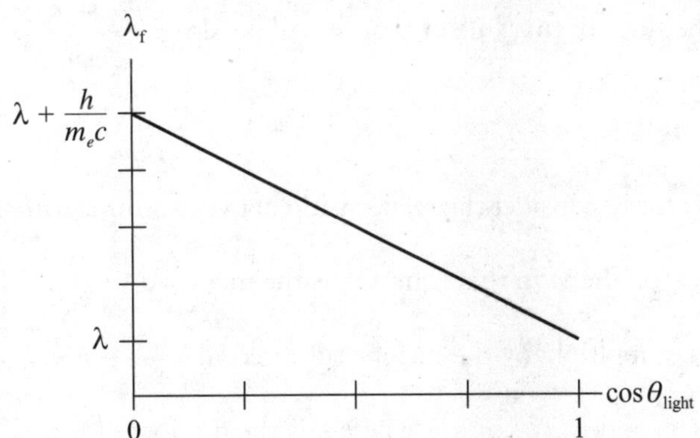

3. A. i. The variables required to find the specific heat are the mass of the metal object, mass of the water, and the change in temperature, of both the water and the metal object. Shown below is one way to list theses variables.

Quantity to be measured	Symbol for Quantity
Mass of the metal object	m_{m}
Mass of the water	m_{w}
Temperature of the water	T_{w}
Temperature of the metal object	T_{m}
Equilibrium temperature	T_{eq}

ii. One possible procedure can be described as follows: Use the scale to find m_{m}, and then place the container on the scale. Subtract m_{c} from the reading of the scale to calculate m_{w}. Next, place the metal object in the oven and turn it on. After waiting a while, remove the metal object from the oven and use the thermometer to measure T_{m}. Note the current temperature of the water T_{w}. Place the metal object in the water. Wait until the temperature stops changing and note the equilibrium temperature T_{eq}. To reduce error, repeat these steps beginning with placing the metal object in the oven.

B. To find c_{m}, you can use the idea that the amount of heat transferred into or out of the metal object must be equal to the amount of heat absorbed by the water. This leads to the following relationship

$$Q_w = -Q_m$$

$$m_w c_w \Delta T_w = -m_m c_m \Delta T_m$$

$$\Delta T_w = \frac{-m_m c_m}{m_w c_w} \Delta T_m$$

Plotting ΔT_w against ΔT_m will result in a linear relationship. Your measurements can be used to calculate ΔT_w and ΔT_m, where $\Delta T_w = T_{eq} - T_w$ and $\Delta T_m = T_{eq} - T_m$. Adding a point to the ΔT_w vs. ΔT_m graph for each set of measurements allows you to create a line of best fit. From the relationship outlined above, the slope of that line of best fit will be slope $= \frac{-m_m c_m}{m_w c_w}$, and so c_m can be calculated as $c_m = -(\text{slope})\frac{m_w c_w}{m_m}$.

C. i. To find molar mass, consider the relationship between temperature and average kinetic energy, $\frac{3}{2}k_B T = \frac{1}{2}mv_{rms}^2$. The m in this equation is the mass of a single molecule, so to transform that into total mass, multiply by the number of molecules N, $\frac{3}{2}Nk_B T = \frac{1}{2}Nmv_{rms}^2 = \frac{1}{2}m_{tot}v_{rms}^2$. By the nature of k_B and R, $Nk_B = nR$, where n is the number of moles. This leads to

$$\frac{3}{2}nRT = \frac{1}{2}m_{tot}v_{rms}^2 \rightarrow T = \frac{m_{tot}}{n}\frac{1}{3R}v_{rms}^2$$

Therefore, plotting T on the vertical axis and $(v_{rms})^2$ on the horizontal axis will give you a straight line.

Vertical Axis: T Horizontal Axis: $(v_{rms})^2$

T (K)	v_{rms} (m/s)	$(v_{rms})^2$ (m^2/s^2)		
300	1362	1.86×10^6		
315	1410	1.99×10^6		
340	1455	2.12×10^6		
360	1501	2.25×10^6		
410	1600	2.56×10^6		
435	1644	2.70×10^6		

ii. & iii.

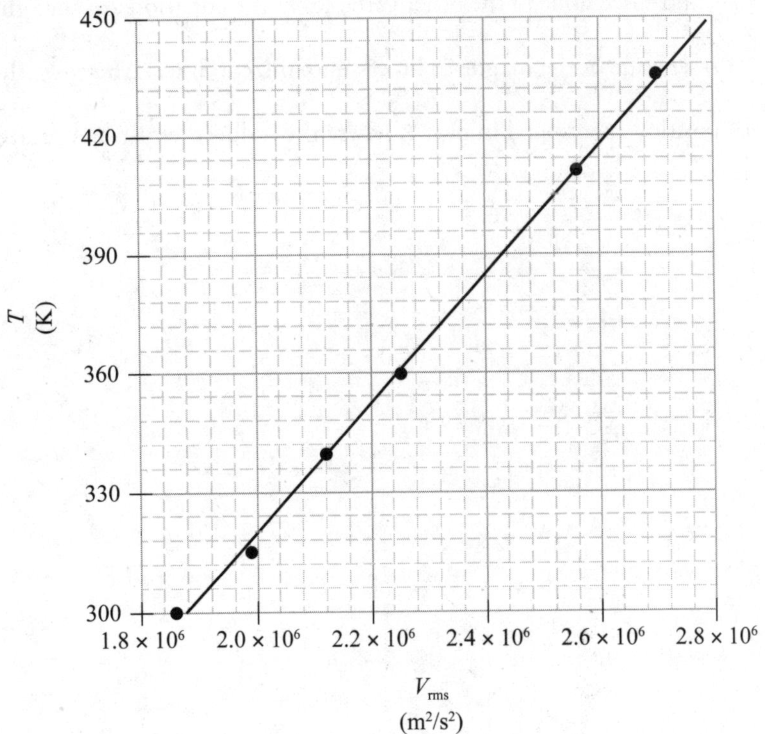

D. The slope of the line of best fit can be found using two points on the line. Choosing $(1.88 \times 10^6, 300)$ and $(2.76 \times 10^6, 444)$ yields

$$\text{slope} = \frac{444 - 300}{2.76\times10^6 - 1.88\times10^6} \approx 1.63\times10^{-4}$$

From the equation in part C (i), the slope is $\frac{m_{tot}}{n}\frac{1}{3R}$. Therefore, the molar mass is

$$M = \frac{m_{tot}}{n} = 3R(\text{slope}) = 3(8.314)(1.63\times10^{-4}) \approx 4.1\times10^{-3}\ ^{\text{kg}}\!/_{\text{mol}}$$

4. A. The induced emf will be at its maximum when the loop is oriented so that the plane of the loop is perpendicular to the page. The emf is proportional to the rate of change of the magnetic flux, and the magnetic flux is given by $\Phi_B = |B|\cos\theta|A|$, where θ is the angle between a vector normal to the area and the direction of the magnetic field. Since neither A nor B are changing in this situation, and $\cos\theta$ changes the fastest when $\theta = 90°$, that is when the induced emf is maximized.

B. Magnetic flux is given by $\Phi_B = |B|\cos\theta|A|$ and for an object rotating at a constant angular velocity, $\Delta\theta = \omega t$. Additionally, the area of a circle is $A = \pi r^2$. Combining these gives you

$$\Phi_B = |B|\cos\theta|A| = B\cos(\omega_0\Delta t)(\pi r^2)$$

C. Yes, the equation agrees with the explanation in part A. The cosine function is changing the fastest when $\omega_0 t = \frac{\pi}{2}$, and since none of the other variables in the equation are changing, this is the point where the rate of change of the magnetic flux is maximized. Since $\Delta\theta = \omega t$, this also corresponds to the $\theta = 90°$ position mentioned in part A where the induced emf is maximized.

PRACTICE TEST 2: FREE-RESPONSE QUESTIONS GRADING RUBRICS

Use the following grading rubrics to get a sense of what you should include in your answers to the four Free Response Questions. These align with what the College Board AP Exam graders will be looking for.

Question 1: Mathematical Routines		**10 points total**
A.	i. For writing some form of Ohm's Law	1 point
	For indicating that the voltage across the resistor will be $V_f - V_i$	1 point
	For correctly substituting an expression for the voltage across the resistor into an expression for current	1 point
	ii. For indicating that the currents will be different	1 point
	For a justification mentioning that the voltages across each resistor is initially the same	1 point
B.	For a graph where V_R starts at V_0 and trends toward 0	1 point
	For a graph where V_C starts at 0 and trends toward V_0	1 point
	For a graph where, throughout the time period, $V_R + V_C \approx V_0$	1 point
C.	For a description that mentions the capacitor will initially discharge and will eventually reach a new steady-state	1 point
	For indicating that the sign of charge on each plate will be reversed in the new steady-state	1 point

Question 2: Translation Between Representations		**12 points total**
A.	For a diagram where both electrons continue traveling in an upward direction after passing through the beam	1 point
	For a diagram where both electrons change direction after passing through the beam	1 point
	For a diagram where the left electron's direction is less affected than the right electron's after passing through the beam	1 point
B.	For a derivation that involves Conservation of Energy or equates the energy lost by the photon to the energy gained by the electron	2 points
	For a correct expression to find the energy of a photon	1 point
	For a correct method to find the wavelength or frequency after scattering	1 point
	For correctly substituting an expression for frequency or wavelength into an expression for change of energy	1 point

C.		For a statement that correctly addresses how the kinetic energy change of the electron might affect its path	1 point
		For a correct justification that links the functional dependence in the equation in part B to the diagram drawn in part A	1 point
D.		For a graph that is linear	1 point
		For a graph that passes through both $\left(0, \lambda + \frac{h}{m_e c}\right)$ and the $(1, \lambda)$	1 point

Question 3: Experimental Design and Analysis — **10 points total**

A.	i.	For a list including all the variables required to eventually calculate c_m	1 point
	ii.	For a valid experimental procedure	1 point
B.		For indicating a correct relationship between temperature change and specific heat	1 point
		For indicating a correct relationship between the change of temperature of the water and the change of temperature of the metal object	1 point
C.	i.	For indicating quantities that would produce a straight-line graph that could be used to determine the specific heat of the metal	1 point
	ii.	For labeling axes correctly (including units) with a linear scale such that the data fills at least half the area of the graph	1 point
		For plotting the data points correctly	1 point
	iii.	For drawing an appropriate line or curve of best fit that approximates the plotted data	1 point
D.		For using an appropriate feature of the best-fit line to calculate the specific heat of the metal	1 point
		For coming to a result that is approximately 4×10^{-3} kg/mol	1 point

Question 4: Qualitative/Quantitative Translation — **8 points total**

A.		For indicating that the induced emf is maximized when the plane of the loop is perpendicular to the page	1 point
		For a justification that involves the rate of change of magnetic flux	2 points
B.		For using a correct equation for magnetic flux	1 point
		For a correct substitution of $\Delta\theta = \omega t$ and $A = \pi r^2$	1 point
		For a correct expression for the magnetic flux involving the relevant variables	1 point
C.		For a statement that addresses the functional dependence of the magnetic flux and the orientation of the loop	1 point
		For a correct justification that links the functional dependence in the equation in part B to the qualitative reasoning in part A	1 point

HOW TO SCORE PRACTICE TEST 2

Section I: Multiple Choice

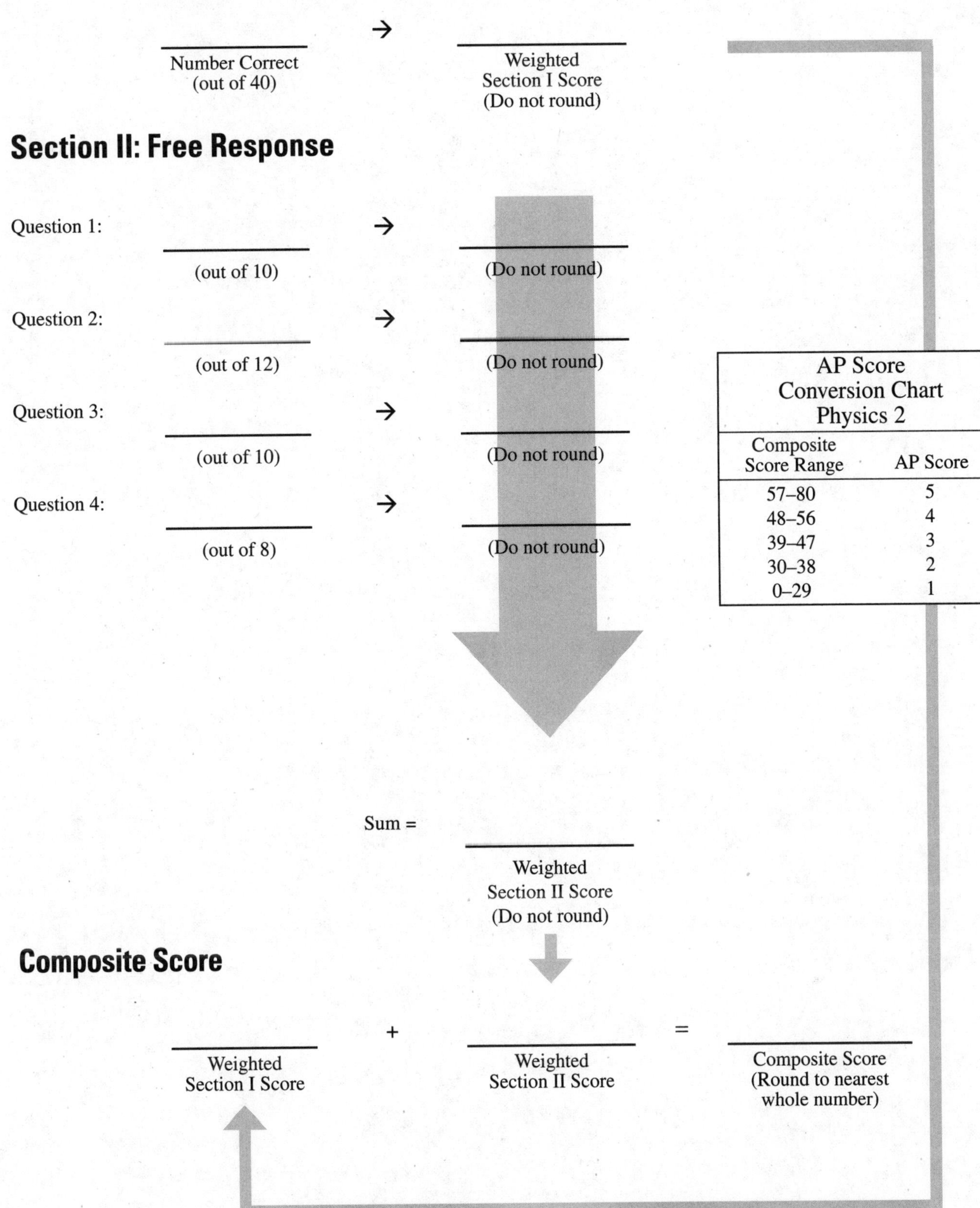

Section II: Free Response

AP Score Conversion Chart Physics 2	
Composite Score Range	AP Score
57–80	5
48–56	4
39–47	3
30–38	2
0–29	1

Composite Score

Completely darken bubbles with a No. 2 pencil. If you make a mistake, be sure to erase mark completely. Erase all stray marks.

1.
YOUR NAME: ____________________ (Print) Last / First / M.I.

SIGNATURE: ____________________ DATE: ___ / ___ / ___

HOME ADDRESS: ____________________ (Print) Number and Street

____________________ City / State / Zip Code

PHONE NO.: ____________________ (Print)

IMPORTANT: Please fill in these boxes exactly as shown on the back cover of your test book.

2. TEST FORM

3. TEST CODE

0	A	J	0	0
1	B	K	1	1
2	C	L	2	2
3	D	M	3	3
4	E	N	4	4
5	F	O	5	5
6	G	P	6	6
7	H	Q	7	7
8	I	R	8	8
9			9	9

4. REGISTRATION NUMBER

0	0	0	0	0	0
1	1	1	1	1	1
2	2	2	2	2	2
3	3	3	3	3	3
4	4	4	4	4	4
5	5	5	5	5	5
6	6	6	6	6	6
7	7	7	7	7	7
8	8	8	8	8	8
9	9	9	9	9	9

6. DATE OF BIRTH

Month	Day		Year	
JAN				
FEB	0	0	0	0
MAR	1	1	1	1
APR	2	2	2	2
MAY	3	3	3	3
JUN		4	4	4
JUL		5	5	5
AUG		6	6	6
SEP		7	7	7
OCT		8	8	8
NOV		9	9	9
DEC				

5. YOUR NAME

First 4 letters of last name				FIRST INIT	MID INIT
A	A	A	A	A	A
B	B	B	B	B	B
C	C	C	C	C	C
D	D	D	D	D	D
E	E	E	E	E	E
F	F	F	F	F	F
G	G	G	G	G	G
H	H	H	H	H	H
I	I	I	I	I	I
J	J	J	J	J	J
K	K	K	K	K	K
L	L	L	L	L	L
M	M	M	M	M	M
N	N	N	N	N	N
O	O	O	O	O	O
P	P	P	P	P	P
Q	Q	Q	Q	Q	Q
R	R	R	R	R	R
S	S	S	S	S	S
T	T	T	T	T	T
U	U	U	U	U	U
V	V	V	V	V	V
W	W	W	W	W	W
X	X	X	X	X	X
Y	Y	Y	Y	Y	Y
Z	Z	Z	Z	Z	Z

The Princeton Review®

1. A B C D	11. A B C D	21. A B C D	31. A B C D
2. A B C D	12. A B C D	22. A B C D	32. A B C D
3. A B C D	13. A B C D	23. A B C D	33. A B C D
4. A B C D	14. A B C D	24. A B C D	34. A B C D
5. A B C D	15. A B C D	25. A B C D	35. A B C D
6. A B C D	16. A B C D	26. A B C D	36. A B C D
7. A B C D	17. A B C D	27. A B C D	37. A B C D
8. A B C D	18. A B C D	28. A B C D	38. A B C D
9. A B C D	19. A B C D	29. A B C D	39. A B C D
10. A B C D	20. A B C D	30. A B C D	40. A B C D

Completely darken bubbles with a No. 2 pencil. If you make a mistake, be sure to erase mark completely. Erase all stray marks.

1.

YOUR NAME: (Print) ______ Last ______ First ______ M.I.

SIGNATURE: ______ DATE: ___/___/___

HOME ADDRESS: (Print) ______ Number and Street

______ City ______ State ______ Zip Code

PHONE NO.: (Print) ______

IMPORTANT: Please fill in these boxes exactly as shown on the back cover of your test book.

2. TEST FORM

3. TEST CODE					4. REGISTRATION NUMBER					
0	A	J	0	0	0	0	0	0	0	0
1	B	K	1	1	1	1	1	1	1	1
2	C	L	2	2	2	2	2	2	2	2
3	D	M	3	3	3	3	3	3	3	3
4	E	N	4	4	4	4	4	4	4	4
5	F	O	5	5	5	5	5	5	5	5
6	G	P	6	6	6	6	6	6	6	6
7	H	Q	7	7	7	7	7	7	7	7
8	I	R	8	8	8	8	8	8	8	8
9			9	9	9	9	9	9	9	9

6. DATE OF BIRTH

Month	Day		Year	
JAN				
FEB	0	0	0	0
MAR	1	1	1	1
APR	2	2	2	2
MAY	3	3	3	3
JUN		4	4	4
JUL		5	5	5
AUG		6	6	6
SEP		7	7	7
OCT		8	8	8
NOV		9	9	9
DEC				

5. YOUR NAME

First 4 letters of last name				FIRST INIT	MID INIT
A	A	A	A	A	A
B	B	B	B	B	B
C	C	C	C	C	C
D	D	D	D	D	D
E	E	E	E	E	E
F	F	F	F	F	F
G	G	G	G	G	G
H	H	H	H	H	H
I	I	I	I	I	I
J	J	J	J	J	J
K	K	K	K	K	K
L	L	L	L	L	L
M	M	M	M	M	M
N	N	N	N	N	N
O	O	O	O	O	O
P	P	P	P	P	P
Q	Q	Q	Q	Q	Q
R	R	R	R	R	R
S	S	S	S	S	S
T	T	T	T	T	T
U	U	U	U	U	U
V	V	V	V	V	V
W	W	W	W	W	W
X	X	X	X	X	X
Y	Y	Y	Y	Y	Y
Z	Z	Z	Z	Z	Z

The Princeton Review

1. A B C D	11. A B C D	21. A B C D	31. A B C D
2. A B C D	12. A B C D	22. A B C D	32. A B C D
3. A B C D	13. A B C D	23. A B C D	33. A B C D
4. A B C D	14. A B C D	24. A B C D	34. A B C D
5. A B C D	15. A B C D	25. A B C D	35. A B C D
6. A B C D	16. A B C D	26. A B C D	36. A B C D
7. A B C D	17. A B C D	27. A B C D	37. A B C D
8. A B C D	18. A B C D	28. A B C D	38. A B C D
9. A B C D	19. A B C D	29. A B C D	39. A B C D
10. A B C D	20. A B C D	30. A B C D	40. A B C D

NOTES

NOTES

NOTES

NOTES

NOTES